O9-CFT-627

A Devotion to Their Science
Pioneer Women of Radioactivity

When the subject of women in atomic science is mentioned, Marie Curie's name immediately leaps to mind. Many other women also worked in the field, but their names have been forgotten. *A Devotion to Their Science* reclaims this first generation of women researchers in radioactivity, providing new insights into the contribution of women to atomic science and dispelling the myth that this field was essentially a male preserve.

A Devotion to Their Science includes biographical essays on twenty-three women who worked in atomic science during the first two decades of the twentieth century, including Marie Curie, Lise Meitner, Irène Joliot-Curie, and a host of lesser-known women scientists whose life stories have never before been told. The biographies highlight the lives and work of these women, noting their contributions and the challenges they faced and overcame. Taken together the essays record their collective experiences, highlighting the support network that developed among them and the reasons women were more predominant in this field than in other sciences in the early part of this century.

By recovering and recording individual and collective histories of the many eminent women in radioactivity whose work had a major impact on the scientific discoveries of the twentieth century, a more complete, gender-integrated view of the history of this fascinating field emerges.

MARELENE F. RAYNER-CANHAM is a laboratory instructor in physics, Sir Wilfred Grenfell College, Memorial University.
GEOFFREY W. RAYNER-CANHAM is professor of chemistry, Sir Wilfred Grenfell College, Memorial University.

A Devotion to Their Science

Pioneer Women of Radioactivity

MARELENE F. RAYNER-CANHAM

AND

GEOFFREY W. RAYNER-CANHAM

SENIOR AUTHORS AND EDITORS

Chemical Heritage Foundation
Philadelphia

McGill-Queen's University Press
Montreal & Kingston • London • Buffalo

ISBN 0-7735-1608-5 (cloth)
ISBN 0-7735-1642-5 (paper)

Legal deposit third quarter 1997
Bibliothèque nationale du Québec

Printed in Canada on acid-free paper

Published simultaneously in the United States by
Chemical Heritage Foundation
315 Chestnut Street
Philadelphia PA 19106-2702
ISBN 0-941901-16-5 (cloth)
ISBN 0-941901-15-7 (paper)

This book has been published with the help of a grant from the Subvention Fund of the Memorial University of Newfoundland.

We acknowledge the support of the Canada Council for the Arts for our publishing program.

Chapter 3 was originally published as "Marie Curie's 'Anti-natural Path': Time Only for Science and Family" in G. Abir-Am and Dorinda Outram, eds., *Uneasy Careers and Intimate Lives* (New Brunswick: Rutgers University Press, 1987). It is published here by permission of Rutgers University Press.

Canadian Cataloguing in Publication Data

Rayner-Canham, Marelene F.
A devotion to their science : pioneer women of radioactivity
Co-published by: Chemical Heritage Foundation.
Includes bibliographical references and index.
ISBN 0-7735-1608-5 (bound) –
ISBN 0-7735-1642-5 (pbk.)
1. Women chemists – Biography. 2. Women physicists – Biography. 3. Nuclear physics – History.
4. Radioactivity – History. I. Rayner-Canham, Geoffrey William, 1944– . II. Chemical Heritage Foundation.
III. Title.
Q141.R29 1997 539.7′52′0922 C97-900199-4

Typeset in Palatino 10/12
by Caractéra inc., Quebec City

Contents

Preface

This compilation contains seventeen full biographies and six briefer accounts of most of the early women pioneers in the study of radioactivity. The fascinating stories of most of these women have never before been told and it is important that their lives be brought to light. It is only just to do so, since most of them contributed to the progress of science without any recognition during their lifetimes or since. We also need to kill the myth that Marie Curie was the one and only woman working in the field – a sort of gender aberration. Just as important, this collection is more than the sum of the essays because it shows, for the first time, that many of the women did not work in isolation but interacted throughout their lives, forming a support network, or "invisible college." In part, the support network resulted from the fact that the researchers, with few exceptions, belonged to one of the three major research schools in radioactivity, and we have organized the book around these three groups rather than dealing with the subjects in alphabetical order.

The difficult part was to decide on some sort of cut-off point, for women entered and left research into radioactivity on a continual basis. For this reason, we chose to include only women born before 1900. This division encompasses what might be referred to as the first generation of women researchers (it could be argued that Marie Curie, having started her research work in the 1890s, actually belonged to the zero generation). However, in the closing chapter, we mention a few of the second generation, such as Marguerite Perey, the discoverer of the element francium, to illustrate that women did not vanish from the field, though they did become more marginalized.

It was our original intent to write the whole of this work ourselves. Yet we wanted this compilation to become the definitive work on the lives of these women, containing the fullest and most comprehensive account possible. However, as we surveyed the quantity of material available on Curie and Meitner, it became apparent that we could not do them justice and that their life histories were best recounted by individuals who had spent years in their study. For some of the women from non-English-speaking countries, such as Gleditsch, we also believed that a thorough account could only be provided by someone fluent in their native language. We were fortunate to find enthusiastic and competent contributors for this purpose. Finally, we were able in two cases (Blau and Noddack) to obtain accounts written by people who had known the women as friends and who could present oral histories to substitute for the lack of biographical material. From our own research, we have edited and added to these solicited contributions to try to provide a consistent style and depth of treatment.

Marelene F. Rayner-Canham
Geoffrey W. Rayner-Canham

Acknowledgments

Apart from Meitner and the Curies, there are no formal archives of the women whose lives are discussed here. They regarded their contributions to science as minimal or even "worthless" and, as a result, kept no record of their lives. In most cases, the only way to piece together their stories has been to obtain copies of their academic records and filed résumés, copies of their publications, obituaries in parochial newspapers, letters to more famous scientists, letters of reference, comments about them in letters between other scientists, and in a few cases, biographical notes from surviving friends and relatives. The research has covered archives in eighteen countries. Through a travel grant from the Canadian Social Sciences and Humanities Research Council, it was possible to search archives at Cambridge University, Oxford University, the Royal Society, the Royal Institution, and the Institut Curie. For all the other archives (more than fifty) we relied upon the dedication of librarians and archivists in those locations to find the particular items for us. Each librarian and archivist is acknowledged in the appropriate chapter, but we wish to mention their efforts at the outset of the book as well. Without their collective contributions, it would have been impossible to assemble these accounts.

Most of the book was written at Sir Wilfred Grenfell College (SWGC) of the Memorial University of Newfoundland. We acknowledge the support and encouragement of SWGC faculty and administrators. In particular, we wish to thank James Greenlee and Dennis and Alice Bartels for critical comments on chapter 2. The remainder of the book was written at New College, University of South Florida (NC), where one of us has had a continuing appointment as visiting research scholar for the summer months.

Elizabeth Behrens, associate university librarian (SWGC), deserves a special thanks for her tireless efforts over the years to track down arcane material and copies of the research papers of these pioneer women. We also acknowledge the similar contribution of Holly Barone, interlibrary loans department (NC), during a half-sabbatical at that institution.

We are most grateful to the Subvention Fund of the Memorial University of Newfoundland and to the Chemical Heritage Foundation, Philadelphia, for their support of this work.

Foreword

For the better part of a decade I have played a game with students in the "women and science" course I teach at Concordia University. During the first class, I ask them to identify three Canadian women scientists and three from other parts of the world, write the six names on a piece of paper, and put the paper into a envelope. I keep the sealed envelopes until two weeks before the end of term; then I read them out for the class and a student writes the names on a board. Unfortunately, the list is brief and repetitious, "Marie Curie" being cited most frequently.

By the second stage of this exercise, students have heard about many other women scientists, including the Canadian Harriet Brooks and the Norwegian Ellen Gleditch. More importantly, they have learned about the factors that have influenced women's scientific opportunities in the Western industrialized world, such as systemic discrimination and the stereotypes of women and science. Readings and discussions have aroused their curiosity about women who have contributed to science at different times in many parts of the world, and who, for a variety of complicated historical reasons, have been overshadowed by the two-time Nobel Prize winner Marie Curie. They want to read about scientists, women and men, to understand gender roles and gender relations in science. From the response of my students it is evident that, in spite of the growing literature on women scientists, there is a great need for books such as *A Devotion to their Science: Pioneer Women of Radioactivity.*

What was the science of radioactivity and who were its forgotten pioneers? What were the factors that shaped their lives? When and for what reasons did the women in this volume decide to enter the new profession, science? How were they treated at various universities as students? Were they encouraged in their endeavours or were they prohibited from pursuing their studies in science? Were there any national and generational variations in their opportunities and experiences? To what extent did they fulfil their potential? How did they deal with obstacles and discrimination? How did they, if they were able to at all, combine a career with a conventional family life – like most of their male colleagues?

These questions need to be answered in order to understand the many similarities and differences in the "careers" of the first generation of women atomic scientists. The essays in this volume show how supportive environments at home, school, and work can encourage creativity and scientific excellence. Many of the authors deal with the challenges women faced because of discrimination based on their ethnic/religious backgrounds as well as on gender. They also deal with the impact of the women's discoveries and contributions to the emerging field of radioactivity.

The Rayner-Canhams have been researching and writing about women in chemistry and physics for well over a decade. Their scientific expertise, their sensitivity to gender issues in science, and their prodigious industry in lecturing and writing about women scientists have enabled them to disseminate information on many previously overlooked women scientists. This collection, with its variety of biographies – some detailed, others (for want of adequate documentation) far less so – will enhance the growing literature on women and science.

The feminist biographical approach of this book will lead to changes in the conventional story of radioactivity. By recovering forgotten women and presenting them as active, creative agents whose work had a major impact on the scientific discoveries of the last century, the authors give us a different, much more complete story that will help in reinterpreting the history of science. This, in turn, should affect the way science is taught in school, lead to fresh insights into the functioning of modern science, and prompt new research into gender relations in what has been known until recently as a largely masculine field. *A Devotion to Their Science* will open the readers' eyes and stimulate their minds. It will answer many questions and raise others. By analyzing the lives of eminent women in radioactivity, this book presents a more realistic, gender-integrated view of the history of this fascinating science.

Marianne Gosztonyi Ainley
University of Northern British Columbia
British Columbia, Canada

Marie Curie in her laboratory. (Archives Pierre et Marie Curie)

Ellen Gleditsch, her graduation photo. (T. Kronnen and A.C. Pappas)

May Sybil Leslie.
(University of Leeds Archives)

Catherine Chamié (right) in the Curie laboratory with Sonia Cotelle.
(Archives Curie et Joliot-Curie)

Harriet Brooks, her graduation photo.
(Notman Collection, McGill University Archives)

Stefania Maracineanu in the Curie laboratory.
(Archives Curie et Joliot-Curie)

Fanny Cook Gates. (Goucher College Archives)

Jadwiga Szmidt (*rear right*). This is possibly a photo of the Ioffe research group. (M. Golonka)

Lise Meitner at a colloquium in Berlin, 1920. (Archiv zur Geschichte der Max-Planck-Gesellschaft, Berlin)

Marietta Blau. (L. Halpern)

Ida Noddack. (Fo Habashi)

PART ONE
The Overview

It is crucial to place our biographical and scientific accounts in context, and this is provided in the first two chapters. They deal in part, of course, with the advances in knowledge of the atom and of subatomic particles that took place in the early decades of the twentieth century. But equally important are the individuals involved in the quest for knowledge. Progress in scientific knowledge is intimately involved with human relations and all advances reflect the preconceptions and prejudices of their era. Hence we will be examining some of the beliefs concerning the atom as well as the relationships between scientists and their research groups. Many of the individual biographies in this book overlap and this is one of its major strengths; for as we show conclusively in chapter 2 – and reiterate in later chapters – the "accepted" historical view that many of these women were "loners" is wrong: they communicated with one another both professionally and socially. It is also crucial to realize that the women researchers belonged to that unique period in history when women were being admitted for the first time to universities. Thus, we review contemporary beliefs concerning women's education and address the question why so many women were attracted into this particular field.

1 Early Years of Radioactivity

MARELENE F. RAYNER-CANHAM
and
GEOFFREY W. RAYNER-CANHAM

In this introductory chapter we summarize the immense progress in our understanding of matter and radiation between about 1880 and the 1940s. This period in scientific history marked the transition in emphasis from individual professional and amateur scientists to the research schools that we recognize today. As well, research was no longer a simple striving for knowledge but a need for priority in discovery, reflecting the intense nationalism of the age.

At the outset of this period the vision of matter was very simple. Ernest Rutherford, who became one of the leading researchers in the field, summarized the situation well: "I was brought up to look at the atom as a nice, hard fellow, red or grey in colour according to taste."[1] The change in our view of matter was brought about by two factors: advances in the knowledge of electricity, light, magnetism, and other physical phenomena and changes in the nature of scientific research. Traditionally, scientific research had been the preserve of the individual professional or amateur scientist, usually employing simple items of equipment. Between the 1860s and the First World War the hierarchical research group appeared, and though important discoveries were still being made by individuals, the work of these groups came more and more to dominate the major advances in science.[2] The formation of groups reflected the increasing complexity of the work, which drew on the combined skills and abilities of scientists of different backgrounds, while the enormous amount of research needed to follow up discoveries in the field required many hands at the research bench. Indeed, the discovery of the phenomenon of radioactivity marked one of the most explosive growths in the history of science; by 1907 more than twelve hundred papers had been published in the field.[3]

Radioactivity was becoming a topic of daily conversation as people became aware that scientific discoveries were changing their lives.[4] Following from this, personal and national prestige became a driving force in the development of the field as well as scientific curiosity. In the study of atomic science and the phenomenon of radioactivity, we find that there were three major groups of researchers. These we will call research schools,[5] though in many ways it is more accurate to refer to them as invisible colleges, a term used to describe groupings of scientists who correspond with each other and exchange results.[6]

THE RESEARCH SCHOOLS OF ATOMIC SCIENCE

Those who worked in this field can be roughly categorized into three sets of collaborators: the French school, the British school, and the Austro-German school.[7] These schools represented not only location of focus but also the style of research. There was a strong perception, embedded in reality, that the approaches of the groups were different. In his discussion of the national differences, Alex Keller notes: "Yet a young Hungarian colleague of Rutherford's wrote to him in 1913 on the contrast between the British Association meeting at Birmingham that September, and a similar congress held at Vienna straight after: 'altogether it was more knowledge in Vienna but far more ingenuity at Birmingham.' A few months later another young colleague, Henry Moseley, described a visit from the eminent French chemist Urbain from whom he learnt that 'the French point of view is essentially different from the English – where we try to find models or analogies, they are quite content with laws.'"[8] Keller comments on the French view of their own science as "the spirit of synthesis, simplicity and clear precision," in contrast to "the haphazard fact-finding sorties of the British, who wanted to turn everything into wheels within wheels, or to the grandiose woolly theorizing, and niggling accumulations of useless data piled up in German theses and treatises."[9] There did seem to be a basis in reality for different scientific styles, Marie Curie, and, later, Irène Joliot-Curie, for example, were often among the first to perform a crucial experiment but, through conceptual conservatism, rarely recognized the paradigm shift[10] necessary for its correct interpretation.

For most of the period in question, the French group was identified exclusively with the name of its leader, Marie Curie.[11] Initially, the "school" consisted essentially of Marie Curie and her husband, Pierre. Following Pierre's death, Marie Curie's research group grew rapidly to the point where there were more applicants than space to house them. Though most of the researchers came and went, André Debierne seemed to have been the day-to-day manager of the research effort with Irène Joliot-Curie, Marie's daughter, taking over in later years. Throughout the continuous history of the French group, it was always Paris-focused.

The British effort largely revolved around Ernest Rutherford. Rutherford had been a student of the famous nineteenth-century physicist J.J. Thomson, at the Cavendish Laboratory, Cambridge University.[12] Enticed by the excellent research facilities at Canada's McGill University, he moved to Montreal in 1898. Chemist Frederick Soddy joined him later and some of Rutherford's major conceptual advances were joint efforts between the two. Rutherford returned to England, first to the University of Manchester and then back to the Cavendish Laboratory. Though many researchers worked with him wherever he was located, he also had a more diffuse network of collaborators and associates. When Rutherford returned to England from Montreal, the study of radioactivity died in North America. Pyenson has shown that research in the field collapsed at McGill with Rutherford's departure,[13] while Badash has commented upon the sterile scene in the United States: "There were no charismatic leaders, and no laboratory seems to have created a vital environment sufficient to establish a tradition with the weak exceptions of Yale and Minnesota."[14]

The situation in Vienna where the Austro-German school originated could more correctly be described as a research circle rather than a research school, as the head of the Institute, Franz S. Exner, favoured collaboration, not a master-student style.[15] Nevertheless, for research into atomic science it was Stefan Meyer who was considered the overall leader. Following the collapse of the Austro-Hungarian Empire, the focus moved to Berlin and the work of Hahn and Meitner, among others. There was still interesting work being done in Vienna, but it was never in the forefront of the major discoveries.

The researchers in Vienna were almost exclusively physics-oriented at a time when interest centred on aspects of radioactivity in the middle ground between chemistry and physics, such as the discovery of new elements and the undertanding of atomic structure. This is, perhaps, part of the reason why the Vienna workers have been little acknowledged in the history of atomic science. By contrast, the Curie group always contained chemists as well as physicists (Marie Curie having had a chemistry background) and Rutherford collaborated with the chemist Soddy for many years. Even though Rutherford thought of himself as a physicist, he was awarded the Nobel Prize in chemistry for his work on radioactivity. Of relevance to this hypothesis, the greatest claim to fame of the Austro-German group was the collaboration between the chemist Otto Hahn and physicist Lise Meitner. Only later did the interdisciplinary science of radioactivity become replaced by that of atomic physics.[16]

Throughout this saga, competition seemed to be a strong driving force between the different groups. Inititially, the principal competition was between the British (Rutherford) and French (Curie) groups, then later between the Austro-German (Hahn and Meitner) and French (Joliot-Curie) groups. The perceived arrogance of the French group antagonized other scientists. Such anti-French views were reinforced by Curie's insistence that the unit of radioactivity be called the curie and that the standard sample be lodged in Paris. Though the First World War severely disrupted scientific relations[17] in general, relations between the British and Austro-German workers in radioactivity seemed to be more amicable, perhaps in part because Otto Hahn had worked with Rutherford in the early days in Montreal. Even so, the superficial congeniality covered a competitive attitude. Otto Frisch, Lise Meitner's later collaborator, remarked of his early years in Vienna that hasty (and erroneous) research was being done for the sake of "beating the English at their own game."[18]

RADIOACTIVITY AND ATOMIC STRUCTURE

If one can possibly place a date on the beginning of the Age of Radioactivity, then it would be the discovery in 1896 that radiation was emitted by a compound of uranium. The discovery that uranium salts darkened photographic plates is associated with the name of the French chemist Henri Becquerel, who observed the phenomenon in 1896.[19] However, Becquerel's research depended upon Konrad Röntgen's discovery of x-rays in 1895[20] and the even earlier discovery of cathode rays. At the time, few scientists pursued investigations into so-called hyperphosphorescence, or Becquerel rays from uranium. Interest rose in 1898 with the simultaneous discovery by Marie Curie in France and G.C. Schmidt in Germany that thorium also emitted Becquerel rays.[21] And so the search was on for the nature of the rays and for evidence of radioactivity among other elements. It was the French group that pursued the isolation of radioactive elements, the Curies extracting polonium and radium from pitchblende in 1898 and Debierne isolating actinium in 1899. Rutherford, on the other hand, was more interested in radiation, showing in 1898 that there were two types of rays, which he called alpha (α) and beta (β), and that they behaved as particles. Later, the French scientist P. Villard reported a third type of radiation. Rutherford suggested that this was very short wavelength electromagnetic radiation, which he named the gamma (γ) ray.[22]

Atomic Transmutation

The long-discredited field of alchemy had claimed that one substance could be transmuted into another, such as lead into gold. Hence the "real" science of chemistry had identified the immutability of chemical elements as one of the cornerstones of its philosophy. J.J. Thomson, one of the orthodox physicists, was convinced that radioactivity was only a simple chemical process and he expended considerable (and fruitless) effort in his laboratory trying to prove this.[23] Even in 1900, the Curies agonized over the apparently unlimited amount of energy being radiated from the radioactive elements without any obvious source and in apparent violation of the laws of thermodynamics.[24]

Thus, it was extremely difficult for atomic scientists to reconcile what was then a fundamental principle of chemistry with the discovery that elements *did* change identity during the release of radiation – the phenomenon that we now call radioactive decay.[25] Rutherford, together with Harriet Brooks, had worked on the identity of "emanation," which was produced from the element thorium. Brooks concluded that emanation was in fact a gaseous element of different atomic weight (what we now call radon). Following from this, Soddy and Rutherford took the bold and somewhat blasphemous step of announcing that one element had been produced from another.[26] The Curies and Debierne had made the same laboratory observations earlier, but they had tried to explain radioactivity purely in terms of energy release. As Marjorie Malley has commented: "Curie's theory was hesitant, conservative, vague, and abstract; Rutherford's hypotheses were bold, ultimately revolutionary, specific, and concrete."[27]

The Concept of Isotopes

Another established principle of chemistry was that of atomic weight: that each chemical element had a unique and fixed value.

Here again philosophical principle was to fall in the face of overwhelming facts. Having recognized that radioactive decay led to the formation of different elements, the identity of those elements came into question. Each of these radioactive products had been given a name, such as thoron or ionium, or the name of its parent followed by a letter, such as thorium X and radium G. In 1906 Bertram Boltwood found it chemically impossible to separate ionium from thorium, while the Hungarian scientist Geörgy Hevesy was unsuccessful in separating radium D from lead. It was Soddy and Kasimir Fajans who simultaneously and independently proposed the group displacement laws: that emission of an alpha particle produced an element two places lower in the periodic table, while emission of a beta particle produced an element one place higher. This proposal meant that it was not previously unknown elements that were being produced but forms of existing elements. The atoms of these elements produced by radioactive decay were predicted to have different atomic masses to atoms of the "ordinary" element. To support this proposal, studies were made of the atomic weight of lead from radioactive ores against the accepted value for "normal" lead. The first major differences were found by Otto Hönigschmid and Stefanie Horovitz, while the most conclusive differences were produced by the Harvard chemist Theodore Richards, based primarily on samples provided by Ellen Gleditsch. As a result of these reports, the existence of isotopes (as Soddy called them) became accepted.[28]

The Modern Structure of the Atom

A whole book could be dedicated to a thorough historical account of the research in this field. For our brief summary we must simplify the history and eliminate the names of many participants and many of the experiments. The first proposal similar to our modern picture was made by the Japanese scientist H. Nagaoka, who had studied in Germany. In 1904 he proposed a "Saturnian model" of the atom in which the electrons swirled around a central core.[29] At the time, scientists were unable to find evidence for this proposal and the model fell into disfavour. In 1906 Rutherford observed that a beam of alpha particles would penetrate a very thin metal foil but the spot produced was highly diffuse. Hans Geiger and Ernest Marsden subsequently showed that the particles were either little deflected or massively deflected. This discovery suggested that the atom was mostly empty space with a "hard" core, leading Rutherford to propose the nuclear model of the atom in 1911.[30]

Nuclear Particles

At that point, to account for the mass and charge, the nucleus of any atom was considered to contain hydrogen nuclei and some electrons. For example, the helium nucleus was considered to contain four hydrogen nuclei *and* two electrons while two additional electrons "circled" the nucleus.[31] This proposal was reinforced by the first artificial synthesis of an element by Rutherford in 1917. He bombarded nitrogen atoms with alpha particles to produce oxygen atoms and hydrogen nuclei (named protons by Rutherford in 1920). The next advance was not to come until 1931 when Irène Joliot-Curie and her husband, Frédéric Joliot, studied the emission of a penetrating radiation when low atomic weight elements were bombarded with alpha particles. The radiation was undeflected by a magnetic field but they were unable to explain its nature. Upon seeing the report of these findings, James Chadwick showed that the radiation was in fact neutral particles, which he named neutrons. It was these neutrons that made up about half (or more) of an atom's mass, and the previous solution of proton-electron pairs was discarded.[32]

Nuclear Fission

Artificial syntheses of nuclei had, with few exceptions, resulted in the formation of a new nucleus higher in the periodic table. The research group formed in Rome and led by Enrico Fermi bombarded uranium atoms to produce new elements of higher atomic weight and atomic number. As they analyzed the results in 1934, they believed they had accomplished their goal.[33] They assumed that any new element would be one or two positions further in the periodic table and it was this belief that determined the tests that they and others used to identify the new elements. According to the periodic law, a chemical element should resemble in its properties the elements in the group above it. Since at that time the uranium series was believed to form a seventh row in the table, element 93 was expected to resemble the manganese group, and this is what Fermi and his colleagues claimed for their new element. Ida Noddack and a few others contested this conclusion, but most scientists accepted Fermi's claimed discovery of element 93. Two groups followed up on this work: Irène Joliot-Curie and Paul Savitch in Paris and Lise Meitner, Otto Hahn, and Fritz Strassmann in Berlin. The Meitner group claimed to identify several more new elements, while Joliot-Curie reported, in addition, a product element with a three-and-a-half-hour half life that behaved like one of the lanthanum-series elements, not like another transition metal. It was easier for the other research groups to dismiss thc Joliot-Curie report than to question the whole basis on which new element status was being assigned. Hence it was not until Hahn and Strassman found one product element that behaved like barium that Meitner and her nephew, Otto Frisch, were forced to question the integrity of the nucleus. The liquid-drop model of the nucleus had been proposed a few years earlier, and Meitner and Frisch realized that it was the key to their conceptual problems – that a nucleus *could* break into two halves, just like a water drop.[34] In other words, the products were not new elements but simply isotopes of elements in the mid-part of the periodic table. So once again, the Paris group had found the evidence for a scientific breakthrough but failed to make the conceptual interpretation.[35]

Hazards of the Work

It is important to consider the health hazards that all of these men and women faced. It is amazing that most lived so long, for nearly all of them were exposed on a regular basis to high levels of radiation with few, if any, safety precautions. The severe skin damage that exposure to radiation could cause was reported as early as 1901,[36] but it was simply regarded as an occupational hazard. May Sybil Leslie had noted, during her time at the Institut Curie, the cavalier attitude towards radiation among the Curie workers.[37] In fact, Marie Curie's cookbooks were still radioactive fifty years later.[38] Rutherford, too, failed to recognize the hazards of radiation, the most vivid example involving a visit to Dartmouth College as part of a lecture tour of the United States. During his demonstrations, Rutherford discarded the paper that he had used to transfer some radium salt into a tube. This paper was saved by his host and used as a radioactive source for a period of forty years.[39]

Though Hahn and Meitner were generally very careful about their exposure to radioactivity, they kept a crate of between 150 and 250 kilograms of radioactive salts under their workbench for many years.[40] Elizabeth Róna had two major accidents involving radioactive materials, one at Paris and the other at Vienna.[41] Perhaps as a result of those close escapes and her frequent exposure to radiation, the last paper that she published described some of the health problems at the institutes in Paris, Vienna, and Berlin. She commented that a number of the participants, such as Sonia Cotelle and Debierne, died from radiation-related diseases. She found it amazing that Marie Curie lived to the age of sixty-seven, considering her continuous exposure to radiation. Even more amazing were the ninety-seven years that Róna herself lived. It is of particular note that, in her review of radiation-related health problems, Róna makes no mention of the potential effect of radiation on a woman's reproductive organs. It is as if she assumed that, for a woman, a life devoted to the study of radioactivity precluded any possibility of a family life. As we shall see, this was largely the case.

Yet it was not just through ignorance or carelessness that radioactive materials were handled in a cavalier fashion. With the burst of optimism at the start of the twentieth century, science was looked upon as the great saviour of humanity. The cancer-killing ability of these materials seemed to demonstrate that the rays of radioactive decay were another wonderful tool that would bring benefit to all.[42] In fact, the use of radium (and later, radon) to cure growths, tumours, and certain forms of cancer was called Curietherapy in France.[43] To admit that radiation had a darker side as a cause of illness and death was unacceptable.

All of the researchers seemed to have held this rosy view but none as firmly as Marie Curie. Even after the illness and subsequent death of Cotelle[44] in her laboratories, Curie still refused to recognize the seriousness of the problem.[45] In part, this was because both she and Irène had handled radioactive materials continually during their careers and had suffered only "minor" problems. Curie recommended plenty of exercise in the fresh air as her personal panacea for radiation-induced illness. Later, Florence Pfaltzgraph, an American journalist, described in a letter to Curie the sufferings of the workers who painted watch dials with radium to make them luminous.[46] Pfaltzgraph asked whether she had "discovered anything which might benefit these women." Curie expressed sympathy and advised the women to eat calves' liver. Irène Joliot-Curie, in turn, did not endear herself to the physicist Cécile DeWitt-Morette by asserting that anyone who was worried about radiation hazards was not a dedicated scientist.[47] It is ironic, then, that Irène, together with Catherine Chamié and another of Curie's later researchers, Marguerite Perey, definitely died of radiation-related causes.[48]

2 Pioneer Women of Radioactivity

MARELENE F. RAYNER-CANHAM
and
GEOFFREY W. RAYNER-CANHAM

Between 1900 and 1910 about thirty women were active researchers in the study of radioactive phenomena. In those days, for a woman to have graduated with a degree in science was a feat in itself. For these particular women, the focus of their lives became the study of the nature of the atom. Many of them depended on their supervisors, mentors, and colleagues for encouragement and support in their work.

When one reads about the history of research in radioactivity, women are rarely mentioned.[1] Yet many women participated in the venture to discover the secrets of the atom. In fact, they seemed to play a disproportionately large share in the research work in radioactivity compared to many other fields of physical science. So why have they been overlooked? As with other aspects of life, it was almost impossible for women to reach the upper ranks of academia to gain the authority, prestige, research workers, and research grants that were necessary to make a name in the scientific community. And it is the "great names" that are passed on from generation to generation in the simplified stories that we refer to as the history of science. As Hubbard commented: "Women have played a very large role in the production of science – as wives, sisters, secretaries, technicians, and students of 'great men' – though usually not as named scientists ... More important, we must understand and describe accurately the roles women have played all along in the process of making science."[2]

In radioactivity studies the research assistants, many of them women, played a particularly important role, as de Solla Price has pointed out: "Thomson and Rutherford were genius experimenters who happened to be rather clumsy, and their assistants were crucial to progress. In fact, much of the apparatus used in the Cavendish Laboratory at Cambridge, where both men worked, was held together by sealing wax and string, not out of poverty, but because the genius experimenters had a dozen pairs of clever hands feverishly tearing down and rebuilding the apparatus as the techniques were pushed in new directions and improvisations."[3]

Three fields in the physical sciences seemed to attract an exceptional number of women during the late nineteenth and early twentieth centuries: astronomy, crystallography, and atomic science. Astronomy is the best-known example and the role of women in this field has been the subject of a number of studies.[4] Less well known is the field of crystallography, the use of x-rays to determine the structure of chemical compounds. Julian has shown that women did not make up the majority of crystallographers, but certainly they made up a higher percentage of the researchers than the average across the physical sciences.[5] The reason for such a high profile in this field was the brilliance of a few of these women. Four in particular, Rosalind Franklin,[6] Dorothy Hodgkin,[7] Isabella Kahle,[8] and Kathleen Lonsdale,[9] attained world renown for their work. Yet it must be remembered that these were merely the four brightest stars among a significant number of women crystallographers.

This book focuses on the third field, atomic science. In this context almost everyone with some science knowledge has heard of Marie Curie, with her two Nobel Prizes. Few people have heard of Lise Meitner and even fewer of Irène Joliot-Curie, the other woman Nobel laureate in the early years of research into radioactivity. However, as in crystallography, these women were just the "tip of the iceberg." Though it has been common practice to focus on the exceptional individuals, both male and female, who attained great fame in their chosen sphere, such an approach to the history of science gives a distorted view. It also does a disservice to those who failed to reach the pinnacle of success, often for reasons beyond their control, such as the "glass ceiling" or family commitments. Even when such individuals have made important discoveries, such as Blau's work on cosmic rays, Brooks's discovery of the recoil of the radioactive atom, or Noddack's proposal of nuclear fission, recognition is not forthcoming. This is an example of the Matthew Effect, which is defined as "the accruing of large increments of peer recognition to scientists of great repute for particular contributions in contrast to the minimizing or withholding of such recognition for scientists who have not yet made their mark."[10] The Matthew Effect has been such a crucial problem for women scientists that Rossiter has renamed it the Matilda Effect.[11]

In those early years, it was a passion for science that fired these women, not an urge for recognition, and many of them regarded their own contributions as minor. Harriet Brooks, one of Ernest Rutherford's researchers, commented: "There are so many other people who can do so much better and in so much less time than I that I do not think my small efforts will ever be missed."[12] As a result, few of the women (except for Curie and Meitner) or their relatives thought it important to keep their papers or correspondence. The few surviving items are found in the archives of their (male) supervisors. Bonta found a similar phenomenon of self-belittlement in her studies of the early women botanists. She remarks: "Because most women naturalists believed that the work was all that mattered, they seemed to feel little or no rivalry toward the more powerful males in their fields and were pleased and grateful for whatever help these men gave them."[13] The comment is equally applicable to the women who were studying radioactivity.

We can obtain a better view of women in a scientific field by looking at the whole picture through accounts of as many women who entered the field as possible. Early research into radioactivity is ideal for this study in that there were a significant proportion of women in the field; yet because it was a completely new area of study, the total number of women was quite small: according to our research, about thirty in the first two decades of this century.

GETTING STARTED

Before we can compare the lives of the individuals in this study, we must first consider the initial hurdle that all of them faced: the struggle for an education. These women received their high school education in an era when the idea of advanced education for women was still being challenged. The powerful arguments of the day have been summarized by Hubbard: "More effective were the extensive treatises, replete with case histories, that 'documented' the drain that menstruation and the maturation of the female reproductive system was said to put on woman's biology and, more importantly, the stress that would fall on these vital capacities if women's intellects were taxed by education. One of the most widely read books of this sort was Edward H. Clarke's *Sex in Education*, published in 1873, which went through seventeen editions in the next thirteen years. Clarke, a former professor at Harvard Medical School and a fellow of the American Academy of Arts and Sciences, details the histories of many girls whose health, he assures us, was severely damaged by education."[14]

At the same time, there were some strong proponents of university education for women, particularly scientific education.[15] Three arguments were given in support of the proposal: that women would become better wives and mothers if they had a scientific background; that simple justice and equal opportunity should permit women to enter scientific careers; and that women had a greater capacity for noting details and superior patience and manual dexterity, enabling them to perform certain scientific work better than men. In fact, the case was made by Becker that women would benefit more than men from training in science: "Prevalent opinions and customs impose on women so much more monotonous and colourless lives, and deprive them of so much of the natural and healthy excitement enjoyed by the other sex in its free intercourse with the world ... many women might be saved from the evil of the life of intellectual vacuity, to which their present position renders them so peculiarly liable, if they had a thorough training in some branch of science, and the opportunity of carrying it on as a serious pursuit."[16] Yet many women themselves questioned the role of higher education for women and, in particular, whether it conflicted with the ultimate goal of domesticity.[17]

But the strongest opposition to women obtaining advanced education came from the academics who used the theory of evolution to support their arguments. For example, the sociologist Herbert Spencer had concluded that the difference between the sexes could best be understood in terms of "a somewhat earlier-arrest of individual evolution in women than men."[18] The prominent chemist Henry Armstrong argued that, as a result of being man's slave, women would, according to the theory of evolution, have an inferior mental development.[19] This was not the rambling of an eccentric but a restatement of Charles Darwin's "scientific" evidence of female inferiority.[20] Nor was Armstrong alone in his opinion, for many scientists of the time had found proof in their research of women's "intellectual inadequacies."[21]

To decide to go to college, then, was a brave act. It required a strong self-image, particularly if the family held to the conventional view that a well-brought-up Victorian or Edwardian girl should stay quietly at home until a suitor appeared on the horizon. Moreover, it was an avenue open only to the daughters of the expanding middle class, the rapidly growing business and professional sectors of society. The daughters of the poor were financially unable to attend university while the daughters of the upper classes were, for the most part, educated with a view to their intended life of leisure rather than intellectual development.[22] In her study of women Nobel Prize winners in science, McGrayne noted two factors that were important for success: sympathetic parents and relatives; and religious values that stressed education (such as Jewish or Quaker).[23] Although the information is incomplete, one or both of these factors are apparent in the lives of many of the women discussed in this compilation.

In continental Europe there was an additional problem: the university entrance examinations offered in high schools were often barred to women. This was true in Germany, where the *Abitur* was only offered in boys' schools,[24] and in France, where the *baccalauréat* was the hurdle preventing women from even applying to a university.[25] The situation in Russia, where as early as 1859 women were admitted to lectures, was initially very promising. However, the outbreak of student demonstrations at which one woman was arrested was used as a pretext to restrict women's access to universities.[26] Accessibility varied with the political regime, and by 1908 women had even been banned as auditors. As a result, Russian women had to travel abroad to obtain a university education.

Self-motivation was a major factor in pursuing further education.[27] M. Carey Thomas described her own feelings, which were obviously coloured by the perceptions of Clarke and Spencer: "The passionate desire of the women of my generation for higher education was accompanied through its course by the awful doubt, felt by the women themselves as well as the men, as to whether women as a sex were physically and mentally fit for it ... I was always wondering whether it could be really true, as everyone thought, that boys were cleverer than girls."[28] In addition, going to college was one of the few ways (nursing and missionary work being among the others) that a woman would escape the family home without having to get married.[29]

Once at university, the woman student's problems were far from over. As one student remarked, she and her sister students "bore the weight of formulated womanhood upon [their] shoulders, although men, even then, were not expected to live to the ideal man."[30] A woman student had to endure the stares and comments of their male peers. Elizabeth Irwin at McGill University, Montreal, recalled that "it required courage in those days to walk from the East Wing to Molson Hall, or the old Library below the Hall. It meant ... running the gauntlet of the men students, who, not yet accustomed to the intrusion of the feminine element, greeted our appearance with the strains, long since forgotten, of 'Hop Along, Sister Mary.'"[31]

Administrators and faculty were often hostile. In 1871 the senate of Heidelberg University described the attendance of ladies at academic lectures as "an unsavoury and disturbing phenomenon" and instructed lecturers "not to tolerate it."[32] The famous physicist Max Planck was of the opinion that even if some women had an aptitude for academic study, it could "not be stressed enough that Nature herself assigned to women the role of mother and housewife."[33]

The problems of the woman student persisted after graduation. What was she to do with her education? In the 1880s, when colleges were starting to open their doors to women, little thought had been given to this question.[34] Antel has shown that most single women graduates actually returned to the parental home, some as "working daughters," while others became "ladies of leisure."[35] A very few – "the independents" – left home to pursue a career. Yet the return home was not necessarily the student's choice. The famous astronomer Annie Jump Cannon, for example, had spent a decade at home and unemployed before receiving the offer of an astronomer-assistant's position.

WHY RADIOACTIVITY?

As we mentioned earlier, there seem to have been three fields of physical science in which women have played a disproportionate role: astronomy, crystallography, and atomic science. Rossiter has shown that the nature of astronomy had changed from the active work of observing (men's work) to the passive role of classifying the thousands of photographic plates (women's work).[36] It was the first of the women astronomer-assistants, Williamina P. Fleming, who commented that women's superior patience, perseverance, and method made such activities particularly suitable for women in science.[37] In their studies of women in astronomy, Lankford and Slavings added: "In evaluating women [for astronomy], male scientists tended to focus on their [women's] ability to do routine work. Indeed, some recommendations read as if they were descriptions of machines."[38]

Portugal and Cohen used the laborious nature of crystallography to explain the number of women in that field: "Since the high speed computer had not yet been invented, the business of calculating data was a very laborious occupation and smart fellows who could find other things to do would generally do them, unless they were absolutely dedicated to the business of x-ray crystallography. Is it possible that these first class women got to be x-ray crystallographers because they were willing to do this work."[39]

It could equally be argued that research in radioactivity also involved tedious, painstaking, and repetitive work and that this was why men avoided the field while women thrived in it. When James Chadwick visited Vienna in 1927, he found that the routine measurements of evidence of radioactive products were being performed exclusively by women, since "they could concentrate on the task more intensely than men, having little on their minds anyway, and by Slavic women because their large, round eyes were better suited to counting."[40]

We would argue more positively, however, that each of these new fields had an aura of excitement for women who were looking for a meaningful career. Astronomy focused on the enormity of the universe, while crystallography and atomic science were both concerned with the investigation of the nature of matter, in one case with the arrangement of atoms in compounds, in the other with the nature of atoms themselves. As Gornick describes in her interviews with modern women scientists: "Each of them had wanted to know how the physical world worked, and each of them had found that discovering how things worked through the exercise of her own mental powers gave her an intensity of pleasure and purpose, a sense of reality nothing else could match."[41]

THE ROLE OF THE RESEARCH SUPERVISORS

There was an additional reason why the study of atomic science was favoured by women: the presence of supportive supervisors who acted as mentors for them. To devote one's time to the tasks of research involves enthusiasm, a significant part of which stems from the support and encouragement of the research supervisor. The crucial role of mentors in the encouragement of women in science is now well established,[42] the clearest example in modern times being the influence of Lee Lorch at Fisk University in encouraging Afro-American women to follow careers in mathematics.[43]

In crystallography, Julian has shown that it was the Braggs, William H. and his son, William L., who welcomed gifted women into their research groups.[44] The field of biochemistry, that link between the physical and biological sciences, also provides some interesting parallels with atomic science in terms of encouragement for women.[45] Mary Creese writes, for example, that "a second and more specific factor in the success of women biochemists associated with Cambridge was the role played by [F. Gowland] Hopkins, whose career there began in 1898. At the time when there were practically no women research workers in any of the other university departments at Cambridge, Hopkins gave them places in his, despite the criticism which this brought him. Even in the 1920s and 1930s, when, as a Nobel laureate with a world-wide reputation he received hundreds of applications for places in his laboratory, nearly half of the posts in his Department went to women scientists."[46]

It was not just a question of admitting women students to the group. Seyre is convinced that, at least in crystallography, it was the non-aggressive and friendly attitudes of the supervisors that were so vital to the encouragement of women: "There is something in the ancient history of crystallography that is hard to isolate but nevertheless was there, that I can best describe as modesty. I have often wondered how much the Braggs were responsible for the unaggressive low-key friendly atmosphere that long prevailed in the field (and no longer seems to very much). Somehow the first and second and a few of the third generation crystallographers consistently conveyed an impression of working for pleasure, for the sheer joy of it – the idea of competition didn't seem to emerge very strongly until the 1960s or so. Uncompetitive societies tend to be good for women."[47]

Creese felt that Hopkins's personal style was also an important influence on the encouragement of women researchers: "He was a great giver and receiver of moral support ... he was open to receive it from the most junior of his research workers, so that they did not feel he was encouraging them like some *deus ex machina,* but as one of themselves; in other words he fully understood and practised the great doctrine of leadership from within and not from above."[48]

The point must be emphasized that a welcoming atmosphere for women researchers was far from common. As an example of the overt discrimination against women, on 27 August 1880 a meeting of the American Chemical Society (ACS) was held in Boston, followed by a dinner and "festivities" that involved the performing of antifemale songs and poems.[49] A subsequent booklet entitled *The Misogynist Dinner of the American Chemical Society* described the events of the evening. The preface of the booklet notes: "The general character of the entertainment ... may be gathered from the fact that not only were no ladies invited by the committee, but even when, through a misunderstanding on the part of some prominent members of the society, several brought their wives with them to the dinner, these ladies were refused admission, and actually turned away from the door." When Rachel L. Bodley, the sole woman member of the ACS, learned of the affair, she resigned and it was eleven years before the next woman member was elected. In Britain, the opposition to women chemists was so vociferous that it was not until 1920 that the Chemical Society would admit them as members.[50]

Discrimination against women scientists could sometimes be quite petty and very hurtful. Otto Hahn recounted how Lise Meitner wrote articles on problems in physics for a review journal edited by a Professor Sklarek. The editor of an encyclopedia wrote to Sklarek asking for "Herr" Meitner's address as, impressed with "his" work, they wanted "him" to write an article for the encyclopedia. In his reply Sklarek noted that it was "Frau" Meitner. The editor responded "immediately and heatedly" that he had no intention of publishing an article written by a woman.[51]

Whereas the Braggs had almost single-handedly founded the science of crystallography and Hopkins, together with Mendel in the United States,[52] had opened the doors to women in biochemistry, radioactivity had a number of early pioneers. A remarkable proportion of these researchers, including Ernest Rutherford, Marie Curie, J.J. Thomson, Frederick Soddy, Kasimir Fajans, and Otto Hönigschmidt, took on women students. The role of the supervisor, we believe, was very important to the early women involved in the study of radioactivity.

Like the Braggs and Hopkins, Rutherford seemed to provide a collaborative, positive working environment for his research workers. His biographer, David Wilson, commented: "It will inevitably be hidden and forgotten that he was a man of exceptional personal kindness. Everyone who remembers Rutherford remembers this – that he was personally kind to them far and away beyond the normal behaviour of a pleasant human being."[53] While Kapitza noted how Rutherford was particularly conscious of the need to boost the morale of new assistants – an important point for women students lacking in self-confidence – "he [Rutherford] was also very particular not to give a beginner technically difficult research work. He reckoned that, even if a man was able, he needed some success to begin with. Otherwise he might be disappointed in his abilities, which could be disastrous for his future. Any success of a young researcher must be duly appreciated and must be duly acknowledged."[54]

During his career Rutherford had many women researchers in his group, of whom Harriet Brooks, Fanny Cook Gates, May Sybil Leslie, and Jadwiga Szmidt were among the earliest. In his two classic works on radioactivity,[55] Rutherford cited research by many of these women scientists. Ellen Gleditsch, though she never worked with Rutherford, seemed to have been influenced by him. She thanked Rutherford in a letter for his "kindly conversations" in Washington in April 1915. After describing her thorough work on the half-life of radium and expressing a wish to work with Rutherford at Manchester, she closed with the remark: "I hope you will forgive me for writing you about my work. As you will understand you yourself gave me the courage in Washington."[56]

Two instances, however, have been cited to show that Rutherford was less than encouraging to women scientists. The first concerns his comment to Lise Meitner, upon first meeting her, that he had thought she was a man.[57] The comment is widely taken to indicate that Rutherford did not expect a woman to have done such research, but there is a less cynical explanation. In Rutherford's two texts the indexed names of women researchers are prefixed by "Miss," "Mrs," "Mlle," "Mme," or "Lady," as appropriate; Meitner is listed simply as "L. Meitner" (the author's name used on her early papers), suggesting that Rutherford may not have realized that she was female.[58] Meitner's own comment on the incident was that "he had not realized that my first name is a girl's name."[59]

Comments by the astronomer Cecelia Payne-Gaposchkin have also been used to cast an unfavourable light on Rutherford's attitude towards women.[60] However, Payne-Gaposchkin's specific evidence for Rutherford's "scorn" towards women was his habit of starting his lectures with the phrase "*Ladies* and Gentlemen," which caused the male students to break into thunderous applause and foot stamping. She, the lone female, was required by university regulations to sit in the "women's row" at the front of the class. Although this emphasis on "ladies" must have been intimidating to as shy a person as young Payne-Gaposchkin, her antipathy towards Rutherford seems to be at odds with the comments of his own women students. In fact, he may have been attempting gallantry, for lecturers at Cambridge at the time commonly addressed their classes as "Gentlemen" even when, as was often the case during the First World War, the audience was exclusively female.[61]

Rutherford has also been criticized for possible condescension towards women, particularly Marie Curie.[62] Yet Curie must have been difficult to deal with since she regarded others as interlopers in "her" field. As Eva Ramstedt recalled in a letter to Curie, "Madame, you have said yourself that radioactivity is your child that you have nourished and educated."[63] Despite Curie's apparent hubris, Rutherford commented in a letter to Boltwood: "I was very glad to see that she [Curie] got a Nobel Prize."[64] For his time, he was willing to express quite vocal support for women's rights. Indeed, Margaret Ashton asked him in 1908 if he would speak to a meeting on joint suffrage in the colonies. She writes: "Can you and would you come in and tell us in a short talk what the effect is and how little it disturbs family life or women's 'womanliness' so called."[65]

Rutherford acted as intermediary for women students other than his own. For example, he wrote to Kasimir Fajans on behalf of a woman physics student at Cambridge, enquiring whether there was any mathematics professor at the University of Munich who would be willing to take on a female student: "One of our Physics students, Miss Taylor, who is attached to Girton College here, wishes to research in the University of Munich, on the Mathematical side, particularly Integral Equations in which she is interested and no doubt she would wish to enter for her PhD. Could you please let me know whether a young English woman would be welcome in Munich and if so which Professor would be prepared to supervise her work. She is quite a fair Physicist and has taken the Mathematical Tripos here in Cambridge as well."[66]

It is interesting that he should have chosen to write to Fajans, who was also supportive of women in science. Elizabeth Róna and Hélène Towara worked with Fajans in the early years. Róna commented[67] that she had chosen Fajans over George Bredig as a supervisor partly because at parties Bredig expected women researchers to join the wives discussing children and recipes, while Fajans treated all his researchers equally at social events as well as in the laboratory.

Frederick Soddy, one of Rutherford's main collaborators, had three women researchers: Ada Hitchens, Ruth Pirret, and his spouse, Winifred Beilby. Soddy made clear his support for women's rights in a speech that he gave on women's suffrage in London, England: "Male suffrage alone is mean. It evokes contempt ... I regard women's cause as won in the fair field of argument and reason. The actual hour alone of victory remains to be fought for."[68]

One of the more complex relationships was that of Otto Hahn and Lise Meitner. In the early days Hahn was very keen to give Meitner her due recognition. In a letter to Ernest Rutherford he commented: "As a matter of fact, we do our work like equal people and not by far like a professor and his assistant. It may have been in the first time, when Miss Meitner came from Vienna to Berlin. But since a number of years this has changed completely, and we do our work independently and she does at least as much as I do."[69] Yet Meitner's biographer, Ruth Lewin Sime, has shown conclusively that, in later years, Hahn denied Meitner recognition of her role in the discovery of nuclear fission, commenting that Hahn's memory was "selective and self-serving."[70]

Hahn, in his autobiography, notes that Meitner was the victim of discrimination in the early years of her work.[71] Emil Fischer, head of the Chemical Institute of the University of Berlin, expressly forbade Meitner to enter the research laboratories where the male students worked. Instead, a wood shop was adapted for her needs. However, this exclusion was no impediment for Meitner as she was extremely strong-willed and determined to become a physicist. This is illustrated by Meitner's reply to a letter from a young science student, Miss Patricia Alison, who asked Meitner why she had taken up her subject of study. Meitner replied: "May be I could add that I never was influenced in my decisions by any question of "career" or personal success. I only was obsessed by my interest in physics and this determined my way."[72]

J.J. Thomson had four women students as researchers during the period of our study: Harriet Brooks, Fanny Cook Gates, Elizabeth Laird, and Jesse Mabel Wilkins Slater. All told, at least fourteen women researchers worked at the Cavendish between 1871 and 1910.[73] Thomson's correspondence suggests that he had a tolerant if patronizing attitude towards women in physics. In 1886 Thomson wrote to a friend, Mrs H.F. Reid, "I think you would be amused if you were here to see my lectures – in my elementary one I have a front row entirely consisting of young women and they take notes in the most praiseworthy and painstaking fashion, but the most extraordinary thing is that I have got one in my advanced lecture."[74]

Of all the senior researchers in early radioactivity, only Bertram Boltwood was openly hostile to women. Boltwood clearly believed that women scientists were little more than potential marriage partners. When Ellen Gleditsch wrote about coming to the United States to work with him, Boltwood wrote to Rutherford that while he was hoping to "ward her [Gleditsch] off," Rutherford should tell his wife "that a silver fruit dish will make a nice wedding present!!!"[75] Mary Rutherford, who had a very traditional view of women's roles, replied: "Are you engaged to the charmer yet, I forget who she was … Wasn't she going to work with you?"[76]

Women scientists from across Europe flocked to work with the famous "Madame Curie" during the early years of atomic discovery. Among those represented in this book are Harriet Brooks, Catherine Chamié, Alicja Dorabialska, Ellen Gleditsch, Irén Götz, May Sybil Leslie, Stefania Maracineanu, Eva Ramstedt, Elizabeth Róna, and Jadwiga Szmidt. There were many others, such as Lucie Blanquies, but unfortunately they left few records except their publications.[77] Marie Curie provided, and still provides, a valuable role model for women interested in science.[78] By all accounts Curie treated the women researchers well, though she did not actively promote the cause of women in science. As one of her biographers, Rosalynd Pflaum, commented, "nothing she did was consciously done to open the doors for women who would follow her example."[79] Of particular relevance, Harriet Brooks and Jadwiga Szmidt worked with both Curie and Rutherford, yet it was to Rutherford that they turned for support and encouragement.

It is unfortunate from a historical perspective that Curie and her biographers[80] ignored the contributions of the workers of the Institut Curie, particularly women such as Chamié, who devoted her life to the institute. The popular image is of a lonely Marie Curie toiling away in a deserted laboratory, when in fact the laboratory was crowded from 1907 on and applicants often had to be turned away. The sole reference by Marie Curie to these other workers is found in her autobiographical notes, which are included in the biography of her husband: "A few scientists and students had already been admitted to work there with my husband and me. With their help [after Pierre's death], I was able to continue the course of research with good success."[81] Yet we must remember the pressures under which Curie worked. Many cynics in France denigrated her contributions, arguing that it was Pierre who had been the brilliant scientist.[82] Marie constantly had to justify her own competence. Rutherford, who is portrayed as sympathetic to Curie by Pflaum, noted this obsession in his comments on Curie's massive *Treatise on Radioactivity*: "It is very amusing in parts to read where she is very anxious to claim priority for French science, or rather for herself and her husband."[83]

As at the Curie Institute, a high proportion of the researchers at the Radium Institute of Vienna – about one-third – were women.[84] In this compilation, women who carried out research in Vienna are represented by Marietta Blau, Stefanie Horovitz, Lise Meitner, Elizaveta Karamihailova, and Elizabeth Róna. According to the list of workers at the Institute there were many more besides, but like the women at the Curie Institute, most vanished into obscurity. The high proportion of women is explained in a review of the cosmic ray research performed at Vienna,[85] where Blau emphasizes the tremendous support given by Stefan Meyer and the "harmonious atmosphere" at the Vienna Institute. In fact, every one of the Vienna women started their careers under his tutelage. Once more, it seems to have been the personality of the mentor that was important. In an obituary Paneth noted: "All those who at any time have worked in Meyer's Institute, as his assistants or as guests, remember with deep gratitude his never-failing kindness,"[86] while Lawson commented on Meyer's "personal charm and good nature, his warm friendship and his inate kindliness."[87]

THE WOMEN RESEARCHERS

As we remarked earlier, those women who received a higher education tended to come from upper-middle-class families. Blau, Chamié, and Meitner were daughters of lawyers, Karamihailova and Róna, of doctors, Curie and Gleditsch of teachers, Horovitz of an artist, Hitchins of a customs official, and Noddack of a businessman. Only Brooks, the daughter of a salesman, and Dorabialska, daughter of a post office clerk, could be said to have lower-middle-class origins.

Of the researchers that are studied in detail here, the majority never married. Brooks and Leslie abandoned their careers after marriage while Curie, Joliot-Curie, Noddack, and Szmidt married fellow atomic scientists. Blau, Chamié, Dorabialska, Gates, Gleditsch, Karamihailova, Maracineanu, Meitner, and Róna remained single. This pattern is similar to that reported by McLauglin and Gilchrist in their study of early women carcinologists.[88] As a dean at Radcliffe College remarked in the 1940s, "In my days, society expected celibacy of women scholars."[89]

The women researchers profiled in this book clearly regarded their work as the purpose of their lives. Thus, in a letter to Marie Curie, Eva Ramstedt commented: "As for me, I have been unfaithful to science, never finding time for research."[90] It was argued at the time that women who excelled academically should regard it an honour and a duty to devote their lives to knowledge as to a religious vocation, thus excluding the possibility of family life: "Civilization rests upon dedicated [women's] lives, lives which acknowledge obligation not to themselves or to other single persons, but to the community, to science, to art, to the cause."[91] Yet there was no equivalent requirement for male researchers to remain single; all five of the main male characters mentioned in this work – Fajans, Hönigschmid, Rutherford, Soddy, and Thomson – were married.

The women, for the most part, were determined travellers in search of a scientific haven. Brooks moved from Canada to England and France, Gates from the United States to Germany and Switzerland, and Gleditsch from her native Norway to France and the United States. Some travelled of necessity: Chamié fled from Russia; Róna and Götz fled from Hungary; later, Blau, Meitner, and Róna fled from Nazi-occupied Europe.

CONTACTS BETWEEN THE WOMEN SCIENTISTS

One of the important factors in any scientific field is contact among the participants. The leading figures in research into atomic science communicated frequently among themselves, exchanging information on the new discoveries of their respective research groups. This personal interaction has been labelled the "invisible college" phenomenon,[92] and it usually occurs through the research supervisors;

the pioneer women scientists in the field of radioactivity interacted even though many of them were subordinates. Their communication was both social and professional, direct and indirect. We shall discuss these contacts in the context of each research group.

Members of the Austro-German group developed very close ties. Elizabeth Róna, Elizaveta Karamihailova, Marietta Blau, and Berta Karlik all corresponded regularly with Lise Meitner[93] and mentioned other members of the group in their letters. Two of the Curie researchers, Ellen Gleditsch and Eva Ramstedt, became "honorary correspondents" of the Austro-German group as their correspondence with Meitner shows. Thus, a letter from Ramstedt to Meitner mentions both Róna and Gleditsch.[94] Such communication continued into later years. Karlik, for example, wrote to Meitner in 1953, mentioning Gleditsch: "I have received a letter from Ellen in which she mentions that she has the intention of sitting on our Commission. She travels to Stockholm at the end of July."[95]

As well as writing to each other, these women occasionally met. Among the meetings that we can document, Meitner in Berlin received visits from Ramstedt during the summer of 1920[96] and Gleditsch during the summer of 1926,[97] while Blau and Róna visited Karamihailova when the latter was in Cambridge.[98] Gleditsch and Ramstedt met from time to time and one such rendezvous was mentioned in a letter from Ramstedt to Curie: "I have heard your news from Mme Gleditsch with whom I spent several days in Brussels."[99] The friendships were of long duration. In 1965, at the age of seventy-five, Róna wrote to Kasimir Fajans, telling him of her European travels (she was living in Miami at the time): "I spent a few days with Ellen Gleditsch in Oslo ... and three days in Vienna with Berta Karlik and Marietta Blau, it was most enjoyable."[100]

One unique feature of the Austro-German group was the degree of mutual collaboration on research projects. Almost all possible permutations of joint publication occurred. Papers were produced by Karlik and Karamihailova,[101] Karamihailova and Róna,[102] Blau and Karamihailova,[103] Blau and Róna,[104] and Karlik and Róna.[105] When Blau and Róna fled from occupied Europe, they first took refuge with Gleditsch. While in Norway Róna published paper with Gleditsch,[106] while Blau published an overview of her work, concluding with the acknowledgment: "I wish to express my sincerest gratitude to Prof. Gleditsch for her kind hospitality that has made it possible for me to continue my investigation."[107]

In contrast, interaction among the women who worked with Rutherford seemed to revolve around Rutherford himself. For example, Gates asked Rutherford in a letter whether Brooks was working with him,[108] while Rutherford noted in a letter to Boltwood: "I understand from Miss Schmidt [*sic*] that Miss Gleditsch is coming to stay with her sometime next week."[109] Communiqués of this sort were a useful conduit for information; in a letter to Rutherford,[110] Szmidt noted that she had received a letter from Gleditsch on the topic of the atomic weight of radium. Gleditsch herself wrote to Rutherford that she "had a letter from Miss Leslie a little time ago, telling me about the British Association Meeting."[111]

The Curie group seemed to function differently again. There appeared to be little social contact between the researchers once they had left the Institut Curie – except for the acquaintances of Gleditsch. Gleditsch maintained a regular correspondence with Leslie and Ramstedt and the trio met at least twice in later years, on one occasion to visit Soddy at Oxford some eleven years after their time together in Paris.[112] Gleditsch must have been in contact with another of Curie's research workers, Alicja Dorabialska, since Dorabialska took on one of Gleditsch's students, Ruth Bakken. Gleditsch also exchanged letters with Catherine Chamié. Curie wrote to Gleditsch in 1924: "Mlle Chamié has told me that you are thinking of bringing a project that specially interests you."[113] And Rutherford's comments to Boltwood suggest that Gleditsch must have corresponded with Szmidt, whom she had first met one summer at the Institut Curie.

Citations are also indicators of personal contact. In one of her publications[114] Lucie Blanquies refers to some of Brooks's unpublished data. One of Slater's articles[115] refers to a paper by Gates, and one of Leslie's papers[116] cites work of Gleditsch. Both of Pirret's papers[117] cite Gleditsch, while a subsequent paper by Gleditsch[118] refers to the work of Pirret. Blau and Róna co-authored a publication on Chamié's photographic method for the study of polonium.[119]

The most overlooked person in this whole saga is Gleditsch. She was the central figure who linked all of the research groups together. She worked or corresponded with most of the key players in the early days of radioactivity and she maintained correspondence with all of the women who are discussed in depth in this work, except Ida Noddack and the two North Americans, Brooks and Gates. Of particular importance was her sheltering of Marietta Blau and Elizabeth Róna when they fled occupied Europe.

CONCLUDING REMARKS

Clearly, to enter university in the late nineteenth century and then to pursue a life in science in the first decades of the twentieth century must have been a tremendous challenge for these women. It was their devotion to science, the support of their supervisors, and, in some cases, their own networks that enabled them to survive. We have laid out the general circumstances of the time, but as the ensuing chapters will show, each woman's life included some unique obstacles and hardships as well as some significant scientific achievements.

PART TWO

The French Group

This group of researchers was led for the most part by Marie Curie (chapter 3), and it is with her life and work that we begin this section of the book. Curie was unique in that she really belonged to an earlier generation than the other women in this compilation. In the first few years of their work Marie and Pierre Curie had few collaborators, but after Pierre's death and the explosive growth of studies in radioactivity, the Paris research group grew. The first woman whose work with Curie is on record was Harriet Brooks (chapter 11), but in view of Brooks's brief research stint in Paris and her protracted time with Rutherford, we have treated her as a member of the British group. The Norwegian chemist Ellen Gleditsch (chapter 4) was the first woman to work with Curie for a significant period, arriving in Paris in 1907 for the first of her many sojourns there. French scientist Lucie Blanquies (mentioned in chapter 10) was the next to join the group (1908–10) but we know little about her. May Sybil Leslie (chapter 5), a British chemist, worked at the Institut Curie from 1909 to 1911 before returning to England to work for a year with Rutherford. As we know much more about her time with Curie than with Rutherford, her account is included in this section. The next arrival was Eva Ramstedt (chapter 10) from Sweden, who worked with the group in 1910–11. Gleditsch, Leslie, and Ramstedt became close friends and had at least two reunions over the following decades. The Russian physicist Jadwiga Szmidt (chapter 13) spent the summer of 1911 working with Curie, but it is her stay with Rutherford about which we have more information, hence she, like

Brooks, is included in the British group. Szmidt was followed by Hungarian scientist Irén Götz (chapter 10), who spent the years 1911 to 1913 in Paris. Of the many women who worked with the Paris group in the 1920s we have included full-length biographies of five: Catherine Chamié (chapter 6), who arrived in Paris from Russia as a refugee and who stayed at the institute from 1921 until her death in 1950; Stefania Maracineanu (chapter 7), who travelled from Romania to work in Paris from 1922 to 1925; Alicja Dorabialska (chapter 8) from Poland, who worked in Paris from 1925 to 1926; Curie's daughter, Irène Joliot-Curie (chapter 9), who was to inherit the Curie "empire"; and Elizabeth Róna (chapter 20), a member of the Austro-German group (hence included in that section) who worked in 1926 with Curie.

3 Marie Curie: Time Only for Science and Family

HELENA M. PYCIOR

Biography seeks not merely to recount the details of a subject's life, but also to grasp the thematic patterns that governed that life. Although by no means a complete biography of Marie Sklodowska Curie,[1] this essay seeks to analyze the thematic patterns most closely associated with her successful combination of the roles of scientist, wife, and mother, and then widow and single parent. The essay is divided into three main chronological parts. The first, covering Curie's early years, emphasizes the importance of her parents as role models, as well as the development of a close relationship, built on shared dreams and mutual support, with her sister Bronia. The next two sections – the core of the essay – are devoted to the scientific collaboration and family life of Marie and Pierre Curie, and finally Marie Curie's twenty-eight years as widow, single parent, and independent woman scientist. The overriding theme of these later periods is the Curies' "anti-natural path" – a simple way of life that allocated the couple time only for science and family. The essay also shows that Curie's marital status and family arrangements were key elements of the sociocultural matrix in which she practiced science.

Marie Curie was born on 7 November 1867, in Warsaw, Poland, which was then under the control of the emperor of Russia. She was the last of five children of Bronislawa (née Boguska) and Wladyslaw Sklodowski, members of the Polish urban intelligentsia with roots in the small landed gentry. Her father taught physics at a gymnasium (secondary school) for boys in Warsaw; and until a few months after Curie's birth, her mother was principal of a private girls' boarding school. Curie was actually born in the boarding school, at the time not only her mother's workplace, but also home for the Sklodowski family.

Despite her auspicious birth to loving and intellectual parents, Curie experienced neither a carefree childhood nor an easy young adulthood. In 1876 she lost a sister to typhus and, at the age of nine, her mother to tuberculosis. "This [latter] catastrophe," Curie wrote late in life, "... threw me into a profound depression."[2] Added to her personal loss, the young girl lived in an atmosphere of political intimidation and oppression. Russian domination of central Poland restricted the professional life and income of Wladyslaw Sklodowski and tainted his children's lives with distrust, hypocrisy, and hatred. These strong emotions intruded even into the private Polish schools of the children's early years and the Russian gymnasia of their adolescence. Since the private schools routinely defied the Russian proscriptions against teaching in the Polish language and about Polish culture, periodic visits by school inspectors became harrowing occasions for conspiracy between teachers and students. While in private school, for example, the precocious Curie was sometimes called upon to mask her feelings and discuss Russian culture in the Russian language with an inspector. Later she enrolled in an imperial gymnasium, the only path to a recognized diploma. Although she graduated first in her class, Curie as an adult remembered most of the gymnasium's teachers as "Russian professors, who, being hostile to the Polish nation, treated their pupils as enemies."[3]

Even with her nationalist objections to the gymnasium, the young Curie wrote a friend: "In spite of everything I like the school ... and even love it."[4] This emotion was a testament to her strong academic bent and also a tribute to her parents' intellectual interests and egalitarian views. Her parents took their daughters' educational needs as seriously as their son's. Following his wife's death, her father "devoted himself entirely to his work and to the care of ... [his children's] education."[5] He infected all his children with his love of science and literature.

The love of learning instilled by his parents led Curie's brother, the leading student of his class at an imperial gymnasium, to the study of medicine at the University of Warsaw. Despite first rankings in their respective classes, Curie and her sister Bronia were unable to follow their brother to the university. The university was closed to women; higher educational opportunities for Polish women were to be found only in foreign universities. But foreign study involved expenses Wladyslaw Sklodowski was unable to afford, despite his sincere support of his daughters' scholarly aspirations.

Abandoning hope of any immediate formal higher education, Curie found employment as a private tutor in Warsaw. This was an important, formative period in her life, during which she and Bronia experimented with Polish positivism and cemented their close relationship. From contact with the positivists, Curie developed an abiding belief that one "cannot hope to build a better world without improving the individuals. To that end each of us must work for his own improvement, and at the same time share a general responsibility for all humanity."[6] This ideology of education as the chief means of social progress cut across gender and class lines; Curie's own voluntary activities of the period included reading to dressmakers and collecting a library of Polish books for other women workers.

But how were the Sklodowska sisters to effect their "own improvement," while Curie was contributing to the family income through her private lessons and Bronia was running the family household? Bronia confided to her sister her ambition to study medicine at the Sorbonne; Curie, in turn, expressed a hope for advanced work in science, sociology, or literature – disciplines given special recognition by the Polish positivist movement. The sisters shared their frustrations as the passing months seemed to bring them no closer to these goals. Finally, Curie evolved a plan: she would enter service as a governess, the only barely lucrative position open to a woman of her social and educational background, and use her earnings to support Bronia's five years of study at the Sorbonne; upon receipt of her medical degree, Bronia would help finance Curie's work at the same university.

From 1885 through 1890, then, Curie worked as a governess for three different families, devoting over three years to the Zorawski family, who lived about 60 miles north of Warsaw. The normal conditions of the life of a governess weighed heavily on Curie. She taught the family's children from seven to nine hours a day. Moreover, as a governess she was obliged to offer social companionship at the Zorawskis' convenience. Thus she often interrupted her evenings of study to make a fourth player at the family card table, play checkers, attend dances, and converse with the family and their friends. Despite such efforts at maintaining the proper decorum of the governess, Curie committed a serious transgression of the position when she fell in love with and agreed to marry the family's eldest son, Casimir. Somewhat naively, Curie had failed to grasp the vast social gap his parents envisioned between them and their governess. Thus she seems to have been surprised by the Zorawskis' violent opposition to the proposed marriage, as well as by Casimir's ready acquiescence to his parents' wishes.

Although she had been rejected as a potential daughter-in-law, Curie's obligation to Bronia kept her with the Zorawskis for two more years. Throughout this awkward period, she hid her sagging spirits behind a brave face. "As for me," she confessed in 1888, "I am very gay – and often I hide my deep lack of gaiety under laughter. This is something I learned to do when I found out that creatures who feel as keenly as I do, and are unable to change this characteristic of their nature, have to dissimulate it at least as much as possible."[7] Other letters reveal that during this period Curie's "lack of gaiety" ran deeply indeed. She was isolated from her family in Warsaw, barely able to afford the postage stamps by which she maintained minimal contact with them, abandoned by her first love, and pushed to the limits of her endurance by the demands of her position. She was led even to question the value of her own life: "My plans for the future? I have none, or rather they are so commonplace and simple that they are not worth talking about. I mean to get through as well as I can, and when I can do no more, say farewell to this base world. The loss will be small, and regret for me will be short – as short as for so many others."[8]

Yet the position as governess did not destroy completely Curie's self-confidence and intellectual drive. Throughout her stay with the Zorawskis, she continued to prepare for formal higher studies, no matter how elusive they came to seem. Even the letter that contained the dark hints of the worthlessness of her life offered a list of textbooks she was studying, with the note: "When I find myself quite unable to read with profit, I work problems of algebra or trigonometry, which allow no lapses of attention and get me back into the right road."[9]

By 1888 Bronia no longer required Curie's financial support. Intent on assisting his daughters' higher studies, Wladyslaw Sklodowski retired with a pension from the gymnasium and took a difficult position as director of a reform school. Now supported by her father, Bronia urged her younger sister to begin to save for her own higher studies. By March 1890 the two sisters were seriously discussing the possibility of Curie joining Bronia at the Sorbonne. Here Bronia, in one of the many crucial roles she played in her sister's life, took the initiative: "You must make something of your life sometime. ... You must take this decision; you have been waiting too long."[10] But Curie hesitated, claiming that, if not stupid, she was unlucky; that she had long ago abandoned the dream of Paris; and finally that familial obligations to her father and another sister, Helena, required that she remain in Warsaw.[11]

During the next year and a half, however, events intervened that cleared the way for Curie's move. Living at home with her father, she rejoined Warsaw's underground "floating university," which now offered her use of a meager laboratory at the Museum of Industry and Commerce. Here, carrying out experiments described in physics and chemistry textbooks, she developed her "taste for experimental research,"[12] a taste she knew could be satisfied only abroad. Also, two obstacles to her escape to Paris were removed: seeing Casimir Zorawski for one last time, she abandoned any remaining hopes of their marriage; and she accumulated the sum required for the trip to Paris and a year's tuition at the Sorbonne.

Fall of 1891, the beginning of the second major stage of Curie's life, found her, at the age of twenty-four, enrolled at the Sorbonne. Initially, the supportive Bronia gave her sister room and board in the small apartment she shared with her husband and new baby. Then Curie, wanting freedom from all distractions, moved into a single room of her own. She could barely afford to pay the monthly rent, had little money or time for food, and sometimes went without heat to conserve her meager resources. Yet she thrived on her solitary life as a student in Paris, and passed the *licence* in physics (comparable to receiving a bachelor of science degree) in 1893 as the first in her class.

Infected by Parisian science, Curie determined to pursue a degree in mathematics. Again there were obstacles to overcome: concern for her father, whom she thought needed her companionship, and lack of money. A careful weighing of her own needs against those of her father resolved the first issue. Her father seemed happy and healthy living near her brother; and as she at this point wrote of her own needs: "It is my whole life that is at stake. It seemed to me, therefore, that I could stay on here [Paris] without having remorse on my conscience."[13] Curie's financial problems were temporarily resolved as a Polish woman mathematician helped her secure an Alexandrovitch scholarship for the next academic year, 1893–94.

The year 1894 was among the most eventful of Curie's life. Not only did she pass the *licence* in mathematics, ranking second among all candidates, but she also met Pierre Curie. At the time of their meeting, Pierre Curie was thirty-four years old and professor at the École de Physique et Chimie in Paris. The son of a physician who had turned his apartment into a hospital for the wounded of the Paris Commune of 1871, Pierre Curie had been raised as a freethinker and educated at home through about the age of sixteen. Following receipt of the *licence* in physical sciences from the Sorbonne in 1877, he had built a solid reputation as one of France's leading young physicists. This early reputation was based on research into the physics of crystals, which was conducted in collaboration with his brother, Jacques, in the late 1870s and early 1880s and resulted in the discovery of piezoelectricity; and independent work on crystallography and magnetism, initiated in 1883 when Jacques left Paris for a professorship at Montpellier.[14]

By 1894, when Marie first approached Pierre (with a request for space in his laboratory so she could conduct the magnetic research that the Society for the Encouragement of National Industry had hired her to do), she and Pierre seem to have decided upon single lives for themselves. Marie planned to return to Warsaw, comfort her father in his advanced years, and fulfil a positivistic mission of teaching science to further the Polish cause. Pierre Curie had concluded that a conventional marriage was not for him. Through 1894, in fact, Pierre seems to have clung to his earlier intense collaboration with Jacques – when the brothers had "lived entirely together," sharing scientific research and recreation and even "arriv[ing] ... at the same opinions about all things, with the result that it was no longer necessary ... to speak in order to understand each other"[15] – as the ideal of a life devoted to science. As Pierre's surviving notes indicate, he was not hopeful of finding a wife who would tolerate his complete absorption in science, let alone participate in it. Thus, prior to meeting Marie, he had written that "women of genius are rare," and speculated that marriage to an ordinary woman would place limits on his anti-natural path of almost complete devotion to science.[16]

After a courtship of over a year, Marie and Pierre Curie married and jointly embraced the "anti-natural path," which permitted the couple time only for science and for their extended family.[17] Traditional to the extent that Marie assumed responsibility for the domestic scene,[18] their marriage was unconventional in many other respects. Marie Curie, who never fussed about her wardrobe, "wore no unusual dress" for their simple civil wedding[19] – no matter to the groom, who later in life amused himself at a banquet by calculating the number of scientific laboratories that could be financed through sale of the jewels adorning the female guests.[20] The newly married couple even refused some furniture offered by the Curie family, as so many more objects to dust. "I am arranging my flat little by little," Marie wrote her brother in 1895, "but I intend to keep it to a style which will give me no worries and will not require attention, as I have very little help: a woman who comes for an hour a day to wash the dishes and do the heavy work."[21]

The decision to concentrate on the essentials of life – with sparse wardrobe and furniture, almost no social ties beyond their extended families and select Parisian scientists, and occasional bike rides into the countryside – fitted the couple's budget and Pierre Curie's simple, eccentric ways. It also took into account Marie Curie's perfectionist streak. What Marie Curie did, she did well: in her early days of marriage, when the couple could afford only limited domestic help, she painstakingly learned to cook;[22] certainly in her laboratory,[23] and probably in her home as well, she enforced high standards of tidiness. Thus a traditional marriage could easily have drained all her energies; instead the Curies calculatedly pared their family life down to the essentials, thus freeing Marie Curie for a scientific career.

The arrival of their first daughter, Irène, in 1897 tested the couple's commitment to the anti-natural path. Around the turn of the century, if marriage did not end a woman's career, motherhood almost certainly did. But a combination of fortuitous conditions, including her husband's moral support, again saved Marie Curie for science. "It became a serious problem how to take care of our little Iréne and of our home without giving up my scientific work. Such a renunciation would have been very painful to me, and my husband would not even think of it; he used to say that he had got a wife made expressly for him to share all his preoccupations. Neither of us would contemplate abandoning what was so precious to both."[24]

Equally fortunate was the availability of child care services. After trying unsuccessfully to nurse Irène, Curie embraced the French custom of turning care of her infant over to a nurse. From that point on, the couple hired a series of nurses and domestic servants to assist with the household chores and to watch Irène and eventually her younger sister, Eve. But even hired help was not the major factor facilitating the Curies' successful (and guiltless) combination of parenthood and first-rate scientific careers. Marie Curie's own explanation emphasized, rather, the willingness of Dr. Eugène Curie, Pierre's father, to assume responsibility for the two Curie girls. Following his wife's death (a few weeks after Irène's birth), Dr. Curie joined his son's household. "While I was in the laboratory," his daughter-in-law later admitted, "she [Irène] was in the care of her grandfather, who loved her tenderly and whose own life was made brighter by her. So the close union of our family enabled me to meet my obligations."[25]

Curie was now free to devote weekdays and some evenings and weekends to study and research. The director of the École de Physique et Chimie, where Pierre Curie was a professor, permitted her to work in the school's laboratory. There, after a year of research on magnetization, she turned to radioactivity, which became not only the subject of her doctoral thesis, but the focus of her research for the rest of her life. Her selection in late 1897 of radioactivity as a research topic – apparently made independently of Pierre – was possibly a stroke of intuitive genius and certainly a brave gamble. Radioactivity had been discovered only the year before, when Henri Becquerel noted that a compound of uranium, put on a photographic plate wrapped in black paper, left an image on the plate. The excitement surrounding Becquerel's discovery aside, "few scientists [of the late nineteenth century] thought radioactivity had much of a future."[26]

Curie soon formulated her own distinctive approach to radioactivity. First she developed a technique for measuring the intensity of radiation (actually measuring the conductivity of the affected air with an apparatus developed earlier by Pierre and Jacques Curie), which she applied to many different compounds. At this early stage, she discovered that thorium was also radioactive,[27] and formulated the hypothesis that radioactivity was an atomic property. She now latched onto this hypothesis – and relentlessly pursued its chemical consequences.

As an atomic property, she reasoned, radioactivity was probably not restricted to two elements. Borrowing materials from various French scientists, she measured the radiation emitted by all standard metals and non-metals, rare elements, and an assortment of rocks and minerals. These measurements produced an anomaly: while only compounds containing uranium or thorium proved radioactive, some of the uranium compounds proved more radioactive than pure uranium. For example, the intensity of the radiation from pitchblende (an ore of uranium oxide) was four times that from metallic uranium. Here Curie made a bold hypothesis: there was a new (undiscovered) element in the pitchblende.[28]

At this exciting point, Pierre Curie joined his wife in the study of radioactivity, a multidisciplinary phenomenon that the couple soon realized involved chemistry, physics, and medicine. Theirs was a fruitful collaboration, in which Pierre served primarily as the physicist, and Marie, although trained especially in physics and mathematics, as the chemist. Together they discovered polonium (named by Marie Curie in honor of her homeland), which proved to be one of three new elements contained in pitchblende. Working with Gustave Bémont, they next discovered radium, while a co-worker, the chemist André Debierne, found actinium.

Having established the existence of radium and polonium by somewhat indirect means, the couple then turned to preparation of radium as a pure salt. This was an exhausting task, involving physical and chemical techniques, whose practical execution seems to have fallen primarily to Marie Curie. As a first step, radium-bearing barium was extracted from pitchblende residue. A ton of residue gave 10 to 20 kilograms of crude barium sulfate. The sulfates were then purified and converted into chlorides. Next there followed fractional crystallization, a complex chemical process designed to eliminate some of the barium from the chlorides. Performed several thousand times, fractional crystallization finally left pure radium chloride. As a result, by 1902 the Curies had established on solid chemical grounds the existence of radium as an element by preparing it in a pure form and calculating its atomic weight.[29]

In this early period of radioactivity research, the Curies did far more than add new elements to the periodic table. Their work "was instrumental in opening up most fruitful avenues of research in a previously untrodden field."[30] Marie Curie herself coined the field's name, "radioactivity." Also, the Curies' astonishing results with radium helped legitimate the field, especially in the eyes of chemists. The honors showered upon the couple, including the Nobel Prize for physics, shared with Becquerel in 1903, were so many victories for the new field. They were also victories for the cause of women in science. In fact, both Pierre and Marie Curie took seemingly calculated steps to ensure that the male-dominated scientific community of the period did not ignore Marie's part in their research. Here her husband's experience as a scientific collaborator, as well as his honest and modest nature, which had been fostered by his early liberal education, proved important. But perhaps more crucial was Curie's self-confidence, which spurred her to publish independently the results of work for which she alone deserved credit and, in her accounts of collaborative efforts, to record carefully those experimental results and ideas that belonged primarily to her. Thus in 1898 she alone published a note containing the announcement of the discovery of the radioactivity of thorium, as well as the bold hypothesis that pitchblende contained a new element.[31] Marie Curie's subsequent accounts of radioactivity never failed to mention these independent contributions, just as they did not fail to note separately the research on radioactivity her husband conducted alone or with collaborators other than herself.[32]

Although concerned about establishing Marie Curie's scientific reputation, the couple did not blatantly challenge the many early twentieth-century scientific and academic mores that favored Pierre over Marie Curie. Thus in 1903 the Royal Institution invited Pierre Curie to lecture on radium; Marie Curie accompanied her husband to London, but merely sat in the audience as he spoke on their joint project. Furthermore, even after receipt of the Nobel Prize, the couple neither expected nor asked that Marie Curie receive such traditional fruits of scientific success as a major professorship and her own laboratory. Instead, the couple concerned themselves with Pierre Curie's academic career. The same honest and modest personality, combined with disdain for convention, that permitted Pierre Curie to collaborate with his wife as an equal also blocked his easy advancement in the French academic world. His professorship at the École de Physique et Chimie afforded him neither prestige, for which he did not care, nor a well-equipped laboratory, which both he and his wife ardently desired. An offer of a professorship to Pierre Curie, and a position in his laboratory for Marie, made in 1900 by the University of Geneva, resulted in new French opportunities for both Curies: for Pierre, an assistant professorship at the Sorbonne; and for Marie, a lectureship at the École Normale Supérieure (for women) at Sèvres. Finally, in 1906, as Marie Curie later emphasized, "because of the awarding of the Nobel prize and the general public recognition, a new chair of physics was created in the Sorbonne, and ... [Pierre] was named as its occupant."[33] Marie Curie, in turn, was to become director (*chef de travaux*) of the laboratory the university promised to create for her husband. Pierre Curie's tenure as a full professor, however, lasted just a few months: in April of the year of his appointment, a horse-drawn cart killed him as he tried to cross a Parisian street.

Pierre Curie's death affected Marie Curie both personally and professionally. In a matter of seconds she went from wife to widow and single parent. The sorrow of Curie was profound. But she chose – as in her earlier school recitations for Russian inspectors and her governess years – to mask her feelings once again. She now assumed an impassive facade, hoping to shield her daughters from her "inner grief, which they were too young to realize."[34]

With her husband's death, Curie also went from female scientific collaborator to the perhaps more difficult position of independent woman scientist. Men scientists could no longer comfortably view her as a scientific wife sharing her husband's laboratory, research, and honors. She was now a single woman scientist, whose Nobel Prize and research entitled her to independent roles in the French academic community and the international community of science. Curie herself, as well as French academics and male scientists, had to adjust to her new position.

The Sorbonne was the first institution forced to deal with the anomalous widow. What was to be her university position in the absence of her husband? This question was one of scientific life or death for Curie, since continuation of her research required a laboratory and therefore some sort of academic association. The French government, however, initially viewed her situation from a different perspective. Treating her as the widow of a great man rather than a scientist in her own right, they offered her a widow's pension. Declining the pension, Curie stated her desire to continue to work. Then staunch supporters – including Jacques Curie, Paul Appell (once her professor and now dean of the Faculty of Science at the

Sorbonne), and Georges Gouy (a close friend of Pierre) – put Marie Curie forward as the French scientist best able to continue her husband's research. Within a few weeks of Pierre's death, the Sorbonne appointed Marie Curie its first woman professor. Years later, attributing her appointment to the emotion surrounding her husband's death, Curie candidly acknowledged: "The University by doing this ... gave me opportunity to pursue the researches which otherwise might have had to be abandoned."[35]

She immediately threw herself back into scientific research and teaching, emphasizing all the time that she was carrying out the dream shared with her husband of a life devoted to science. Her major research following Pierre Curie's death – much of it done in collaboration with André Debierne[36] – centered on completing the proofs that radium and polonium were true chemical elements. She turned to this work because of Lord Kelvin's doubts about the elemental nature of radium, published in a letter of 1906 to the London *Times*, and her own earlier doubts concerning polonium, the salt of which (unlike radium) had never been isolated. She again performed the complex and tedious extraction processes on pitchblende. She produced "perfectly pure" radium salt and, with Debierne, polonium salt as well. Then she and Debierne went a step further. In 1910, in a paper written jointly for the *Comptes rendus*, a weekly publication of the Académie des Sciences, they announced that they had clinched the proof of radium's elemental nature by preparing radium as a pure metal.[37]

The accomplishments of her early widowhood were crowned by a Nobel Prize in chemistry, awarded in 1911 to Curie alone for her long-standing work on radium and polonium. This unprecedented second Nobel Prize was a testament not only to her scientific genius, but also to her success at beginning to carve her own niche as an independent woman scientist in the male-dominated scientific community of her period. Of course, even in her early widowhood, Curie enjoyed advantages unknown to any other contemporary woman scientist. At the time of Pierre's death she was already a Nobel laureate. Moreover, besides permitting her to continue her research, her Sorbonne professorship established her as France's official expert in the field of radioactivity. As a result, the scientific community could not easily ignore her.

Still, there were undercurrents of tension and paternalism as male scientists of international stature were compelled to deal directly with Marie Curie without the intermediary of a husband or other male collaborators. For example, prior to the International Congress of Radiology and Electricity of 1910, at which Curie and Debierne were

France's major representatives, Ernest Rutherford (1871–1937) and Otto Hann (1879–1968) "wondered what Madame Curie would request, and they prepared, with their friends, to soften her by a little flattery, in that they all agreed that whatever new unit of radioactivity should be set up it would be called a 'Curie'."[38] But flattery did not prevent Curie from insisting on a definition of the new unit (named in honor of her late husband) that differed from that established at an early session of the congress. Feeling ill, Curie left the session before its end and, later that evening, in the privacy of her hotel room, prepared a statement on the issue. The statement, delivered to Rutherford, declared that she "desired" a change in the propositions adopted at the session and, in particular, acceptance of her definition of the "curie."[39] "She had her way," Robert Reid has noted, "her bald dictatorial statement was accepted. But she created her critics, some of whom were becoming less tolerant of her mistressly attitudes."[40]

The claim of Curie's "mistressly" behavior seems to rest primarily on her failure to use normal conference procedures to argue her position on the new unit. The decisions of the radiology congress, however, were not generally reached through public debate. Furthermore, the nature of Curie's participation in the congress differed from that of her male peers even before her retreat to her hotel room. As David Wilson's biography of Rutherford shows, "The Congress was called to confirm what the inner circle of experts had already decided,"[41] and this circle – consisting at least of Rutherford, Hahn, Bertram Boltwood (1870–1927), and Stefan Meyer (1872–1949) – had decided to handle Curie with flattery rather than frank and open discussion. Denied an equal voice in the inner circle and shunning public debate, Curie simply defined her own mode of expression at the congress – direct communication by letter with the inner circle.

The strained relationship with Curie was symptomatic of the general inability of male scientists to adjust quickly to independent women scientists. For example, in 1913 Boltwood raised a fuss over the arrival of a woman – Ellen Gleditsch, one of the first of many women scientists trained by Marie Curie[42] – to work in his laboratory at Yale. Writing that he had tried unsuccessfully to dissuade Gleditsch from coming to Yale, Boltwood added in a letter to Rutherford: "Tell Mrs. Rutherford that a silver fruit dish will make a very nice wedding present!!!" Mary Rutherford's response was equal to the situation: "Are you engaged to the charmer yet, I forget who she was?"[43]

Even men generally regarded as supportive of women scientists – including Curie's peers, Rutherford[44] and Hahn – related differently to female and male colleagues. Hahn collaborated on radioactivity with Lise Meitner (1878–1968) for over thirty years. They published results jointly, and Hahn sometimes took additional, informal steps to assure that Meitner's contributions were not overlooked. In 1912, for example, Hahn wrote to a reluctant "Rutherford begging that due credit should be given to his collaborators von Baeyer and Lise Meitner."[45] Yet Meitner did not participate fully in all the special occasions of Hahn's laboratory. For example, Rutherford's visit of 1908 to the laboratory started inauspiciously for Meitner. As Meitner recalled years later: "When he [Rutherford] saw me, he said in great astonishment: 'Oh, I thought you were a man!'"[46] Her gender exposed, Meitner then fell at least partially into the role of hostess for Mary Rutherford. "While Mrs Rutherford did her Christmas shopping, sometimes accompanied by Lise Meitner," Hahn recounted, "Rutherford and I had long talks."[47] Following the visit, Rutherford wrote to Boltwood about Meitner – but he misspelled her name and, describing her as "a young lady but not beautiful," noted that she was no threat to Hahn's bachelorhood.[48]

Given these patterns of gender differentiation evidenced by key male scientists of the radiology congress, it is not surprising that tension between Curie and the inner circle persisted at least through 1912. In his account of the congress, Rutherford stated that Curie had agreed to prepare the international radium standard.[49] But after Stefan Meyer and his Austrian colleagues prepared a radium standard by late 1911, Rutherford, Boltwood, and Meyer plotted the best strategy for assuring that Meyer's standard was accepted along with Curie's. In a letter to Boltwood, Rutherford explained that Curie was being "very obstinate" about her standard and predicted that Meyer's standard would make Curie "far more reasonable." Rutherford then proposed an informal meeting to compare the two standards. But making an issue of Curie's gender, he warned Boltwood: "It is going to be a ticklish business to get the matter arranged satisfactorily, as Mme. Curie is rather a difficult person to deal with. She has the advantages and at the same time the disadvantages of being a woman."[50] The potentially awkward confrontation between Curie and the inner circle of radioactivity, to which this correspondence seems to have been leading, did not materialize. The inner circle arranged to have the two standards compared in April 1912 while Curie was recovering from surgery. Debierne represented Curie in the affair, after which Rutherford noted: "I think we perhaps got through matters very much quicker without Mme. Curie, for you know that she is inclined to raise difficulties."[51]

Actually, the international radium standard was one of the least of Curie's worries during the academic year 1911–12. She was now already beginning to suffer serious side effects from her fourteen years of research with highly radioactive material, and she had personal problems. With the passing years, responsibilities as a single parent weighed more heavily on her. Honoring the wishes of Eugène Curie (who remained with her following his son's death), Curie had moved her extended family to the Parisian suburbs, and she spent at least an hour a day commuting back and forth to her laboratory. Concern for her daughters' education also led her to design a special elementary school, run for two years by herself and other Sorbonne professors. Then, at the beginning of 1910, she lost her father-in-law. His death was painful and costly in time as she "passed all her free moments at the bedside of ... [the] sick man who was both difficult and impatient."[52] The death left Curie without a special caretaker for her two daughters, then aged twelve and five; although employing Polish governesses (the first of whom had been a distant relative) for the girls, she now sometimes had to take time away from her outside work to meet their needs. As late as 1919, for example, she admitted to Rutherford that a serious illness afflicting Eve had kept her away from her laboratory and caused her delay in responding to one of his earlier requests.[53]

In addition to the death of her father-in-law, Curie endured a scandal concerning her relationship with Paul Langevin (1872–1946). Langevin was a leading French physicist who had studied under Pierre Curie and, as one of his many scientific projects, continued Curie's work on magnetism.[54] He and Marie Curie had taught together at Sèvres, and both before and after Pierre's death he was a select member of the small social group with which Marie Curie's anti-natural path permitted her contact. Many factors promoted intellectual sympathy, possibly leading to physical intimacy, between the two scientists. Langevin, however, was a married man with four young children – and a wife who supplied Parisian newspapers with letters supposedly exchanged between her husband and Curie. The letters, whose authenticity seems never to have been completely verified, implied that the two scientists shared an apartment and showed that Curie had urged the unhappily married Langevin to force his wife to accept a divorce.[55]

The Curie-Langevin scandal, as developed in the international press, impugned not only Curie's personal morals, but her scientific reputation as well. Writing in *World Today*, Henry Smith Williams used the Curie-Langevin relationship to resurrect the question of whether or not a woman was capable of independent, creative scientific research. He noted that "discriminating critics" had always believed major credit for the Curies' work on radioactivity was due Pierre, but that these same critics had been confused in 1910 when the widowed Marie Curie announced the isolation of the metal radium. Thus it had seemed for a while, Williams admitted, that Curie's "work could not be impugned as shining in the light of any man's reflected glory." But, he continued, the Curie-Langevin correspondence – with "passages that seem pretty clearly to suggest an intellectual dominance of the man over his woman colleague, such as scholars of the first rank are not supposed to accept" – reopened the question of the scientific talents of Marie Curie and women in general. Although declining to take a definite stand on the question, Williams related the opinions of "one of the most famous of European men of science" that Marie Curie was "an average plodder" and that, in general, a woman could excel in science only while "working under guidance and inspiration of a profoundly imaginative man ... [and while] she was in love with that man."[56] The article thus pointed to the conclusion that Curie's scientific accomplishments were to be attributed to the love and inspiration of Pierre Curie, and later Langevin.

A much more subtly constructed account of the scandal, leading by innuendo to a similar conclusion, appeared in the *New York Times*. Ignoring the fact that the widowed Curie had collaborated primarily with Debierne, the *Times* article described Langevin as Curie's "co-worker in scientific research" and referred to their "constant close association in their scientific researches." The article used extracts from the Curie-Langevin correspondence that had appeared in Parisian newspapers not only to elaborate on such personal details as the couple's apartment, but also to explore the nature of their intellectual sympathy. Here Curie was subtly portrayed as Langevin's subordinate, even in the area of radioactivity:

> The bond of interest Prof. Langevin and Mme. Curie found in their scientific work is shown in another letter cited. "I have been thinking of your letter," writes Prof. Langevin, "and will tell you what to say on the consequences of the discovery of radium." ... Mme. Marie Sklowdowska [*sic*] Curie is the most prominent woman to-day in the scientific world. To her has always been given the joint credit with Prof. Curie of the great discovery of thirteen years ago.[57]

Curie's self-confidence and talent for masking her feelings, combined with the assistance of two key women, helped her survive the scandal. Just a month after news of the correspondence with Langevin broke, she claimed her Nobel Prize in chemistry. At her side was her faithful sister Bronia, who periodically returned from Poland (where she had lived since 1898) to support her younger sister through difficult times, including Eve's birth, Pierre's death, and now the scandal. The Nobel Prize and Marie Curie's acceptance address – in which she continued her practice of delineating her independent contributions to the study of radioactivity[58] – at least partially neutralized the adverse professional effects of the scandal. But still the scandal exacted a heavy personal toll. It seems to have assured that for the rest of her life Curie had no close personal relationship with any man and kept an impassive face turned towards the outside world. Combined with serious medical problems, the scandal kept her away from research for over a year and made her more dependent than ever on other women. During this period, besides accompanying her to Stockholm, Bronia saw her through an operation on her kidneys; and Hertha Ayrton, the English mathematician and physicist, gave her privacy and care in a house on the Hampshire coast.[59]

Curie resumed scientific work only in late 1912, and even then not the driven, prolific research of her early widowhood. She now took time away from research to travel abroad for honors, and at home to oversee construction of the new laboratory promised Pierre in 1905. The foreign honors and conferences, as well as her new laboratory, seemed to help her regain her self-esteem in the wake of the Langevin scandal and to solidify her position in the international community of scientists. Rutherford had first planned to bring Curie to England in early 1912 for receipt of an honorary degree as one of the "little things [that] ... help to smooth matters over."[60] Curie's illness, of course, intervened. But Rutherford extended a similar invitation for 1913, and Curie traveled to Birmingham, where she received an honorary degree from the local university and participated in the annual meeting of the British Association for the Advancement of Science, along with Rutherford, Niels Bohr, and J.W.S. Rayleigh. At Birmingham there were hints of the diplomatic skills Curie was honing for survival in the male scientific community. In late 1911 she had written to Rutherford to thank him for the kindnesses he had shown her during the first Solvay Congress, and now at Birmingham, perhaps deliberately repaying Rutherford's flattery in kind, she described him to reporters as "the one man living who promises to confer some inestimable boon on mankind as a result of the discovery of radium."[61] Following completion of her laboratory, moreover, Curie wrote to Rutherford on equal terms. Thus, in 1919, armed with increased confidence and the official stationery of the laboratory, she (unsuccessfully) suggested to Rutherford that the two of them simply decide on names for the emanations of polonium, radium, and actinium and impose the names on the field.[62]

By 1918, as World War I ended and the Laboratoire Curie finally opened, Marie Curie was over 50 years old. Serious illness, scandal, depression, the distractions of building a new laboratory, and an international war had kept her away from intensive research for about seven years. In the postwar period she resumed her scientific research, while throwing her major energies into fashioning the Laboratoire Curie into one of the world's leading research institutes. Work for the laboratory was professional, but it also had a personal element, for there were early indications that Irène Curie would follow her parents into the field of radioactivity. Whereas the younger daughter, Eve, had an artistic rather than scientific bent, Irène participated in Marie Curie's efforts to bring X-ray equipment and technicians to the French battlefields and later in 1918 joined her mother's laboratory as a *préparateur*. Indeed, until Irène's marriage in 1926 to Frédéric Joliot, she and her mother traveled the anti-natural path together, sometimes to Eve's dismay.

Neither the financially strapped postwar French government nor the French industrial and private sources that Curie resourcefully tapped were able to provide all the radioactive materials and scientific equipment desired by Curie. Quite remarkably, however, Curie's familial status now helped attract additional funds for her laboratory. In this later stage of her fundraising activities – as in so many other crucial moments of her life – Curie was assisted by a woman, this time the American journalist Marie Mattingly Meloney. The two women first met during an interview in 1920. As Curie described the deficiencies of her laboratory, Meloney quickly formulated an American fund-raising campaign for the double Nobel laureate. Curie was promised a gram of radium, worth about $100,000; and in return Curie agreed to visit the United States as a model for American women. As her published articles and surviving correspondence indicate, Meloney objected to the direction the first feminist movement had taken in the United States and especially to the willingness of some of the feminist pioneers to forego marriage and motherhood for careers. She determined to bring Curie to the United States as the perfect role model of a woman who had successfully combined career and family.[63]

When Meloney's private appeal to her friends failed to raise enough money for a gram of radium, she established the Marie Curie Radium Fund, which was charged with collecting the required sum through an appeal to all American women. At this point American women – as diverse as academic women. wives of male philanthropists, Polish-American women. women afflicted with cancer (who had a special interest in the therapeutic potential of radium), and

Girl Scouts – became Curie's benefactors, The fund raised over $150,000, and Curie made a triumphant tour of the United States during which she was hailed variously as a working mother, a female scholar and scientist, and the healer of cancer.

The warm reception accorded Curie by American women contrasted with the behind-the-scenes objections raised at Harvard and Yale to proposals for honorary degrees for her. Although Curie's visit conveyed a generally optimistic view of women in science,[64] these objections told a different story of at least a few male scientists and scholars at major universities who would deny an honorary degree to a woman scientist even though she was a double Nobel laureate. Thus, in December 1920 Charles Eliot, then president emeritus of Harvard, informed Meloney that the university's Physics Department was "not in favor of conferring such a degree" on Curie. The department included William Duane, who had worked in Curie's laboratory for six years on his way to a professorship at Harvard in 1913. According to Eliot, Duane favored awarding the degree, telling his Harvard colleagues that Marie Curie "had a large share in the details of the researches" on radium. Yet, Eliot added, she had done no work of great importance following her husband's death.[65]

These remarks infuriated Meloney. Responding to the insinuation that Pierre Curie's contributions to radioactivity were greater than his wife's, Meloney reminded Eliot that Curie's two Nobel prizes had already settled this issue in her favor. Then Meloney criticized Eliot's claim of the insignificance of Curie's activities during her widowhood.[66] This claim affected Meloney to the quick, for it overlooked not only Curie's independent scientific work, but also her responsibilities as a single mother. Meloney noted the irony of the situation: she was bringing Curie to the United States as an example of a working mother, but Eliot, who had written a series of articles extolling motherhood and large families as the highest "social and political service" women could contribute to mankind,[67] refused to give Curie any credit for the time and energy she had devoted to her family. Harvard's refusal of the honorary degree stood, and it was announced in June 1921 that the touring Curie, if forced by ill health to choose between acceptance of an honorary degree at Wellesley's commencement exercises and attendance at a special Harvard reception in her honor, would go to Wellesley.[68]

Unlike Harvard, Yale gave Curie an honorary degree – apparently at the instigation of medical doctors, however, and without prior consultation with the university scientists. When informed of Yale's intentions, Boltwood objected to the degree. He also initially refused to serve on a special committee of the American Chemical Society set up to welcome Curie, and to open his laboratory for her inspection, even though, as he reported to Rutherford, "the Madame ... [has] expressed a particular desire to visit New Haven and call on me." Later, upon discovering that his "action was likely to be misunderstood and to cause some hard feeling," he joined the committee and honored Curie's request to see his laboratory.[69]

These slights, which Curie may not have known about, mattered little as the touring scientist met thousands of admiring Americans at scientific and medical banquets, on major college and university campuses, and in small and large cities around the nation. Her self-confidence seems to have blossomed with the almost uniformly favorable public reception. Even the visit to Boltwood's laboratory went off well, with the Yale scientist admitting that he "was quite pleasantly surprised to find that she was quite keen about scientific matters and in an unusually amiable mood."[70] Boltwood's admission of surprise at finding a double Nobel laureate "keen about scientific matters" seems strange. But it is possible that the tour of 1921 was a unique occasion that forced Boltwood to deal with an especially confident Curie on equal terms. Significant reinforcement of Curie's positive self-image was, in fact, one of the lasting legacies of her American tour. As Eve Curie later explained, the tour convinced her mother that "at fifty-five [she] was something other than a student or a research worker: Marie was responsible for a new science and a new system of therapeutics. The prestige of her name was such that by a simple gesture, by the mere act of being present, she could assure the success of some project of general interest that was dear to her."[71]

Even as Curie now lent her name and time to scientific conferences, committee work, and medical causes, her life continued to center around the Laboratoire Curie. When she drained French sources, she used annual income from the American Radium Fund and her new American connections to acquire additional radioactive materials and the latest scientific equipment, essential for survival in the highly competitive field of atomic and nuclear physics of the 1920s and 1930s. Her international reputation so attracted research students and colleagues that seventeen different nations were represented at the laboratory in one year of her directorship. The combination of good apparatus, strong radioactive sources, and talented personnel resulted in a continuous string of publications from the Laboratoire Curie and at least two major discoveries of the late 1920s and early 1930s: the spectral analysis of alpha rays by Salomon Rosenblum in 1929 and, even more important, the discovery of artificial radioactivity by Irène and Frédéric Joliot-Curie in mid-January 1934, just half a year before Marie Curie's death from the effects of years of close work with highly radioactive materials.

The Joliot-Curies' discovery was a fitting culmination of Marie Curie's life, in which science and family were the most important elements. "Believe me," Curie reflected late in life, "family solidarity is after all the only good thing."[72] Curie's appreciation of family was a product of the supportive Sklodowski environment, which fostered her dreams and talents from childhood through young adulthood. Still, she waited until the age of twenty-eight before finding in the eccentric and modest Pierre Curie a husband with whom she could blend science and family. Even then her successful combination of the two necessitated sacrifice – in particular, a life that embraced the anti-natural path.

As the demands of science conditioned the Curies' family life, so Marie Curie's marital and familial status affected her scientific career. Her marriage to Pierre Curie assured her laboratory space and contact with France's leading male scientists. It also set in motion a scientific collaboration that produced a joint Nobel Prize. Her husband's death in 1906, followed by that of her father-in-law a few years later, meant additional domestic responsibilities for Curie and somewhat less time for science. Moreover, Pierre's death altered her professional status: no longer a scientific wife sharing her husband's laboratory, research, and honors, she was now forced to carve out a niche as an independent woman scientist. In a period still tainted by gender distinctions, the change in Curie's marital status from wife to widow thus forced adjustments not only in her small family unit, but also in her own professional life – at the Sorbonne and within the inner circle of radioactivity. As the Langevin scandal of 1911 showed, there was special pressure on her as a widow to prove that she was a creative scientist in her own right. Finally, the Laboratoire Curie flourished at least partially because of Curie's successful blending of science and family – which, after all, formed the basis of her invitation to the United States, where she found supplementary funding to keep the laboratory on the cutting edge of atomic and nuclear physics and thereby facilitate work such as that completed by the Joliot-Curies.

Thus, Marie Curie's very practice of science – including her early scientific opportunities, collaborative style, relationships with professional colleagues, scientific reputation, and effectiveness as director of the Laboratoire Curie – was fundamentally conditioned by her marital and familial status. In short, while the thematic patterns associated with Curie's combination of the roles of scientist, wife, mother, and then widow and single parent are not the only interesting patterns in her life, they may be the dominant ones.

4 Ellen Gleditsch: Professor and Humanist

ANNE-MARIE WEIDLER KUBANEK
and GRETE P. GRZEGOREK

Ellen Gleditsch was one of the few women in this compilation to reach the top of her field, achieving a professorship at a university. Her life was full of travel, interaction with other atomic scientists, and work for international organizations. During the Second World War she was active with the Norwegian resistance. Yet despite her contributions to radioactivity research and her major role as a Norwegian scientist and humanist, her name is almost unknown, even in her homeland.

Gleditsch was born on 29 December 1879 in Mandal, a small town on the North Sea in southern Norway.[1] She was the eldest of ten children all but the youngest having been born in rapid succession. Her mother, Petra Birgitte Hansen, was of Norwegian ancestry and the daughter of a sea captain; her father, Karl Kristian, was a teacher and the son of a Lutheran minister. The Gleditsch family had come to Norway about a century before Ellen's birth from Yugoslavia via Germany and Denmark, where her ancestor, Carl August Ludwig von Gleditsch from Sachsen-Weimar, had taken employment as a corporal in the Danish army and had married a Danish woman. The couple later moved and settled in Norway, which in 1790 was part of Denmark.

Ellen's parents had to support their growing family on a teacher's salary, but while poor the family's life was intellectually stimulating. Ellen described her mother as a very intelligent woman with a particular gift for languages: "She was quick in her work and quick with her answers."[2] With a keen power of observation, her mother taught the children to study and learn from nature, its plant and animal life and the stars above them. Ellen also acquired a life-long interest in folk music and songs from her mother's love of music. Despite her large family, Petra Gleditsch found time to involve herself in the political debates of the time and took part in the fight for women's right to vote.

Karl Gleditsch, Ellen's father, was also considered a liberal. Like many Scandinavians, he loved nature and often took his family hiking in the mountains and sailing on the North Sea around Mandal. When Ellen was eight years old, her father was appointed principal in the school district of Tromso, on an island far to the north of the Arctic Circle. The family found that, while isolated, the town had a vibrant cultural life that boasted lecture series, growing libraries, amateur theater, and musical societies.

As the eldest of ten children in a happy and caring family, Ellen learned at an early age the importance of compassion and regard for others. Typical of large families, the overworked mother would turn to Ellen for help with the younger children. These early years instilled in Gleditsch the work ethic, unselfishness, modesty, concern for others, and deep humanitarian values that would dominate her personality and stay with her throughout a long life.

Ellen attended a private coeducational school from which she graduated in 1895 at the top of her class. Although she studied Latin, German, and later, English, her main interest was in the sciences and she achieved the highest mark in mathematics. Had she been a boy she would undoubtedly have continued her studies towards the *examen artium* (matriculation exam) that would have opened the door to university studies. However, at the time the examination was reserved for boys, so it was arranged that Ellen would start working at the pharmacy in Tromso where she would have some opportunity to pursue the study of botany. On nature walks with her father she had acquired his enthusiasm for everything growing in the Norwegian countryside. The local pharmacist, himself an avid botanist, became her first mentor and encouraged her in studies of the natural sciences.

In 1897, at the age of eighteen, Ellen began working as an apprentice apothecary. In those days, pharmacy work involved a fair amount of mixing and preparing of chemicals, usually under the supervision of the head pharmacist. In this setting Ellen received her first training in chemistry, but she also studied pharmacology and in 1902 was awarded a non-academic degree with highest standing in the subject.[3]

That same year she moved to Kristiania[4] (Oslo) to continue her studies. In an interview in Paris some ten years later, she described her struggles to continue her education: "After the pharmacology exam, I was advised by Dr Bodtker to continue my studies at the university laboratory under Professor Hiortdahl. I wanted so badly to advance my education, but I had no money, and one has to survive, so I started to work as a tutor on the side. It went all right, but I did not have as much time for my studies as I had wished. It was a matter of bread first and then science. But actually, it went better than one would think. In 1903, I became an assistant at the chemical laboratory of the university, and in 1905 I passed the matriculation exam, and in 1906 the qualifying university entrance exam."[5] Eyvind Bodtker became Ellen's long-time friend and advisor, and she worked as his teaching assistant in 1906–07. This year marked a turning point in her life, for she had written her first scientific paper, which, on Bodtker's advice, was translated into French and published in a French journal.[6] The paper was to be her only venture into organic chemistry.

Impressed by French cultural and scientific life during his periodic visits to France, Bodtker undoubtedly influenced Gleditsch and awakened in her a wish to branch out beyond the rather confining scientific community in Oslo. She told him of her dream to study with Marie Curie in Paris and he offered to help. While Gleditsch looked after the laboratory in Oslo, Bodtker travelled to France and visited Curie's laboratory. He did not find Curie on his first visit, so he left a message that "my talented assistant wishes to work for you, solely out of love for science, not to gain a degree."[7] When he returned a few days later and met the famous scientist, his request was initially turned down; there was no more room in the laboratory for assistants. The persistent Bodtker showed Gleditsch's published article to Curie. Curie became interested but still lamented that there was no room. When Bodtker retorted that "Mademoiselle Gleditsch is so small and slight that she will not take up much room," Curie relented.[8] The deciding factor was probably that most of the researchers then at the laboratory were physicists and Curie badly needed another chemist, a position for which Gleditsch was ideally suited.

Bodtker immediately sent a letter to Gleditsch notifying her of his successful visit to the Curie laboratory: "Mme Curie is expecting you already in the beginning of October. You should come down here soon, in the beginning of September, to polish up your French and make yourself acquainted with the city." Marie Curie made an extremely sympathetic impression on Bodtker: "I believe you will find her very pleasant… You should write Mme Curie right away. As your French probably is not yet that good, I suggest the wording shown on the other side."[9]

To finance her studies Gleditsch was awarded a grant of four hundred Norwegian kroners (sixty dollars) from the legacy of Josephine, dowager queen of Norway and Sweden. Students were expected to pay for the privilege of working at the Curie laboratory, but Curie told Gleditsch in a letter that "if you would take on this work [recrystallization of barium and radium salts], which will only take up part of your time, and which is of general benefit for the laboratory, I could exempt you from the fees because of these favours. You can at the same time work on another problem of greater interest, which could lead to new results."[10] Thus, in September 1907 Gleditsch arrived in Paris to start work at Curie's laboratory. During her early days in Paris she shared living accommodations with one of her brothers, Adler, a topographer who was in Paris for a short period of study. Later she rented a small apartment behind the Panthéon, from which she could easily reach the Curie laboratory and the Sorbonne, where she studied.

The Curie laboratory was located at 12 rue Cuvier. Its cramped space was shared by ten people. Besides Curie, there were her nephew and personal assistant Maurice Curie, the chemist André Debierne, and seven students – some French but most from abroad. It was Debierne who helped Gleditsch to adjust to the unfamiliar surroundings. As she later described them: "The rooms Marie Curie had at her disposal were all spread out. At the entrance there was a large office with an adjoining dark-room, and close by a room which served as a library. You had to cross the yard to get to the large laboratory, where most of the scientific equipment could be found. Next door was Marie Curie's small office and another room used by André Debierne."[11] The group of young students in the laboratory, many of them women, provided Gleditsch with valuable contacts, including May Sybil Leslie and Eva Ramstedt. She became especially close to the Swedish Ramstedt, who arrived in Paris from Stockholm in 1910.

As Curie wished, Gleditsch worked at the tedious task of separating and recrystallizing barium salts that contained some radium, with the purpose of producing pure radium salts. Gleditsch was given sole responsibility for extracting these valuable substances and she was promoted from student to Curie's personal assistant after only one year. In a later interview she described how she used to lock the doors to the laboratory where she was handling minerals worth 100,000 francs: "The smallest mistake would be disastrous."[12]

The minerals with which she worked came from a small radium production plant situated in Nogent-sur-Marne outside Paris. The plant has its own peculiar history. Marie and Pierre Curie had declined to participate in any economic exploitation of their discovery of radium and they did not take out any patents. The laboratory found it difficult to obtain sufficient amounts of radium-containing minerals for the production of radium salts. Help was obtained from a chemist, Armet de Lisle, who realized the significance of the Curie's discovery of radium. On the site of his family's quinine production plant, he opened the first radium factory in the world. He planned to use French radium-containing minerals, but he was forced to look elsewhere as the available ores were scarce and had a low radium content. Armet de Lisle kept the commodity at a reasonable price. The quality of the radium was tested in the Curie laboratory and Curie herself signed the certificates.[13] At times when their own facilities were inadequate, her assistants were invited to carry out some of their work in de Lisle's factory.

The more interesting scientific problem that Curie entrusted to Gleditsch followed from Cameron and Ramsay's claim that they had observed the transformation of copper to lithium and other alkali metals when solutions of copper salts were exposed to radium "emanation" (radon). Curie was sceptical of this transmutation of elements, and with Gleditsch's help she set out to reproduce the experiment. Their negative results led them to discount the possibility of transformation.[14]

The new field of radiochemistry was expanding rapidly. The concept of isotopes was unknown and the discovery of all those seemingly new, radioactive elements caused confusion as to their placement in the periodic table. Gleditsch chose to work with uranium- and thorium-containing minerals.[15] This turned out to be a fruitful choice because therein lay the key to the phenomenon of radioactive decay and the relationships between elements as they undergo transmutation.

Curie, described by Gleditsch as somewhat reserved, took a keen interest in the work of her students and she visited the laboratory daily. She also gave frequent lectures at the Sorbonne, which Gleditsch attended. "The students had the sense that here was not just a great scientist, but also a great personality."[16] At the time of Gleditsch's arrival, Curie was forty years old and had been widowed for one year. Gleditsch, at twenty-eight, was far less experienced both in her personal life and as a scientist. Despite their differences, a unique and lasting friendship developed between the two women. Gleditsch was often invited to the Curie home in Sceaux on Sundays where she befriended Marie's two daughters. Irène, then ten years old, also became a lifelong friend and, later, a colleague.

Gleditsch went home to Norway during the summer holidays from the end of July to the middle of September, while Curie took her family and a few friends to Brittany. While on vacation in

Norway and Sweden, Gleditsch helped locate radium-containing minerals. These ores were sent to the Curie laboratory and the radium factory in Nogent-sur-Marne. During one of her brief visits to Norway, Gleditsch became engaged to a young officer. It was said that the two were very different and that intellectually she was his superior. The engagement, apparently the only relationship Gleditsch had that might have led to marriage, was soon broken off. In a speech to the International Federation of University Women (IFUW) in 1926, Gleditsch gave her views on what happens when women attempt to combine a research career with family life: "Their [women's] presence in the home is often required. A woman who wants to be a researcher has to reconcile two opposing demands. The research requires first and foremost a tranquil atmosphere, opportunity to think in peace and quiet, and to concentrate on a particular problem. Material worries, concern for a husband or children who are left at home without adequate help or care, will kill all chances of a first rate effort."[17]

Before leaving Paris in 1912, Gleditsch received a *Licencíee ès Sciences* (the equivalent of a BSC) on the basis of her studies at the Sorbonne. Returning to Norway, she was granted a modest fellowship at the University of Oslo, which required her to lecture on radioactivity and supervise work in the laboratory. In the fall of 1912 her first article about Marie Sklodowska Curie was published in *Tidsskrift for Kemi.*[18]

Gleditsch tried to continue the research started in Paris on the determination of the half-life of radium but she was hampered by a lack of proper equipment. The small scientific community at the University of Oslo, where she was by far the most experienced radiochemist, must have felt especially confining and isolated after her five years in Paris. It was not surprising, then, that she began to explore further opportunities to study abroad, and she turned her eyes towards the United States.

Before she was able to realize her plans, Gleditsch was struck by a series of personal tragedies. In the early part of 1913 she lost both her parents and one of her brothers within the space of two months. It was Ellen who took care of the necessary arrangements, and it was to her that the family instinctively turned for comfort and support. Despite her extremely heavy workload, she set up house soon afterwards for herself and two of her brothers. One brother, Adler, lived with her until her death some fifty years later, while the other, twelve-year-old Kristian, received his education through Ellen's support. About ten years later when their financial situation started to improve, she was able to hire a housekeeper and free herself from the drudgery of housework.

In 1913 Gleditsch was awarded a sizeable scholarship from the newly formed American-Scandinavian Foundation, and she contacted Bertram Boltwood at Yale (the leading authority on radiochemistry in the United States) and Theodore Lyman at Harvard for permission to work in their laboratories. The events and correspondence leading up to Gleditsch's arrival at Yale and her experiences in the United States shed light on the bias of the time against women doing science alongside men and the inability or unwillingness of most scientists, including women, to acknowledge the existence of those barriers.

In his answering letter, Lyman pointed out that no woman thus far had ever set foot in the physics laboratory at Harvard, a fact that Gleditsch may not have known. Boltwood's letter of 11 September 1913 was equally typical of contemporary attitudes. He politely cautioned her against making a hasty decision about Yale before she could see for herself whether "the facilities I can offer are sufficient to make you feel it is worth your while."[19] Writing to Rutherford the following day, his tone was very different: "I have a piece of news that will interest you. Mlle Gleditsch has written that she has a fellowship from the American-Scandinavian Foundation (I never heard of it before!) and wishes to come and work with me in New Haven!! What do you think of that? I have written to her and tried to ward her off, but as the letter was unnecessarily delayed in being forwarded to me, I am afraid she will be in New York before I get there. Tell Mrs Rutherford that a silver fruit dish will make a very nice wedding present!!!"[20]

Gleditsch did indeed arrive in New York before Boltwood was able to "ward her off." Gleditsch's study of the relationship between radium and other radioactive elements in minerals led to her most fruitful research – the determination of the precise half-life of radium. Boltwood had previously studied the rate of decay of radium and he had derived an approximate half-life of about 2,000 years while Rutherford and Geiger had found a value of 1,760 years (later corrected to 1,690 years). This was crucial work, as Badash explains: "Probably the most significant constant in radioactivity studies was the half-life of radium, the time required for half its amount to transform into emanation [radon]. As the only highly active radioelement that could be prepared in pure and reasonably large quantities, with a half-life long enough to consider those quantities constant for most practical purposes, radium was regarded as the standard substance in this field."[21] Gleditsch determined the radium half-life to be 1,686 years, a number widely accepted by the scientific community and only much later adjusted to the current value of 1,620 years.

With her work on the determination of the half-life of radium, Gleditsch gained the respect of other scientists. The president of the American Chemical Society, Theodore W. Richards, invited her to visit him at Harvard, and in the spring of 1914 she was awarded an honorary degree, Doctor of Science, at Smith College in Massachusetts.[22] Lyman also changed his views and even offered her a place in his laboratory at Harvard as a guest.

During the year that Gleditsch worked in the laboratory at Yale, Boltwood came to respect her as a person and a scientist. In a letter delivered to the steamer as she was returning to Norway, he wrote: "I want to tell you how much I enjoyed having you working here and how satisfying it was to have this opportunity of learning to know you and finding that we have so many scientific interests in common. I wish you every possible success in the future."[23] The respect appeared mutual. In a letter from Oslo, Gleditsch wrote: "I would have liked spending a few hours in New Haven [prior to leaving for Europe], talking to you, mainly about radioactivity. It's a pity Norway is so far away."[24]

Many years later, Gleditsch was asked in an interview how she felt about "men as colleagues." She replied, "I have worked as the only woman in a laboratory without any reason to complain"; nonetheless, "some men on principle are against women as scientific researchers." Obviously referring to Boltwood, though without naming him, she described how she had once worked with a learned man who "was reputed to hate women." She related how she had heard from a co-worker that the individual had referred to Gleditsch as a rare exception because unlike other women "she does not scream." It is telling of the attitude towards women of the time that Gleditsch called this statement "the biggest compliment of my scientific career."[25]

Still, Gleditsch's praise for men as colleagues is surprising. After all, she was in Paris when the press speculated about the Curie-Langevin relationship. This had resurrected the question whether or not a woman was capable of independent and creative scientific research. An interview with Gleditsch in the *New York Press* six weeks after she arrived at Yale reflected the same prejudice. The reporter introduced her as a "remarkable woman scientific researcher" who was searching for an inexpensive method for preparing radium. He then expressed disbelief at finding that this beautiful, petite, and charming woman was actually a famous scientist: "Impossible ... what do words like radioactivity and gamma rays have to do with such sweet lips." Eventually he recovered from his surprise and related how Harvard University had opened its laboratory to her, the first woman to be so honoured. The journalist noted that although Miss Gleditsch came from the country that first adopted universal suffrage,[26] she had little time for feminist interests. When asked if she had ever considered marriage, Gleditsch replied, "Due to the amount of work I just do not have time; my interest is research and that takes up all my thoughts."[27]

On her way home from the American continent she visited London and Manchester. Rutherford was not in his laboratory and she wrote to Boltwood: "I have a strong feeling that the laboratory is completely different when Mr Rutherford himself is there, and I am looking forward to a new visit, when I can find him there."[28]

Realizing the importance of being visible in the international science community, Gleditsch wrote to Boltwood upon her return to Norway to seek his advice. She asked if they could publish the results together. "I would be very flattered, and the first part of the work was done in your laboratory."[29] With Boltwood's help the paper was published in English in the *American Journal of Science*.[30] Boltwood wrote to Gleditsch: "I have also taken the liberty of having your paper published as a contribution from the Sloane Laboratory since the work had its inception in this laboratory, and particularly because it will make it possible to send out about 150 copies of your reprint with our regular laboratory packages. These will go to all the leading laboratories and workers in various parts of the world, a very desirable condition when one publishes in the American Journal since this has a very limited circulation."[31] The paper appeared under her name alone; "I will publish my confirmatory results later," Boltwood remarked.[32] Gleditsch had now established herself as a leading specialist on the isolation and separation of radioactive substances from minerals. Boltwood later acknowledged the accuracy of her measurement in a letter to Rutherford, admitting that his higher value of the half-life was due to impurities of ionium (thorium-230).[33]

Gleditsch now undertook a lecture series on atomic theory, discussing the emerging quantum theory and the recent work by Bohr, Rutherford, and Moseley. She expanded the work with radioactive minerals and her interest in this area would continue through her life as a research scientist. She developed analytical procedures for the isolation of numerous radioactive substances found in Norwegian minerals, and she continued to supply the Curie laboratory with samples.

With the concept of radioactive series, isotopes, and half-lives of radioactive elements, scientists could use the ratio of lead to uranium in a mineral to measure its age. However, various stable isotopes of lead from different sources are present in minerals, and Gleditsch was one of the first to point out the importance of knowing the relationship between these lead isotopes in order to get accurate age determinations. She isolated lead from bröggerite (Norwegian uranite) and had its atomic mass determined by Richards at Harvard.[34]

The First World War brought isolation. In November 1915 Gleditsch wrote to Rutherford about her continued work on American and Norwegian minerals and added at the end of the letter: "I had a letter from Miss Leslie a little time ago, telling me about the British Association Meeting. I wanted very much to go to England for the meeting, and if I had been alone in the world, I guess I would have gone in spite of war and mines. But my parents are both dead, and I have the charge of a young brother and did not like to leave him. I hope, though, that it may soon be possible for me to cross and to take up some research work in Manchester. I hope you will forgive me for writing you about my work. As you will understand, you yourself gave me the courage in Washington."[35]

Ten days later she sent a letter to Curie again expressing her wish to break out of her isolation: "A long time has passed since I had some news from you. I hope you are faring as well as times like these allow ... As you might know, I met Mme Ramstedt this summer. Together we have studied radioactivity in springs and water, and electricity in the atmosphere. We hope to be able to continue next year. It felt good to talk about Paris, about the Curie laboratory and of memories from our days as students. And it was useful for me at least to speak with somebody who is working on radioactivity. I have friends here, chemists and physicists, but none of them has done much with radioactivity ... How I wish I could travel to Paris and visit you! If only the conditions would soon allow it!"[36]

At last the University of Oslo recognized the immense importance of radiochemistry and realized that a native Norwegian was a leading expert in the field. When a non-tenured university position (a "dosentur") was opened up at the university, Gleditsch was the only applicant. In 1916 she was appointed "dosent" in radiochemistry. After many years as a "research fellow" she had at last received recognition for her work. The new appointment also meant that she could now benefit from a decent salary and a more challenging and responsible position at the university.

The years abroad made Ellen Gleditsch a proponent for international collaboration. For the rest of her life she would encourage students, especially women, to seek experiences abroad as a necessary part of their education. Much later, she recalled:

> In that atmosphere, I learnt how a problem arises in one laboratory, is taken up in another and then perhaps gets its solution in a third. I got to know

how competition and the race towards a solution is closely connected to the advancement of science. A kind of brotherhood is established between those who work in the same laboratory, where various problems are aired, discussed and perhaps result in publication of a solution. And this fellowship eventually reaches laboratories in other countries where the same problems are being investigated. And eventually one realizes – perhaps only through small personal disappointments or successes, that whoever takes that crucial, final step towards a solution, if it is taken from this or that side of the border, is of secondary importance.[37]

During World War I, there was desperate need for someone with expertise to supervise the production of radium, which had declined dramatically. Curie turned to Gleditsch for help. In a letter of 22 June 1916 she wrote: "I'm writing you today to ask if you could possibly come to France to work in the factory of Armet de Lisle. Mr Armet would be very happy to have you here so that the work at the factory can be resumed, and I do think he will offer favourable conditions." She ended the letter in a personal tone: "As for myself, I am overjoyed at the thought of seeing you again, because I have quite warm memories of you and I hope that sometime you would be able to come to my new laboratory as well as visit me in my home."[38] In response, Gleditsch applied for leave from the university and travelled to Paris via England. A letter from Rutherford to Gleditsch at that time indicates that he had helped with security clearance for her stopover in England.[39] She worked in the radium factory until close to Christmas, when she made the same perilous journey back to Oslo, this time facing not only enemy warships in the North Sea but also winter storms.

In 1917 Gleditsch published two textbooks. The first, a *Brief Textbook of Inorganic Chemistry*,[40] caused her much grief. Thorstein Hiortdahl, professor of chemistry since 1872, originally wrote the book at the end of the nineteenth century, and after five subsequent revisions a major overhaul was needed. Gleditsch believed she would have a free hand, but her seventy-seven-year-old co-author had different ideas. In a letter to Boltwood she confided her doubt that she would ever co-author a textbook again. She commented: "He [Hiortdahl] is very interested in the book and nothing can be put on paper without his consent. He is a very old man, so I find it difficult to meet him half-way."[41] Although much of the correspondence between the two has been lost, the following passage from one of Hiortdahl's letters to Gleditsch is indicative of their struggle: "I have lived with chemistry for well over 50 years. When I started, I learnt nothing of molecules and atoms, one only talked about equivalents. But time has changed. I have seen one and the same substance, not only acetic acid [they had argued over the formula of acetic acid, which Hiortdahl wanted to write as $C_2H_4O_2$], but also others, dressed up in different clothing according to the chemical fashion of the time. But although they often changed their appearance, the essential body stayed the same."[42] The book was nevertheless published. When a seventh revision was eventually needed, Gleditsch undertook the work alone, and it appeared two years after Hiortdahl's death.

The second book published by Gleditsch in 1917 was also coauthored, but this time with a close friend and contemporary, the Swedish scientist Eva Ramstedt. *Radium og de radioaktive processer* was published simultaneously in Norway and Sweden and it became the first advanced textbook in radiochemistry written by Scandinavian scientists.[43]

The two friends had worked all summer in Stockholm to complete the book, and a copy was immediately shipped off to Curie. Curie wrote to acknowledge receipt of the book and expressed her delight that "they both were doing something worthwhile."[44] Her brief reference to the book may be excused considering that it was published in Norwegian and Swedish. Rutherford's response had a more personal tone: "I regret to say that I do not know your language sufficiently to read it, but I know the subject so well that I almost persuade myself that I can. The work should be very useful in giving a clear account of radioactive work."[45]

Gleditsch often chose to write in her native language, which isolated her from the international scientific community. Some of her works were published in French or English, but most were written for the local audience, that is, Norwegian scientists and students. For this she received recognition at home. In 1917 she was elected a member of the Academy of Science in Oslo, only the second woman to be so honoured after the biologist Kristine Bonnevie. It was five years earlier that the Norwegian Parliament had passed a bill concerning women's access to public office (including appointments to professorial chairs). Shortly thereafter, Bonnevie had been made professor of biology and was later given a seat on the Academy of Science, but not without opposition from other academicians as well as from the public. She paved the way for women's membership in the academy and the appointment of Gleditsch went through without incident.

It was probably the sense of isolation that drove Gleditsch to apply for leave from her position at the university, and starting in October 1919 she spent nine months in Paris. With her she brought a young women, Randi Holwech, who had been promised a place in the

Curie laboratory, which at the time held as many as thirty to forty men and women of different nationalities.

From Paris Gleditsch travelled to England in the summer of 1920 to visit Rutherford at Cambridge and Soddy at Oxford. The laboratories had just resumed their operations after the war, and Soddy wrote: "I should be glad to see you if you come to Oxford. I am afraid, however, I have little or nothing to show you here. My own work has been interrupted by the war and my move here. I am afraid it will be a long time before I get anything very serious started again."[46] Her friends Eva Ramstedt and May Sybil Leslie accompanied her on these visits. From Oxford the three women sent a postcard to Marie Curie: "This is the first time we are together since 1911 when we worked in your laboratory. We are thinking of you, talking about you and the good memories."[47] Soddy maintained a friendly relationship with Gleditsch. When he was awarded the Nobel Prize for chemistry in 1921 he wrote to her: "Very many thanks for your kind and congratulatory note on the Nobel award and for what you say about it. As one of the few radiochemists in the world, I value your appreciation of my work especially."[48]

After her visit to England Gleditsch returned to Paris. Curie had left on a long journey to South America and she had assigned Gleditsch to take over supervision of the Curie laboratory during her absence. Curie sent a postcard to Gleditsch at the Institut du Radium from the ship *en route* to Rio de Janeiro: "I leave everything in your capable hands."[49] With her duty completed, Gleditsch headed back to Norway, this time via Strasbourg where she had been invited to give a lecture at the university about the dating of minerals from nuclear decay.[50] When she finally reached Oslo, she was awarded the prestigious Nansen prize by the Norwegian Academy of Science for her dissertation "Minéraux Radioactifs I. La Broeggerite."

During the 1920s Gleditsch travelled several times from Norway to France to work in the radium factory at Nogent with Armet de Lisle and André Debierne. In a letter written to Curie in November 1922, Gleditsch referred to a diary she had left behind in the Curie laboratory that had been returned to her in Oslo.[51] Unfortunately, attempts to locate this valuable source of documentation after her death have been unsuccessful.[52] Her correspondence tells of a visit with the Curie family at their vacation place at l'Arcouest in August 1921.[53] Gleditsch also made a sidetrip to Cornwall to investigate a controversy about the uranium content of minerals supplied by a mine there. She was apparently asked to undertake this mission because of her fluency in English.[54]

It was at this time that Gleditsch first mentioned in a letter to Curie that she was suffering from anemia, which resulted in frequent colds during the winter months.[55] One can speculate that her anemia was caused by radioactive exposure. However, as she went on to live to the impressive age of eighty-eight her ability to fight off any adverse effect of radiation must have been remarkable. When in 1922 Gleditsch brought up the subject of radiation damage in a letter to Curie, it was obviously a sensitive issue.[56] The frequent correspondence between the two women continued through the last decade of Curie's life. The topics of discussion were now mainly technical as Gleditsch continued to provide Curie with a constant supply of radioactive minerals from Norway.

Curie often requested Gleditsch's help at the laboratory, and she appeared happy to comply as long as she could arrange leave from her duties in Oslo. No doubt the long, dark Scandinavian winters coupled with her love for Paris and the French culture made Gleditsch more than willing to travel south whenever the opportunity arose. In December 1924 she wrote to Curie: "I was very upset when I realized that I would not be able to escape from here this winter. I had already made plans for some quiet time in your laboratory. Anyhow, I hope I shall be able to come later."[57] In the same letter she asked Curie a favour. On the next visit to Paris, she wanted to bring one of her young students, Sonja Dedichen. "Would there be room in the Curie laboratory for her? She is intelligent and serious, very kind, maybe too kind, and very childish." Curie replied that unfortunately Dedichen could not be accommodated during the next semester, but she arranged a place for her with another scientist[58] and promised a place in her laboratory the following year, should Dedichen still be interested.

The story of Sonja Dedichen deserves special mention as it illustrates how both Curie and Gleditsch intervened directly in the lives of their young protégés to assist them in their pursuit of science. Dedichen was eventually given a place in the Curie laboratory and arrived in Paris in October 1925. In May 1926 Curie wrote to Gleditsch:

> Today I want to write a bit about your protégée, Miss Dedichen, who is eager to return to Paris for the next academic year, but so far has not obtained permission of her parents. She somehow counts on you to convince them to allow her to do so. As far as I am concerned, I would very much like to give her permission to return. Since she arrived here, she has been busy working with ionium-thorium. At first, she appeared rather inexperienced and had several accidents due to lack of attention, which you can imagine had very

serious consequences. However, she has made strong efforts to improve her performance and my opinion of her is now higher, and at the same time we love her pleasant and kind manners. I believe strongly she would benefit from a second year of research here. She could carry through a project which would give her greater satisfaction.[59]

Gleditsch replied a few weeks later, expressing appreciation for Curie's active interest and conveying the permission of Dedichen's parents.[60] But events did not develop as Sonja or her two mentors had expected. On 1 October 1926, Sonja Dedichen wrote to Curie from Oslo: "It is not necessary to reserve a space for me in the laboratory this year, because I am not returning. The fact is I have become engaged and will be getting married perhaps as soon as in three months to a Norwegian medical student in Oslo. Therefore it is not sensible to go to Paris and study chemistry now."[61] Curie responded that Mme Cotelle could complete Dedichen's work and that she wished Dedichen much happiness in her marriage and for the future.[62]

During the interwar period, Gledisch focused on the study of isotopes in minerals. Some of her work was performed jointly with her younger sister, Liv, who was also a chemist.[63] Gleditsch's book *Radioaktivitet og grundstofforvandling*[64] (*Radioactivity and the Transmutation of Elements*), published in 1924, had to be reprinted the following year due to high demand. In January 1925 she wrote again to Curie about applying for leave to work in Paris on a series of topics, including "the age of minerals" and "penetrating radiation and its intensity related to the amount of radium present." However, her duties at the university were demanding and she complained of little time for research with a teaching load of four courses per week.

In 1920 the League of Nations was created to promote world peace and disarmament, and numerous interest groups grew up as offshoots. Gleditsch became heavily involved in one such group, the International Federation of University Women, founded in 1919. By this time there were many women graduating from universities, but it was extremely difficult for them to attain academic positions, especially in research. Women academics in many countries worked under adverse conditions, and it became apparent they had to band together; the IFUW answered this need. "Its function was to act as a chemical catalyst in international conditions, serve as a point of contact between individuals and between nations, to bring them together through the power of understanding and friendship."[65] Personal contact would do away with national and international prejudices, which after the war seemed especially necessary.

Corresponding national associations were set up in many countries. The Norwegian Women Academics' Association (NKAL) was formed in 1919, and Gleditsch, one of its founders, was elected president in 1924. She felt that the association should not concern itself with women's rights issues but should focus on science as a means to peace and on the conditions under which women scientists worked: "It's totally unimportant if work is carried out by a short woman in Bulgaria or by a tall man in the United States if it's done well. And that is what we have to do: to work so well that no one dares to say: it's good work to be done by a woman, but that everybody would say: that's good work. Races and gender do not matter in science."[66] Gleditsch continued to serve as president of NKAL until 1928, when she was replaced by her sister Liv.

Only three months after she took over the presidency of the local association, the world congress of the IFUW was held in Oslo under the auspices of the five nordic associations. At the congress Gleditsch was elected to the executive of the IFUW and in 1926 she was chosen as its president. She had excellent qualifications for this important post: knowledge of university life in other countries, fluency in four languages, a deep conviction of the importance of intellectual cooperation across borders, and belief in research free of political interference and equality for women and men.

In the spring of 1929 Gleditsch set out on a six-week lecture tour of the United States as president of the IFUW. She hoped to stimulate a more active American participation in the international exchange of women academics, create interest for providing scholarships for women scientists, and lecture on her own work at some American universities. Upon her arrival in New York in March, the *New York Times* ran an article describing Gleditsch as "a scientist of distinction as well as a student of world affairs."[67]

Her extensive trip took her first to New Orleans for the National Convention of the American Association of University Women. From there she went to the West Coast as a guest of Mills College in San Francisco. She spoke to women students at Stanford University and visited Berkeley before returning east for a final visit to Yale, where she had studied some fifteen years earlier under Boltwood. This was the end of her leadership of the IFUW. When she stepped down in 1929 due to work pressures she had served for five years, including three years as president. By this time, the IFUW had grown in size and importance to a membership of 40,000 in thirty-one countries.

On her way back to Norway she visited Paris to see Marie Curie, not only to describe her experiences in North America but also to tell her friend and former teacher about the debate that was raging in

Oslo concerning her nomination to a chair of chemistry, and to seek her advice on the matter. Curie subsequently sent a letter to Professor Kristine Bonnevie, *pro-dekanus* (associate dean) of the university. Curie made it clear that she felt it was improper to involve herself in this appointment and that it should be decided solely from the point of view of the general interest of the university. She added, however, that she had a high regard for Ellen Gleditsch's intelligence and for her dedication to scientific research: "For many years she worked in my laboratory, where she made herself well respected as a good chemist." She praised Gleditsch's subsequent contributions to research and teaching and reiterated her respect and friendship for her long-time collaborator.[68]

It is not known whether the letter helped secure Gleditsch's appointment. Without doubt, she needed all the help she could get. It was true that the ground had been broken for women academics, for the University of Oslo already had a woman professor, Kristine Bonnevie, the chair of biology. But the debate surrounding Gleditsch's appointment reached the daily press, and there were those who seriously questioned the wisdom of appointing a woman to such a position. A planned move of the Chemical Institute from the old campus in central Oslo to the new site at Blindern on the city's outskirts was considered an additional challenge to the new appointee.

In her reply to Curie, Bonnevie did not hide the fact that the departing occupant of the chemistry position, Heinrich Goldschmidt, was strongly opposed to Gleditsch's appointment.[69] It was not only that she was a woman; Goldschmidt, with his strong German connections, favoured one of the other major candidates, O. Hassel, whose sympathies were pro-German, while Gleditsch throughout her life had strong ties with France and its scientific community.

The university's president, physicist Sem Saeland was also opposed to Ellen Gleditsch's appointment. His influence, felt throughout the academic community, was pointedly satirized in the student newspaper, *The Fig Leaf*,[70] in an article entitled "Will Little Ellen Become Professor?" The title referred to Gleditsch's diminutive stature – she stood all of 154 centimetres tall and was at times referred to as "the Petite Scholar." Among the students, whose affection for this unusual professorial candidate was clear, it was understood that "if you have a problem – go to Ellen."

Although the voices against Gleditsch's appointment grew weaker, the debate over "the potential danger of having a woman in such an important position" raged on in the popular press. No longer just an academic issue, it had become a problem of general social concern.

Gleditsch stayed out of the debate, and when the appointment finally came through on 23 June 1929, she reacted with her usual calm. In a letter to Curie dated 26 June she wrote: "It is done. Three days ago I was appointed professor. I sincerely hope I will be able to do a good job; and of work, there will be no shortage. I also hope I can visit you from time to time; those visits will for me be the source of inspiration and encouragement."[71]

On 5 May 1930, Ellen Gleditsch was formally inaugurated as professor of chemistry in a solemn ceremony. Some months earlier she had turned fifty and she now stood at the height of her career. In an interview for the newspaper *Tidens Tegn* the day before the appointment became official, she made it clear that she was looking forward to the change: "I will get better working conditions, can move up from the basement laboratory where I have spent the last years, I will have a more interesting job. I will lecture on inorganic chemistry to students in pharmacology, science and medicine, and I will supervise the laboratory research." She also talked about her wish to continue her research on radioactive minerals in Norway. The interviewer was impressed by this energetic woman: "That little dainty professor with the intelligent high forehead and the lilting voice seems to be able to manage it all."[72]

A new laboratory, which Gleditsch helped to plan, opened at Blindern outside Oslo in the fall of 1934. On 17 September she started a lecture series there on the history of chemistry. The early 1930s were productive years for Gleditsch. She was inspired by a visit to the Curie laboratory in 1931, and publications started to appear in rapid succession.[73] Some of the research was undertaken with her assistant, Ernst Föyn, and her students Ruth Bakken (who later worked with Alicja Dorabialska), Bergljot Qviller, Sonja Haneborg, and Th.F. Egidius. Despite her accomplishments she was often introduced as a student of Marie Curie, and it appears that she lived in the shadow of her mentor. There is no indication that she ever resented this; on the contrary, she took every opportunity to express her great admiration for her famous teacher.

Gleditsch held a strong conviction that education should be accessible to all, especially to young women. When asked once what young women ought to fight for she replied, "That their parents will understand that it is just as important that girls get an education as boys, and this regardless of whether the girls marry or not."[74] Despite her prominent stature, Gleditsch was never too busy to help her students. She believed that "to work with students is to work with the future. It is the student today who will continue our work tomorrow. It is youth working together, intellectually and internationally, we must build on."[75]

One of her former students, Bergljot Qviller Werenskiold,[76] talked in 1984 about the influence that Gleditsch had on her:

I started my studies at a time when Ellen Gleditsch was "dosent" and her lectures were not compulsory for everyone. Boys had to economize, and make sure they finished [their studies quickly]. There were therefore many female science students who flocked around Ellen, girls who were fascinated by Mme Curie and Ellen as a spokesperson for her findings. The girls – well, there were not that many – could relax more about their studies than the boys. I started in 1922, took my "embetseksamen" [first university degree] in 1930 as the first one with a major in radiochemistry. During the last part of my studies I was her personal assistant. After I graduated, I continued several years as her assistant. Thus I had almost daily contact at work with Ellen Gleditsch up to 1936 when I myself took charge of an institute.

No, we did not just know each other as colleagues. Already in 1924, she asked me to join her on a hiking tour in the mountains. Ellen was in a great mood and never seemed to tire. Year around, both at work and during official celebrations, Ellen meant an awful lot to her students. She helped them and supported them in every way. Her youngest sister Liv who later became a "lektor" [lecturer], received such good support that she would always remember it. Ellen had an unbelievable amount of energy. She was not a "workaholic" but was always busy with one thing or another. Feminist? Both yes and no. She was not a feminist as we see it today. But she was very clear on one point, that women should have the same access to education as men. Yes, she influenced me a lot, not only professionally; I too, became president of the Norwegian Women Academics Association.

During my first university years I often wondered why Ellen had not married. She was so charming. We women students talked much about that. After a while we realized that for a woman who had such high goals as Ellen, it had been difficult to think of marriage. Such was the situation in those days.[77]

To popularize science, Gleditsch had begun to give public radio lectures in the 1930s on "The Chemical Elements" and "Radium and Radioactivity." She was a prolific writer, producing articles in French, English, German, and Norwegian. As well as research papers, she wrote biographies of numerous scientists who had been personal acquaintances. She felt it was her duty to inform the world of what was happening in her own discipline, and to keep the scientific community in other countries abreast of the latest developments. She argued frequently that "research shall be free, independent of national borders, and for the good of mankind."[78]

The news of Marie Curie's death on 4 July 1934 reached Gleditsch by telegraph within a few hours, and she immediately sent telegrams to Irène Joliot-Curie and Marie Curie's secretary. She could not reach Paris in time for the private funeral two days later, but her contact with the Curie family persisted after the death of Marie Curie. Gleditsch corresponded regularly with Irène. When the Joliot-Curies received the Nobel Prize the following year, they turned down an invitation to visit Norway. However, in 1946, some ten years later, Gleditsch had the pleasure of seeing Irène in Oslo when the latter received an honorary doctorate from the university. Irène returned again in 1953 to go trekking through the Norwegian mountains with Gleditsch, then seventy-four years old but probably in better physical condition than her younger friend, who suffered from a radiation-related illness. Irène died three years later of leukemia at the age of fifty-nine, and with her death regular contact between Gleditsch and the Curie family was broken after almost fifty years.

When fascism and nazism surfaced in earnest in the thirties, Gleditsch was ready to do her part for its political and racial victims. After the *anschluss* of Austria in 1938, she was instrumental in finding a safe haven in Norway for refugees primarily from the university community. In this she was helped by Ernst Föyn, who at the time was working at the Vienna Institute, and by Irène and Frédéric Joliot-Curie, who were involved in similar endeavours in France. Places were found in her laboratory in Oslo and other parts of the country for fleeing scientists, among them Tibor Graf, Elizabeth Róna, and Marietta Blau from Austria, all well-known researchers who became her co-workers during their stay. After the discovery of artificially induced radioactivity, Gleditsch pioneered work with Róna and Föyn on radioactive tracers and their use in the study of dynamic equilibria in saturated salt solutions.[79]

At the outbreak of war in 1939 it must have been devastating for Gleditsch, now sixty-one, to see her long struggle for peace suffer such a serious setback. But she kept her belief and hopes. Throughout the dark years of 1940–45 when the Germans occupied Norway, Gleditsch never waivered in her efforts to defy the occupiers and the puppet regime. On the very first day of the occupation, the Germans took over part of the University of Oslo. Gleditsch and her colleagues tried to continue their usual activities to ensure that they were in a position to act if necessary. A clandestine group of professors, Gleditsch included, banded together to keep up the spirit of resistance and to be ready for expected attacks on academic and personal freedom. These six to eight people worked in secrecy from the rest of the university population but were nevertheless able to influence decisions within the departments and the opinions of professors and students. They communicated with the "Homefront," the

Norwegian-organized underground movement, and set up a relief structure to provide economic aid to professors at risk of losing their jobs.

Gleditsch's brother Adler spent several years as a prisoner of war, two of them in a notorious concentration camp near Oslo, for passing maps of German installations to the Allies. Her sister Liv was arrested during the last year of the war. Gleditsch continued to show her disdain for the occupying forces by working to aid students, academics, and others who found themselves in danger. On several occasions she concealed wanted underground fighters in her apartment. In 1943 there was a big German raid on Blindern, one of the campuses of the university, and all the men at the Chemical Institute were arrested. The women quickly saved what they could from the laboratories. Gleditsch collected all the precious metals she could lay her hands on and put them in a suitcase, which she hid under her bed in her apartment.

Shortly afterwards she was arrested and interrogated about her activities by a well-educated German officer. She talked her way out of danger by turning the conversation to the officer and to science. When she switched to German he was won over and released her, thanking her for a pleasant meeting and expressing the hope that they would meet again under more pleasant circumstances. She was very proud of her "charm offensive." At one point, the secret police, in an attempt to curtail her freedom of movement, ordered her and her longtime friend Bonnevie to be exiled from Oslo. The two were accused of political activities at the university and were not allowed to return without special permission.

Food was becoming scarce as the war progressed. In order to augment their supplies, Gleditsch and others kept and cultivated individual parcels of land outside the university at Blindern. They grew vegetables and potatoes and at night took turns keeping watch over their plots to discourage thieves – an unusual assignment for a university professor. On the pretext of demonstrating needlework and the making of national costumes, she travelled around the countryside relaying messages to people on behalf of the underground resistance.

During the war, Gleditsch had assumed yet another responsibility. Her youngest brother, Kristian, and his wife needed a home for their nine-year-old daughter, Chris, while they spent the war years in England. Working against Franco's regime during the Spanish Civil War, the two had organized emergency help for children. They found Chris in a hospital after a bomb attack and adopted her. Gleditsch loved children, and the arrangement turned into a happy one for both of them. She became like a mother to Chris, and Chris recalled the years spent with "aunt" Ellen as the happiest time of her life.

No sooner had the war ended than Gleditsch found another cause that would occupy much of her time and energy for several years. This was the newly established United Nations organization UNESCO, which had commissions in many countries. She was appointed to an eight-member working group on the Norwegian national committee, bringing to the job her usual knowledge, experience, and capacity for hard work. At UNESCO's first general assembly held at the Sorbonne in November 1946, Gleditsch's speech, which stressed the importance of fighting illiteracy and ignorance, made a deep impression on the audience.[80]

Gleditsch continued to support young chemists after her effective retirement on 1 January 1946. She kept an office at the Chemical Institute and was actively involved in reviving its activities after the occupation ended in 1945. She stayed in touch with friends and colleagues such as Ramstedt and Meitner, whom she visited in Stockholm,[81] and she accepted an invitation to Paris in 1947 from Frédéric Joliot-Curie to take part in ceremonies on the tenth anniversary of Rutherford's death.[82]

At the next UNESCO congress (1949) Gleditsch was re-elected to the working committee. In 1952 she was named to the Norwegian commission for scientific coordination as the government representative, where she became an effective spokesperson for international control of the atom bomb. Her work for UNESCO gave her great pleasure, as it offered an opportunity to work for the very things that had always interested her: to transform and educate people to create a better world. Unfortunately, her connection with the organization was broken abruptly in 1952 when she resigned in protest over the admittance of Franco's fascist Spain as a member. Unwilling to compromise her convictions, she had no choice but to take this drastic step, and others followed her lead. For six years, she had been a prominent and highly respected member and her contribution was missed.

Gleditsch had resumed research after the war, her last scientific paper appearing in 1952.[83] She continued writing, however, focusing on an interest in the history of science. Her last article, about the Swedish chemist Carl Wilhelm Scheele, was published in 1968, the year of her death.[84] Gleditsch's list of publications has around 150 entries, including scientific papers, monographs on current topics and famous scientists, and the textbooks on radiochemistry and inorganic chemistry.[85]

During her long life Gleditsch received many honours and distinctions. The one that she probably treasured most was an honorary doctorate from the Sorbonne bestowed, in 1962. The first woman to be so honoured, she was eight-three years old and fifty-five years had passed since she first started her studies at that university. A few years earlier, she had been made an honorary citizen of Paris and received the medal of the City of Paris. She was granted an honorary doctorate by the University of Strasbourg and was given honorary membership in a number of chemical societies at home and abroad. It was symbolic that her last public lecture, given at the University of Oslo in November 1967, was a memorial lecture about Marie Curie. At the age of eighty-eight Gleditsch could still hold the capacity audience spellbound as she gave a lively description of her teacher.

Less than a week before her death she gave a dinner party for some of her old students, all women, and expressed delight that their education and training had made them valuable citizens from Norway to Tanzania.[86] She was spending the Whitsunday weekend at her small country house in Enebakk outside Oslo when she was suddenly taken ill. She died of a stroke a few days later on 5 June 1968, in her eighty-ninth year.

Initiatives had already been taken in 1964 to establish a scholarship in Ellen Gleditsch's name. It took many years to build up enough capital, but by 1986 a stipend of 30,000 Norwegian kroners was finally in place, and the call went out for nomination of recipients of the Ellen Gleditsch Scholarship Foundation.

The detailed biography by Kronen and Pappas is unquestionably the best account that we will ever obtain of this truly outstanding person. However, the picture they paint is somewhat two-dimensional. The official person and the admired teacher are seen through the eyes of colleagues and former students. There are many questions that could, perhaps, have been answered by Gleditsch's missing diary.[87] It would certainly have given her personal reflections on events and people. So many of her letters to famous scientists of the time are missing. Sadly, so is much of her correspondence with some of her close women friends and family. These letters might have told whether she was ever discouraged by the hurdles that she faced; ever enraged by biased attitude towards women scientists in academia, among male colleagues, or in society at large; whether at times she felt exploited by Marie Curie with her frequent demands, disappointed by her rather formal communications, and hurt by being kept at a distance. The "official" Ellen Gleditsch, full of humour and energy, is always understanding, cheerful, and encouraging, refusing to become involved in the politics around her. Only occasionally do we catch a glimpse of something more complex.

Despite her impressive list of scientific publications, Gleditsch's greatest contribution to science was probably as an educator: of the public at large through numerous articles, books, and popular lectures on radio and at the university, and as mentor to her many students and assistants. She was a role model for the many women who were drawn to science, possibly inspired by tales of Marie Curie. But what stands out more than anything else in our limited picture of Gleditsch is her compassion for her fellow human beings and her dedication in fighting for humanitarian causes. It was particularly as a peace activist and a proponent for higher education for woman that she made her greatest contributions outside her work as a scientist.

It has been suggested that women such as Gleditsch provide better role models for young women aspiring to a career in science than Marie Curie, who has been described as a "driven and probably obsessive personality"[88] as well as "very dull and boring."[89] We would argue that Gleditsch provided a *different* kind of role model, not a better one. Her strength lay in her role as mentor. She actively promoted the careers of her students, especially young women, by securing places for them in laboratories where they could gain valuable experience. Her single status allowed her to take an active part in their lives; they became her extended family.

One would expect a small country like Norway to nurture the memory of someone like Gleditsch, with her international stature and long list of accomplishments, as a model for young women and men. But despite her tremendous achievements, little was known about her before the biography by Kronen and Pappas appeared in 1987.[90] Even today, if you ask a Norwegian who Ellen Gleditsch was, the reaction will most likely be puzzlement. Some might have heard of Henry Gleditsch, the actor, or Jens Gleditsch, her granduncle, a controversial liberal bishop, but not of Ellen Gleditsch herself. Should they be asked to name the country's first woman professor in chemistry and told that she was also an internationally famous nuclear scientist, peace activist, and lifelong friend and co-worker of Marie Curie, the reaction would probably be the same.

Gleditsch remains invisible[91] to the great majority of the population – overlooked and forgotten in her own country. Kronen and Pappas commented: "The Norwegian public knew little or nothing before 1929 of Gleditsch's international stature and contacts, and even many years later her male Norwegian colleagues showed her little respect."[92] Norwegian reference books mostly ignore her. One such work covering the reign of King Haakon VII (1905–57) has listings of the first female PhD in the country, the first woman member of Parliament, the first woman lawyer to defend a case before the superior court, and various opera singers, actresses, writers, and musicians. Ellen Gleditsch, the first woman professor in chemistry and the pioneer in Norway in the field of radioactivity, is excluded.[93]

Gleditsch's fate of being "written out of history" is not unique, of course. It is shared by most of the women whose lives and contributions are related in this anthology. Gleditsch is not the only woman scientist whose "male colleagues showed them little respect," even as they rose in stature and fame abroad. We hope that this book will help to right a few of the wrongs that these women experienced during their lives as scientists.

5 May Sybil Leslie: From Radioactivity to Industrial Chemistry

MARELENE F. RAYNER-CANHAM
and
GEOFFREY W. RAYNER-CANHAM

May Sybil Leslie worked with both Curie and Rutherford. While her own research into radioactivity contributed no major findings, she is known for the industrial chemistry research that she performed during the First World War. During this period, she rose to the position of Chemist in Charge of an industrial chemistry laboratory. Later, she attained the rank of lecturer at the University of Leeds – a major accomplishment for a women until quite recently, let alone in Leslie's time.

Leslie's life began with great promise. Born on 14 August 1887 at Woodlesford, Yorkshire, she was awarded a West Riding County Major Scholarship in June 1905 to study at the University of Leeds.[1] Majoring in chemistry, she graduated with first-class honours in 1908. The following year she was awarded an MSc for research under H.M. Dawson on the kinetics of the iodination of acetone. This work, which showed that the reaction rate was independent of the iodine concentration, was the subject of her first publication,[2] now a classic in its field.[3]

In 1909 she was also awarded an 1851 Exhibition Scholarship, which she decided to use to work with Marie Curie in Paris. Arthur Smithalls, professor of chemistry at the University of Leeds, wrote to Curie: "This lady – May Sybil Leslie – is of very high personal character and of exceptionally good intellectual abilities. She studied chemistry for three years and distinguished herself greatly. In the last year she has been involved in physico-chemical research with my colleague Dr H.M. Dawson and I am sure you would find her in every way an earnest and excellent worker."[4] A positive response must have been received, for Leslie herself wrote to Curie[5] asking the exact date on which she was expected to arrive and seeking advice on "suitable lodging of moderate terms."

We are fortunate that, while in Paris, Leslie wrote four letters to Smithalls telling him about her work and environment. She noted that the day-to-day supervision of the laboratory was actually performed by André Debierne, the discoverer of actinium.[6] For Leslie he was an excellent *chef du laboratoire,* "listening patiently while I murder his language and tie myself into linguistic knots." She thought very highly of Marie Curie, though Curie rarely visited the laboratory: "She does not appear to come around much to the students but receives them very kindly when they seek her. She does not speak English at all, nor does she appear to understand spoken English except a few scientific terms. She speaks very quickly and to the point and is very quiet in manner but by no means languid."[7]

As well as doing research, Leslie attended Curie's lectures at the Sorbonne. She found the presentations quite difficult to follow, partially because of the mathematical content but also because of the speed with which Curie covered the material. She commented that a throng of curious spectators came to see Curie teach but the numbers diminished rapidly after the first two or three lectures, once the novelty had worn off and the technical nature of the presentations had become apparent.

According to Leslie, there were now thirteen to fourteen students working in the laboratory, the majority of Polish origin. "There are only two ladies besides myself, Norwegian Mlle Gleditsch, and French, Mlle Blanquies. Of the French lady I see very little because she does not spend all her time here, but of Mlle Gleditsch I see much since she lives in the same *pension*. She has been exceedingly good to me and has prevented me from feeling lonely."[8] Later, Eva Ramstedt travelled from Sweden to join the group and Leslie's friendship with the two Scandinavian women continued for the rest of her life.

During her sojourn in Paris, the Seine overflowed its banks and devastated the city. This made life in the laboratory really miserable and for some weeks there was neither electricity nor heat. The former was endurable as some laboratories had gas jets, though others had to rely on candles. The cold, however, made the laboratories intolerable for any length of time. Moreover, the electroscopes that Leslie used for her measurements of radioactivity became inoperable in the damp conditions.[9]

Leslie's work revolved around the extraction of new elements from thorium. For a chemist used to working with grams of pure chemicals in beakers, the manipulation of kilogram quantities of minerals in huge jars and earthenware bowls must have been a completely new experience. One very real problem was the scant concern at the Institut Curie for the hazards of radioactivity. Upon her arrival, Leslie had remarked that the chemistry room was highly radioactive,[10] and she expanded upon the point in a later letter to Smithalls: "A number of people seem to be employing radium emanation [radon] at present and my electroscope is disgraceably sensitive to the influence of anyone entering from the salle active so that I spend half my time in keeping dangerous people out and in airing the room. Formerly more care was taken to prevent the distribution of activity all over the laboratory, but as the foundations for a new Institute of Radioactivity for Mme Curie are now laid, all precautions seem to have been abandoned."[11] In spite of these problems (which probably contributed to her early death), Leslie's research had gone very well. Her work, published later in three articles, established the molecular weight of the emanation from thorium (an isotope of radon) and discussed the decomposition products from thorium.[12]

Leslie spent the two years from 1909 to 1911 at the Curie Institute, though she returned to England during the summer vacations of 1910 and 1911 to help Smithalls with his domestic science class at Scarborough as an honorary demonstrator.[13] As she had yet to publish any of her work in Paris, she was ineligible for a third year of scholarship support and it became necessary to look for another position. In the last of her letters to Smithalls, Leslie asked for a letter of recommendation for entry to a teachers' training college. "The West Riding scholarship people have remembered my existence and have sent me an application form for a post in their new Teachers Training College at Bingley and I am thinking of making use of it. My lack of experience will probably be an insurmountable difficulty in the way of my obtaining such a post but I can at any rate make the attempt."[14]

The imminent publication of her work in Paris must have persuaded the committee to extend her scholarship, so she did not pursue the teaching position. Instead of staying a third year with Curie, however, she applied to work with Ernest Rutherford at the Physical Laboratory of Victoria University, Manchester. Rutherford accepted her thanks to a positive reference from Marie Curie, who called her "an assiduous and intelligent worker."[15] At Manchester, she continued her work on thorium and extended her studies to actinium in 1911–12, getting another publication under her belt in the process.[16] Rutherford commented on her work in a letter to Bertram Boltwood: "Miss Leslie is comparing accurately the diffusion constants of the thorium and actinium emanations. She has found, as I long ago anticipated, that the first values I found for the diffusion constants are much nearer right than all the later values. I think she will also find that the diffusion constants of thorium and actinium are not very different."[17] Indeed, her results showed that the "emanations" from the different decay series were probably the same substance (radon).

After leaving Manchester, she spent two years as a science teacher at the Municipal High School for Girls in West Hartlepool. During this time, she still managed to continue research with Dawson, this work being on ionization in non-aqueous solvents.[18] From 1914 to 1915, Leslie held a position as assistant lecturer and demonstrator in chemistry at University College in Bangor, Wales. It was in 1915 that Leslie entered the world of industrial chemistry, having been hired to work at His Majesty's Factory in Litherland, Liverpool. She obtained this position because the male research chemists were being called up for military duty, and it is possible that the position at Bangor was offered her for the same reason. Her initial rank was that of research chemist, but in 1916 she was promoted to Chemist in Charge of Laboratory, a very high position for a woman at the time. Her research, elucidating the chemical reactions involved in the formation of nitric acid and determining the optimum industrial conditions for the process, was later published in part.[19] This work was vital for the munitions industry, which required massive quantities of nitric acid for explosives production.

One of her young assistants at Litherland, Edward Rogans, recalled Leslie's pleasant nature.[20] Leslie had her own office attached to the laboratory and a young woman graduate, Miss Dicks, as her assistant whose task was to collect details of all work performed in the laboratory by the six staff members. The laboratory was responsible for analysis of the different chemicals used in the production of explosives, but the most dangerous work was the collection of nitric acid samples from the nitric acid absorption towers. In June 1917, the Litherland factory closed[21] and Leslie was transferred with the same rank to His Majesty's Factory in Penrhyndeudraeth, North Wales.

Leslie was awarded a DSc degree in 1918 by the University of Leeds, mainly in recognition of her contribution to the war effort. The referee for her doctoral application reported on her work with Dawson, Curie, and Rutherford and then described her industrial research: "The remaining set [of work] comprises two joint papers[22] and four independent inquiries and were carried out at HM Factory at Litherland. The confidential nature of these investigations precludes any reference to the subject matter in this report; but they appear to me to constitute the most weighty claim in Miss Leslie's application for the doctorate. The problems she has had to solve are not only of the first importance at the present time, but have been attacked in a manner showing unusual resource as well as novel methods of procedure."[23] With the return of the (surviving) male chemists at the end of the First World War, Leslie lost her position – a common fate among women workers at the end of the war.[24]

Leslie accepted a position as demonstrator in the Department of Chemistry at the University of Leeds in 1918 and was promoted in the following year to assistant lecturer. In 1924 she moved to the Department of Physical Chemistry where she was promoted to lecturer (akin to associate professor in the North American system) in 1928. Two of her former students, Charles Whewell and Emma Stott, recalled: "She was certainly highly regarded as a chemist ... Mrs Burr [Leslie] was outstanding not only because of her skill as a scientist, but also because she was one of the few women on the University staff at the time."[25]

In 1923 Leslie married Alfred Hamilton Burr, a lecturer in chemistry at the Royal Technical College, Salford.[26] Burr had also worked at the factory in Litherland in 1916 and presumably it was there that they first met. After marriage, Leslie continued to be an active researcher at Leeds, publishing an article on solvate formation in 1926 in which she determined a formula to represent the solubility of 2,4-dinitrotoluene in sulphuric acid-water mixtures.[27] Leslie also ventured into textbook writing, contributing one of the volumes of J. Newton Friend's classic *A Textbook of Inorganic Chemistry*[28] and co-authoring another.[29]

According to university records, Leslie resigned her position at Leeds in 1929. Whether she did so because of the difficulty of working so far from Burr, for health reasons, or because of pressure from her spouse is unknown. In 1931, when Burr was appointed head of the chemistry department at Coatbridge Technical College, Scotland, she moved with him.

From time to time, Leslie met her friends Ellen Gleditsch and Eva Ramstedt from the Paris days. Their first reunion was in 1920[30] when the three of them visited Soddy in Oxford[31] and Rutherford in Cambridge.[32] In 1932 the trio sent their mentor, Marie Curie, a postcard from Scotland.[33] Gleditsch also used Leslie as an intermediary when she wished to contact Frederick Soddy about his visit to Norway to receive the Nobel Prize.[34]

After A.H. Burr died in 1933, Leslie moved back to Leeds, resuming her research work at the university. Her first project was to complete her deceased husband's research on wool dyes by examining the nature of the dyeing process in cellulose acetate rayon using partition data.[35] She also worked with her former supervisor, H.M. Dawson, on hydrolysis mechanisms of sodium bromoacetate.[36] In addition to her research work, she was employed as subwarden of a women's residence, Weetwood Hall, at the university from 1935 to 1937. This was one of the types of "women's work" in science identified by Rossiter.[37] In fact, Leslie followed "scientific women's tradition" in three ways: she married a scientist, gave up work after marriage and then took up a woman's role in science.

Leslie died at Bardsey, near Leeds, on 3 July 1937, having given up research only a month earlier. While the cause of death was not recorded, it was quite possibly radiation-related, considering her young age and her experiences in Paris.

It is sad that Leslie's life and work have been overlooked for so long. Her chemical abilities were certainly acknowledged during her lifetime. She had been elected an associate of the Institute of Chemistry in 1918 and fellow of the Chemical Society in 1920.[38] Newton Friend must have thought highly of her abilities to have made her a contributing author in his definitive series of monographs on inorganic chemistry. Her obituary in the *Yorkshire Post* noted that Leslie was "one of the University's most distinguished women graduates."[39] H.M. Dawson commented that her reputation as a researcher was "deservedly high" and that as a teacher she was "exceptionally gifted." He went on: "To her intimate friends she was known as a woman of the highest ideals, of wide human sympathies and of great earnestness of purpose. Her reticence and inate modesty limited the circle of her acquaintances, but such restriction would doubtless count for very little in comparison with the respect and sincere regard of those who were privileged to enjoy her confidence."[40]

6 Catherine Chamié: Devoted Researcher of the Institut de Radium

MARELENE F. RAYNER-CANHAM
and
GEOFFREY W. RAYNER-CANHAM

Most of the women researchers at the Laboratoire Curie stayed for only one or two years. Apart from Irène Joliot-Curie, the one other exception was Catherine Chamié, who spent about thirty years there, becoming a significant figure in the operations of the institute.

Chamié's early life is a reflection of the turmoil that engulfed Europe around the beginning of the twentieth century. She was born into an affluent family on 13 November 1888 in Odessa, Russia.[1] Her father, Antoine Chamié, was a Franco-Syrian notary from Damascus while her mother, Hélène Golovkine, was Russian. Catherine completed her school education in Odessa in 1907 and enrolled at the University of Geneva in the same year. Like Sofia Kovalevskaia,[2] Yulya Lermontova,[3] and other Russian women, she had to travel across Europe for a university education. This must have been quite an adventure for a nineteen-year-old woman in those days – much more so than it would be today. By 1913 she had obtained a doctor of science degree in electrical physics. Her thesis work, carried out with Professor A. Schidlof, was on the influence of the rapidity of variations of the magnetic field on the alternating hysteresis.[4]

After completing her degree, Chamié returned to Russia. In 1913–14 she studied voltages in gas discharge tubes at the physics laboratory of the University of Petrograd. This work was brought to a halt by the First World War and for the next two years she nursed war casualties at the University of Odessa clinic. She stayed on at the University of Odessa after the war as an assistant to Professor Pimtchenko, the mathematician, performing research on the singular solutions to differential equations. During that time she was a member of the Society of Mathematicians of the University of Odessa. Chamié showed a great breadth of intellect, for she was simultaneously working on a dissertation on the methodology of science for the Philosophy of Science division of the arts faculty.

On 4 April 1919, a pogrom forced the entire French colony of Odessa, including the Chamié family to flee the city. The family travelled across Europe until they reached a refugee settlement at Vareppe, near Grenoble in Switzerland. In her rapid departure from Russia, Chamié noted that she had had to leave behind all the notes of her research. After five months in the refugee camp, Catherine took it upon herself to search for employment to support her family. Thus, she arrived in Paris in September 1919, possessionless.

Apart from finding employment, her primary desire was to improve her scientific knowledge. To this end, she enrolled in a series of courses on radioactivity offered by the Collège de France during the 1919–20 academic year. Following completion of the courses in May 1920, she was successful in obtaining a position as a professor at the École Secondaire Russe, where she gave private lessons to students preparing for the *baccalauréat* examination.

Having gained some experience in the field of radioactivity, Chamié applied to work with Marie Curie in January 1921. In her letter to Curie, she comments: "I am enclosing my curriculum vitae which shows that, for the last seven years, there has not been any possibility of working in the field of my speciality – physics – and that I have lost my laboratory skills. Several years more of such a life and I will lose the moral right to the title of doctor of science. It is for this reason that I wish to devote the spare hours in the week to research work in your laboratory."[5]

Chamié's references were glowing: a confirmation of her degree status from the University of Geneva, extolling her dedicated work and her experimental skills in the laboratory;[6] a reference from the Groupe Académique Russe commenting upon Chamié's scientific knowledge, high intelligence, and "profound analytical skills";[7] and an appreciation from the École Secondaire Russe of her zeal and intellect as a teacher.[8] On the basis of these testimonials, Curie allowed Chamié to work in the Laboratoire Curie, starting 15 April 1921. She worked on her own and in collaboration with others of the Curie group. Her first publication, a joint effort with D.K. Yovanovitch, was on the preparation of a radium salt that could be used as a standard radioactive reference.[9] Over the following years, she conducted research with Ellen Gleditsch[10] and Irène Joliot-Curie,[11] among others. Her own research started with a more detailed investigation of the ionization of quinine sulfate. This phenomenon had been investigated much earlier by Fanny Cook Gates.[12] Gates had shown that the ionization was not the result of some new form of radiation but was simply a physical process. Chamié repeated the work and extended it to show the close relationship between the potential produced and the hydration of the crystals.[13] Over the years at the institute, she published research work in many areas of radioactivity.

One particular interest of hers was the phenomenon of the grouping of radioactive atoms in different media – a form of colloidal effect.[14] Very small quantities of radioactive substances were employed in chemical reactions, thus it was often difficult to know whether the radioactive compound was soluble or insoluble in the solvent. This method, later called the Chamié Effect by Otto Hahn and Irène Joliot-Curie,[15] involved exposing photographic film to the radioactive solution. If the film was uniformly fogged, then the radioactive substance was dispersed as ions or molecules in the solution. If, however, there were distinct dark patches, then the radioactive substance must be coagulated into colloidal particles, hence insoluble in the solvent. This technique became quite well known among nuclear researchers; it was used, for instance, by the Vienna scientists Marietta Blau and Elizabeth Róna in their studies on the chemistry of polonium.[16]

As well as her research work, Chamié was responsible for the classification of radioactive minerals and the administration of the Measurement Service of the Institut du Radium. The latter task, for which she received a small salary, involved being accountable for the radium sources. Her friend and co-worker Elizabeth Róna commented: "The Institute was highly contaminated. The staff was more concerned with the safety of the radium sources than with their own. Mlle C. Chamié, who became my friend, was the custodian of the radium preparations. It was her duty to get the preparations out of a safety box in the morning and return them in the evening. A small cart with some (but not enough) lead bricks around the radium preparation was used, which she pushed to and from the safe. We left together at the end of the work day because we did not live far from each other. But each evening I had to wait outside, because she felt the need to go back to see whether she had really returned the radium preparation."[17]

Chamié seems to have played quite a major administrative role at the institute, judging from a reference that Marie Curie wrote for her. Curie noted that "she has given me, from the beginning, an excellent impression as a serious person dedicated to her scientific work. This impression has been confirmed to such a degree that I refer to Mlle Chamié for advice on the running of the laboratory."[18]

As a further indication of Chamié's status, she wrote Curie during the latter's absence, commenting upon the state of things in the laboratory: "I am most appreciative of your best wishes and I thank you sincerely for the postcard. Everything is going well in the laboratory. Several researchers have already gone on vacation [as] it is very hot in Paris. Mlle Galabert left on vacation on 12 July and I have replaced her in the Measurement Service until 20 August. There are a lot of measurements at the moment. Mlle Gleditsch has just left for a conference of university women in Brussels, we have not accomplished much work."[19] And it was to Chamié that Curie delegated the task of writing a letter of condolence on the accidental death of one of the researchers (François Raymond)[20] and the responsibility for reading and correcting the errors in a biography of Curie.[21]

For many of the women researchers, their enthusiasm had to compensate for the minimal stipends (if any) that they received. This case is clearly true for Chamié. In 1929 Curie wrote to the dean of the Sorbonne Faculty of Sciences requesting that Chamié be awarded the Jacques Vignon bursary:[22] "This bursary will be of great help to Mlle Chamié to add to the very modest salary that she receives in my laboratory and it will permit her to take a much-needed vacation." Fortunately for Chamié, the award was given in her favour.[23]

Two years later, Chamié's financial situation was just as precarious and Curie had to write another request for financial assistance. This memorandum implied that Chamié was only being paid by the institute for her work for the Measurement Service, her hours in the research laboratory being strictly a labour of love. Curie explains Chamié's circumstances in some detail: "Mlle Chamié has worked for several years in my laboratory where she has been in charge of a section of the Measurement Service. For this work, she receives a stipend which, this year, is 9000 francs. As well, she has an external position at the Russian school. The family situation is extremely difficult. Of Syrian nationality, she lived in Russia before the revolution. Her very rich family was ruined and forced to emigrate. Through her work, Mlle Chamié is the sole breadwinner of the family. If it is possible to grant her a half-load of research funding, she would be able to terminate the poorly-paying teaching work and devote more time to the Laboratory."[24]

Unfortunately, at this point the "paper trail" ceases. The only evidence of Chamié's later activities comes through her ongoing stream of publications from the Institut de Radium. These continued up to

1950, the later ones mostly being joint publications with Mlle Hélène Filçakova[25] and Mme Henriette Faraggi,[26] but the very last was a solo effort.[27] As well as journal articles, Chamié co-authored a book of radioactivity data[28] and she wrote two books on the subject of psychology; she must have maintained her interest in philosophy and social sciences from her days at Odessa. In the first work on psychology,[29] published in 1937, she gave a broad perspective of her views on personality and learning, while in the second,[30] published in 1950, she focused on the psychology of knowledge. This latter study discussed the nature of knowledge and the way it was taught and assimilated at the secondary and tertiary educational levels.

A thorough search of the Archives Curie et Joliot-Curie produced no definitive documentation on the later life of Chamié, not even an obituary.[31] However, it would appear that she worked until 1949 and died sometime in 1950.[32] Róna, in her review of laboratory contamination during the early history of radioactivity, noted that Chamié died from overexposure to radiation. This, in Róna's opinion, was due to the radiation received as Chamié transported the poorly shielded radioactive samples to and from the safe every day.[33]

Chamié's life, then, is one of total devotion to science and it is this persistence through life with little, if any, reward for which she should be remembered. She seemed to play a crucial role in the running of the Institut de Radium. Unlike Curie, but more typical of the women researchers, she did not marry; nor, it seems safe to say, did she attain any position of financial security or significant authority. Her research work was interesting but none of it proved to be of great significance. Yet we should not only focus on the "greats." The faithful, pioneering women foot-soldiers of the laboratory, like Chamié, deserve recognition for their role in those early years as well.

7 Stefania Maracineanu: Ignored Romanian Scientist

MIRUNA POPESCU,
MARELENE F. RAYNER-CANHAM,
and
GEOFFREY W. RAYNER-CANHAM

As we look back in time, it is difficult to appreciate the challenge that faced the young women scientists who decided to leave their homelands to pursue research with one of the stars of radioactivity. Stefania Maracineanu travelled west from Romania all the way to Paris to work with Curie. Her work was outside the mainstream of research in radioacitivity, but during the 1930s her results were quite controversial.

Most of the women who devoted their lives to the study of radioactivity relied on correspondence and visits for mutual support and encouragement. Maracineanu, however, seems to have been isolated from the group of women scholars. Stefania (Stéphanie) Maracineanu was born in Bucharest on 18 June 1882.[1] She had a bleak childhood about which she did not like to talk. After obtaining a degree in the physical and chemical sciences in 1910, she became a teacher at the Central School for Girls in Bucharest.

It was in 1922, with the financial support of the Romanian minister of Science, that she arrived in Paris to pursue graduate research with Marie Curie. Her first task was to determine the precise half-life of polonium,[2] the element discovered by Curie herself. Maracineanu found that the half-life of polonium seemed to depend upon the identity of the metal on which the polonium layer had been deposited. She considered that the alpha rays from the polonium had partially transformed some atoms of the metal into radioactive isotopes. Had this observation been verified, it would have been the first example of artificial radioactivity, a discovery for which the Joliot-

Curies would later receive credit. As part of her study, she devised a method for the measurement of the intensity of strong α-ray emitters.[3] On the basis of such promising research work, she was awarded a doctorate degree in 1924.[4]

Maracineanu then applied for a position in her homeland to run the proposed Romanian Laboratory for the measurement of radioactivity. Curie wrote a reference for her: "I have great esteem for the work that she has accomplished. In particular, she has acquired a perfect knowledge of precise electrometric measurements."[5] It is not clear whether she was unsuccessful or whether the position did not materialize, for she continued to work in Paris with Curie for one more year, and then at the Meudon and Paris Astronomical Observatories with H. Deslandres until 1930.

Maracineanu became convinced that solar radiation had an effect on the radioactivity of substances, and this became the focus of her research. In particular, she noted that the half-life of polonium was different when the polonium was deposited on lead, if the deposition was perfomed in sunlight.[6] She argued that the sun's rays had caused the lead, then known to be the end of the radioactive decay series, itself to become radioactive. In 1925 while at Meudron, she reported that lead exposed to sunlight produced scintillations on a zinc sulfide screen similar to that of alphaparticles.[7] This radiation, she noted, lasted as long as a month. It is important to note that Maracineanu was not alone in her convictions that sunlight could induce materials to become radioactive. Three years earlier, A. Nodon had proposed that the sun and the upper atmosphere emitted penetrating radiation, enough to discharge electroscopes and darken sealed photographic film (this may have been the effect of cosmic rays).[8] The following year, he claimed that the rate of radioactive decay was affected by the sun.[9]

Maracineanu then turned her attention to the measurement of the level of radioactivity in ancient lead roofs. Her studies suggested that south-facing roofs showed higher levels of radioactivity than north-facing roofs.[10] H. Deslandres, who presented her results, supported her findings, himself concluding that the induced radioactivity must be due to some unknown form of radiation from the sun that he designated ultra x.[11] German scientists became involved in the controversy over the source of this radiation from the sun, atmosphere, or both. One of her former colleagues at the Institut de Radium, Franz Behounek, claimed that it was simply natural radioactivity that was being found,[12] while W. Kolhörster argued in support of Maracineanu.[13]

Elizabeth Róna[14] became involved in the debate, finding no evidence that the polonium in Maracineanu's earlier experiments could rapidly have diffused into the lead to cause the supposed radioactivity of the lead itself.[15] However, Maracineanu may have been observing decay fragments that had become imbedded in the lead as a result of the recoil from particle emission, the phenomenon first observed by Harriet Brooks.[16]

Maracineanu had thought that the solar radiation had converted lead back to polonium, but by 1928 she had decided that radium D, the radioactive isotope of lead, lead-210, was actually being formed.[17] In 1929 she found evidence of mercury, gold, and helium in the sun-exposed lead, suggesting that the lead had been transmuted into the two other metals.[18] A flurry of papers on this subject appeared in 1930, most, but not all, attacking Maracineanu's claims. Fabry and Debreuil reported that they had found no evidence of nuclear transformations in old lead roofing as Maracineanu had claimed.[19] Maracineanu responded by suggesting that they had not obtained proper samples.[20] She had some supporters: the German scientist G.I. Pokrowsky had reported that X-rays and gamma rays had caused elements including lead to become radioactive,[21] while the French scientist Reboul maintained that even non-radioactive metals actually emitted some form of radiation.[22]

Three papers upheld the contention that the sun-exposed lead roofs showed evidence of radioactivity. However, the three pairs of authors – Smits and Mlle MacGillavry,[23] Boutaric and Roy,[24] and Lepape and Geslin[25] – were circumspect about the cause of the radiation. Moreover, Smits and MacGillavry objected to Maracineanu citing their preliminary and confidential results. Boutaric and Roy decided subsequently that the traces of radioactivity originated with exposure to rainwater rather than sun.[26] Maracineanu wrote a strong response to these papers, challenging their doubts about the source of the radioactivity.[27] Meanwhile, in the German literature Maracineanu and Behounek had a lively exchange on the topic of the quality and interpretation of Maracineanu's measurements.[28]

This marked the end of Maracineanu's ill-fated series of publications on the possibility of the sun's rays causing radioactive transformations. Nevertheless, Maracineanu was convinced that she had been the first to observe the phenomenon of artificial radioactivity. That is, through the bombardment of a stable isotope with radiation she had generated a different, radioactive substance. It is possible in her early experiments, when a layer of radioactive polonium was placed on different metal surfaces, some of the nuclei of the underlying metal atoms could have been affected by the alpha rays in a nuclear reaction. However, the findings described by Maracineanu more likely resulted from the recoil of the decaying radioactive atom.

There is certainly no contemporary evidence to support her theory of light-induced radioactivity.

Convinced of the validity of her arguments, Maracineanu wrote to Lise Meitner in 1936 expressing her dismay that Irène Joliot-Curie, without acknowledging Maracineanu, had used much of her work – particularly that dealing with the phenomenon of artificial radioactivity – in her publications.[29] Irène Joliot-Curie had commented in 1934, "We recall that the Romanian researcher [Maracineanu] made a public announcement of the discovery of the artifical radioactivity."[30] Maracineanu nonetheless made her disagreement with the Joliot-Curies public in a spirited statement of her claims to the prior discovery of artificial radioactivity during her work in Paris.[31] In her own country she received greater recognition. At a meeting of the Romanian Academy in 1936, the president, Alex Lapedatu, congratulated her on her research work in the field of artificial radioactivity: "This work places her among the ranks of the famous scientists who have studied the problem."[32]

In 1930, the year she returned to Romania, Maracineanu took her research in a new direction, publishing a paper in which she posited a link between radioactivity and induced rainfall. She recounted how her experiments on radioactivity induced by solar activity seemed to cause localized rain.[33] In the wet climate of Bucharest, it was difficult to prove her hypothesis that artificial rain could be generated by the presence of radioactive sources. To provide a more challenging test, Maracineanu obtained permission from the French authorities to travel to the desert regions of the Touggourt territory of Algeria.[34] According to the authorities, the results of the expedition were not promising,[35] but the representative of the governor general of Algeria sent Maracineanu an encouraging letter expressing his appreciation for her tenacity in pursuing the research in spite of the debilitating heat.[36] Upon her return to Bucharest, and as a result of her observations of the weather, she claimed to find a correlation between earthquake activity and rainfall.[37] She argued that the link was electrical purturbations in the atmosphere and the ground.

Maracineanu died in 1944. The exact date was not recorded. In August of that year Romania was invaded by the Russian army; whether her death was in some way related to this event is not known.

Although we might dismiss some of Maracineanu's proposals as unacceptable, we tend to forget that in the early part of the century, the nature of radioactivity was not well understood. Many famous scientists, including J.J. Thomson, held some beliefs that were later discredited. Maracineanu was obviously a dedicated scientist who, had she picked a different research topic, might have become a household name. Instead, she became involved in a fierce controversy of the late 1920s – on the losing side. Maracineanu did not become part of the network of women scientists. As we discussed in chapter 2, most women researchers at the Institut Curie never seemed to develop strong personal relationships with each other. This was detrimental to her career as contact with others might have influenced the direction of her work. The path she forged for herself as a scientist was isolated from the mainstream work of the time. In Romania's scientific community, however, she became a recognized figure through her work in the field of radioactivity.

8 Alicja Dorabialska: Polish Chemist

STEPHANIE WEINSBERG-TEKEL

Alicja Dorabialska was another of Curie's protégées. Her contributions were recognized in her native country of Poland, but she was completely unknown in the rest of the world as almost all of her eighty-one publications appeared only in Polish journals. Dorabialska's life was remarkably similar to that of Ellen Gleditsch; both were promoted to professorships against strong opposition and both were active in the resistance during the Second World War.

Dorabialska was another of the women who devoted their whole lives to science. She was born on 14 October 1897 in Sosnowiec, Poland,[1] a small mining town with copper, iron, and coal mines in its suburbs. At the time of Dorabialska's birth Poland was partitioned into three sectors: Russian, German, and Austrian. Sosnowiec was in the Russian sector. Although it was an ugly, smelly, neglected town, in Dorabialska's eyes it was the most exciting city in the world.

Dorabialska's father worked as a clerk at the post office while her mother was the daughter of a Polish patriot who had been arrested and sent to Siberia for his political activities. The atmosphere at home was coloured by memories of Polish heroes, rebellions against Russian oppression, Polish patriotic songs, and nostalgia. Polluted Sosnowiec was not a desirable place to spend the summer, so Alicja's parents bought a small piece of land in the suburb of Wotomin, seventeen kilometres from Warsaw, where in 1903 they built the house in which they subsequently spent their summer vacations.

There were no Polish schools in the Russian-occupied sector of Poland, so Alicja was taught at home by her mother, thus escaping the tension that Marie Curie encountered at her Russian-speaking school. In 1908 Dorabialska was immediately enrolled in a new private technical school for girls, and she loved it from the very first day, considering it her second home. Five years later, she and her mother moved to Warsaw, where her sister, Lily, was finishing high school and taking piano lessons at the Musical Conservatory. Alicja Dorabialska completed her high school program at the Technical School in Warsaw, graduating in 1914.

Her strongest desire was to study mathematics at the University of Warsaw. However, when war broke out in June 1914, it was questionable whether the university would open in the fall. To Dorabialska's delight, the German army was retreating by the end of October 1914 and the university term opened on schedule. The following May it was the turn of the Russian army to retreat, and in June 1915 her father was ordered to go east. The family moved to Moscow. At that time, Russian universities did not accept women as students, but there were so-called Higher Women's Courses with equivalency to university programs. The registrar's office in the University of Moscow's Department of Physics and Mathematics accepted Dorabialska's marks and it was there that she spent her second, third, and fourth years of university.

In 1916 Dorabialska met Professor Wojciech Swientoslawski, who became her teacher, mentor, advisor, and above all, friend. Upon completion of her degree, she left her family and returned to Poland with Swientoslawski. He had accepted the position of chairman of the Department of Physical Chemistry at the University of Warsaw and he asked Dorabialska to join the department. She started her job in August 1918. Though the position was supposed to be for two years, she stayed for sixteen, until 1934. Initially, she was also working towards a doctorate, so during those early years she was a graduate student and an assistant professor at the same time. As a physical chemist, Dorabialska studied heat changes in organic chemical reactions. Her first two research publications, recording her as sole author, appeared in 1921.[2]

Marie Curie had been Dorabialska's idol since childhood, but at that time she had no idea what Curie's research into radioactivity actually involved. After receiving her doctorate in 1922, she remedied the situation by enrolling in the radiology program at the Radiology School. Dorabialska took part in a wide range of radiochemical experiments and learned the use of radiochemical instruments. In the spring of 1925 Marie Curie came to Warsaw, having been invited by the Polish government to the opening of the Radiological Institute. The institute was a gift of the Polish people to Curie on the occasion of the twenty-fifth anniversary of the discovery of polonium and radium.

At a party given by the Polish Chemical Institute, of which Dorabialska was then secretary, she was introduced to the famous scientist. After a short conversation Curie invited her to Paris to work at the Institut Curie for the 1925–26 academic year. Thus, on 1 October 1925 Dorabialska arrived in Paris with her mother and sister. Her studies were carried out in collaboration with one of Curie's senior researchers, D.K. Yovanovitch.[3] She combined her previous study of heat changes in chemical reactions with her new interest in radioactivity to investigate heat release during radioactive changes. These were among the earliest investigations of the enormous energy released during the decay of radioactive elements.

Curie and Dorabialska became good friends and working colleagues. Curie had become almost blind from the effects of radium radiation on her pupils and her daughter Irène had to accompany her on walks. After Irène's marriage to Frederick Joliot, Curie asked Dorabialska to accompany her on these excursions. The two held long conversations in Polish on their strolls through the streets of Paris, sometimes reminiscing about their lives in Warsaw. One of Curie's biographers wrote of these walks: "One Polish girl who came to work in the new laboratory, Alicja Dorabialska, would on dark winter evenings often see Marie Curie safely from the laboratory to the Quai de Béthune. As they walked hand in hand down to the Seine, Marie would confess that she did not fully understand radium's effects on the human body; she suspected that radium was the real cause of her cateracts, and was the reason why she had to stumble so uncertainly through these streets."[4]

On her return to Poland in late 1926, Dorabialska resumed her work with Swientoslawski. In particular, the two of them designed a microcalorimeter for the measurement of heat changes during the transformation of one element to another.[5] Most of Dorabialska's publications from this time until the commencement of World War II concerned the study of heat changes accompanying nuclear reactions. The energy released in nuclear reactions became a matter of intense interest both as a power source and in terms of its destructive potential, yet Dorabialska's research, published in Polish journals, was completely overlooked. Almost all of these publications were authored solely by herself. She was also concerned about the human side of science and she wrote ten biographical articles on scientists, including one on her friend and mentor Swientoslawski.[6]

Dorabialska visited Paris again in 1929. Her sojourn was mentioned in letters from Irène Joliot-Curie reporting on the activities in the laboratory to an absent Marie. Irène commented that "things are not going too badly in the laboratory: Mlle Dorabialska is working and I have already spoken with her"; two weeks later she added, "In the laboratory, Mlle Dorabialska is measuring her polonium. There are several little problems, probably caused by chemical reactions."[7]

In 1931 Dorabialska accepted an invitation from Karol's University in Prague to study there for one year, after which she returned to the University in Warsaw. Shortly after her return, the professorship in physical chemistry at the University of Lwow (Lemberg) became vacant. Her selection as the successful candidate caused quite a furor at the university: was it acceptable to have a woman in such a high and responsible position? For Dorabialska, the debate was of no consequence: she was happy at the University of Warsaw and she was not eager to leave it. However, in September 1934 she agreed to take the position at Lwow. After a few years she obtained an assistant, Ruth Bakken, who had studied with Ellen Gleditsch in Oslo. This is our only direct indication that Dorabialska was acquainted with Gleditsch. It is quite probable that they first met during Dorabialska's 1925–26 year in Paris, for Gleditsch was there at the time.[8] Unfortunately, none of the correspondence between Dorabialska and Gleditsch has survived.

While Dorabialska viewed her years in Lwow as a time of great success and achievements, the experience was not always pleasant. She had been in Lwow for two years when anti-semitic fighting broke out among the students. Even as advisor to the Polish Youth Association Dorabialska felt helpless, and she recalled this episode with sadness and shame.

War again! Every year Dorabialska travelled to Wotomin to spend the summer with her family. In 1939 her visit came to an abrupt end when, on 1 September, the family awoke to the sound of exploding bombs – Germany had attacked Poland. Dorabialska decided to return to Lwow, a journey that took her three days and nights. A week later the Germans embarked on a siege and eventually occupied the city. On 17 September they gave Lwow to the Russians, their allies at the time. Dorabialska called the war "the six year long night of horror." Throughout this time she was a member of the underground forces, but with characteristic reticence and humility she does not elaborate on her activities. It is noted in a Polish encyclopedia that she spent the war teaching at the secret study sessions held by the Warsaw Polytechnic as a representative of the Polish underground.[9]

In February 1945, Dorabialska was appointed professor of chemistry in the reborn Warsaw Polytechnic. She stayed until the fall of the same year, when she moved to a new polytechnic with better facilities in the nearby city of Lodz. Within a few years, the school in Lodz opened its own Department of Physical Chemistry and Dorabialska was invited to become its head. During the post-war years her interest shifted to radiochemistry, the study of chemical processes using radioisotopes as tracers. In contrast to the pre-war years, she now had a regular supply of research students to help with the variety of projects that she undertook.[10] From the honorariums for her articles, she funded a symbolic memorial in the Warsaw cemetery to honour her many colleagues who died during the Second World War. The inscription reads:

> Here the dead lie alive
> In this grave of chemists, whose ashes
> were scattered by the enemy during the years 1939–45
> and did not find a place of silence in a Polish cemetery.

Dorabialska remained in Lodz until her retirement in 1968, when she returned to her favourite city, Warsaw. She died in 1975 and was buried in the cemetery containing the memorial.[11]

There are remarkable parallels between the lives of Alicja Dorabialska and Ellen Gleditsch. Both worked with Curie and became especially close to her. Later, both managed to climb the normal career ladder but against bitter opposition. During the Second World War, both did what they could to support the Allied cause. Finally, and most regrettably, neither received any recognition for their pioneering work outside their own countries, and very little within.

9 Irène Joliot-Curie: Following in Her Mother's Footsteps

E. TINA CROSSFIELD

Until Irène's marriage in 1926, the lives of Irène Joliot-Curie and Marie Curie were so tightly entwined that to tell the story of one is to recount the life of the other. Both women achieved a high level of education and shared a Nobel Prize with their scientific husbands. A competent scientist in her own right, Irène's brilliant collaborations with Frédéric Joliot contributed to developments in atomic fission, nuclear medicine, and France's post-war energy self-sufficiency. Her personal role as mother, teacher, cabinet minister, and peace advocate enabled Irène to lead an exceptional life despite her courageous battle against radiation disease.

Irène was born in Paris on 12 September 1897 under fortuitous circumstances. She had been conceived during one of Marie and Pierre's most productive scientific years. Largely unaware of the effects of radiation on human cells, Marie suffered abnormally from fatigue and morning sickness.[1] Her father, Vladislav Sklodowski, insisted she recuperate at a seaside cottage in the fishing village of Port-Blanc. When Marie went into premature labour after setting off for a bicycle ride with Pierre, she was rushed back to Paris. Irène was delivered one month early with the assistance of Pierre's father, Dr Eugène Curie. In the midst of this happy event, Pierre's mother died of cancer and the family was plunged into grief. For Irène, losing her grandmother meant that Eugène Curie would soon move in with them. He was to become her closest childhood companion, and he profoundly influenced her sense of world justice.

While Irène brought a certain amount of joy and amazement to her mother, their early relationship was inconsistent. In the grey, linen-covered journal where her mother noted household expenses, Irène's first words were jotted down, along with the dates of her first tooth and first steps. Her favourite names for her mother and father were Mé and Pé, affectionately shortened from *mère* and *père*.

Marie, however, had problems regaining her strength. The family doctor worried about a possible lung lesion and advised her (unsuccessfully) to rest in a sanitorium. A wet nurse was soon needed for Irène, while a helper was engaged for the housework. Irène began to miss the comfort and security of being close to her mother, a yearning that stayed with her throughout her young life. Unknown to Irène, part of the reason for her mother's distancing lay in the past; Marie's own mother had died of tuberculosis when she was seven, and although Bronislawa Sklodowska loved her children deeply, the common practice was to avoid excessive physical contact lest the bacteria spread. As a result, Marie refrained from hugging her own children because to her it seemed unnatural.

While Irène was loved and well cared for, her mother's attentions were divided. Pitchblende, the raw material from which Marie would extract radium, demanded long hours of toil and put an enormous strain on her health. Pierre, overburdened with teaching commitments, often returned to a curiously luminescent lab in the evening with an exhausted Marie. In the morning the couple would depart for work as usual, while Irène ached for attention and reassurance.[2] Irène learned early on that her main competition was the lab and resented every moment her parents spent there. It is more than likely that she foresaw participation in science as a means to establish closer emotional ties with her parents.

There were few periods when Irène's parents were relatively healthy. Her father complained of chronic pain in his legs, which the family doctor diagnosed as rheumatism, prescribing strychnine to ease the symptoms. Her mother was tested periodically for tuberculosis, but the results remained inconclusive. Like Pierre's discomfort in his joints, Marie's colds and frequent depressions were blamed on their hectic working schedule. Even though radiation burns were common, they were lightly regarded by the Curies – nothing that a weekend at the seashore couldn't remedy. Concerned over the state of the Curies' health, a colleague, Georges Sagnac, wrote a ten-page letter to Pierre urging them to take better care of themselves. Sagnac also recognized that something basic was amiss in Irène's life and ended his letter on a sympathetic note: "Don't you love Irène? It seems to me that I wouldn't prefer the idea of reading a paper by

Rutherford to getting what my body needs and of looking at such an agreeable little girl. Give her a kiss for me. If she were a bit older, she would think as I do and she would tell you all of this. Think of her a little."[3]

It is highly probable that Irène was exposed to varying amounts of radiation as a child. The clothing worn in the lab by her mother was the very same skirt and sleeve that Irène clung to at the end of the day. Contaminated notebooks were brought home for evening study, while vials of radioactive material were casually transported in shirt pockets. It was not yet known that radium was damaging to bone marrow and that radon (the gas produced from radium) caused fibrosis of the lungs and other respiratory diseases. Despite her mother's philosophy of fresh air and exercise, Irène was often unwell. In a letter to Henriette Perrin, a close family friend, Marie complains: "I am not very pleased with Irène, who has a lot of trouble getting over her whooping cough; from time to time she begins to cough again, and yet she had been in the country for three months."[4]

Eugène Curie's role as caregiver was of special significance to Irène's development. Grandpé, as she fondly nicknamed him, satisfied the emotional gap left by her often absent parents. Taking her on long walks as he had taken his own sons, Eugène was a patient teacher and intriguing storyteller. At the dinner table he often passionately discussed current events, and by the time she was twelve Irène had adopted many of his democratic and socialistic ideals. Much as they enjoyed talking politics, however, it was clear that Grandpé considered science to be *la vie royale*. He had practised medicine for a living but preferred to do research in the Museum of Natural History. Interested in many things, his private studies included a series of articles on tubercular infections.[5] Like Marie's side of the family, he had little use for organized religion but respected culture and was highly committed to family and community. In sharp contrast to the Spartan interior of their house, Grandpé's room was cluttered with fascinating memorabilia, his prize possession being a large bookcase filled to capacity with literature and sheet music. Irène loved every moment they spent together.

Aside from Grandpé, there was little extended family and the Curies had few social outlets. Pierre's brother, Jacques, had married and taken a university appointment in Montpellier. Marie's closest sister, Bronia, had returned to Poland with her husband, Casimir Dluskis, to establish a sanatorium. On Sunday afternoons in good weather, a collection of young scientists and students often assembled in the Curie garden to discuss the latest happenings, among them, André Debierne, Jean and Henriette Perrin, Georges Urbain, Paul Langevin, Aimé Cotton, Georges Sagnac, and a few students from the school for women at Sèvres where Marie taught physics. This élite group would remain close, professionally and personally, throughout their lives. For all the politics, literature, and music, the word radium took on a persona all its own. While Henriette Perrin recited fairy stories to her children, Irène's attention drifted towards the language of physics and the coveted world of adults.[6]

While company on Sundays was acceptable to Irène, on weekday afternoons it was not. Lonely for her mother's attention, Irène regarded any guests as unwelcome intruders. When Marie brought four of her Sèvres students home for tea one afternoon, Irène hid behind the living-room door and refused to be introduced. After being coaxed out by one of the students, Eugènie Feytis, Irène scurried behind her mother's long skirt. In a tone of voice that reminded the students of her father's, Irène repeatedly demanded of her mother, "You must take notice of me."[7] Irène slowly became friends with Eugènie, whom she treated as an older sister.

Until she went to primary school, Irène had little contact with children her own age. When her sister Eve was born on 6 December 1904, Irène soon became impatient with the noisy infant that her mother constantly carried. Eager to distinguish herself from her sibling, Irène pointed out that Eve's dark hair and blue eyes did not resemble her own fairer features. Fortunately for Irène, Jean and Henriette Perrin moved into the house next door with their two children, Aline and Francis. The rose-coloured fence that separated their gardens became a favourite play area through which the children passed chocolates, toys, and secret messages. Irène spent as much time in the Perrin household as in her own, and like their parents, the children remained friends into adulthood.

Irène was six years old when the Nobel Prize in physics was awarded jointly to Marie and Pierre Curie and Henri Becquerel. Their peaceful house was quickly invaded by journalists and photographers who hounded the family for interviews. Irène first ran from these intrusions, then discovered it was more effective if she just ignored them. Even so, she never forgot the damage and distress that all the media attention caused her parents. When Irène and her husband, Frédéric Joliot, received a similar telegram in 1935, she pulled him away for the afternoon to a local department store with the excuse of having to buy a new tablecloth.[8]

Little is known about Irène's relationship with her father. He was a kind man who worked hard and wished his daughters to have a proper education, one in which science would play a large part. When Pierre was tragically killed while crossing a busy Paris street, eight-year-old Irène could only sense that something was terribly wrong. A message was quickly dispatched to the Perrin household where she was playing to ask if she could remain there for a few days. Two of Marie's siblings, Bronia and Josef, arrived shortly from Poland, while Pierre's brother Jacques came from Montpellier. In addition to her uncles and aunt, a great many well-known people appeared at the house to offer their condolences, but Irène never saw them; she was not told of her father's death until after the funeral.

Deprived of the right to mourn for her father, Irène turned stoically silent. Pé's memory lingered painfully in the house on Rue Kellermann, and she half expected him to reappear suddenly at the dinner table, or by her bedside. Marie would never speak to her children about their father, nor about their relationship together. Although Eugène was greatly affected by the loss of his son, he believed it best to move beyond sadness and deal with the living. His positive outlook had a beneficial effect on Marie and the children. Without their blue-eyed jovial grandfather, Irène and Eve would have had little emotional release from the tragedy.

Eugène Curie assumed responsibility for managing the household, and a succession of Polish governesses came to look after the girls. The first was a friendly woman named Marya Kamienska, who was the sister-in-law of Josef, Marie's older brother.[9] From her Irène and Eve learned to speak Polish. Other housekeepers came and went, but it was Marya who made the greatest impression. The resumption of routine was critical to Marie's career as it enabled her to accept Pierre's chair in physics at the Sorbonne. In her biography of her mother, Eve mentions their constant struggle against sorrow despite Marie's best efforts and admits that their childhood years were mostly unhappy.[10]

Irène's general education was quite unique and in some ways revolutionary. Her mother believed in the value of sports and manual labour and fully expected Irène and Eve to be self-supporting as adults. Their bodies were toughened by swimming, cycling, acrobatics, and long walks in inclement weather. They also learned how to speak foreign languages, cook, sew, ride horseback, and play the piano. One Christmas, Marie and the girls vacationed in the Juras where Irène skied for the first time. Like many other activities, winter sports were still a rarity for women. Irène would later boast that "she was one of the oldest women skiers in France."[11]

Unwilling to trust the public school system with Irène and Eve's formal education, Marie persuaded her colleagues to participate in a private cooperative. The concept of an alternate school was supported by Eugène Curie, who had taught his sons in a similar fashion. The children would receive one formal lesson every day and draw from the staff and facilities at the Sorbonne and Collège de France. Jean Perrin, Paul Langevin, and Marie Curie taught physics; Henriette Perrin and Mme Chavannes were in charge of French literature, history, and visits to the Louvre; and Henri Mouton and the sculptor Magrou looked after natural science, drawing, and modelling. The plan was effective, although there was some public opposition. One gossip columnist wrote: "This little group, which hardly knows how to read or write, is given complete freedom to perform experiments, construct apparatuses, test reactions ... The Sorbonne and the building in Rue Cuvier haven't blown up yet, but all hope isn't lost!"[12]

Despite what journalists thought, the school gave many of these students an impressive advantage and many would later enter scientific careers. Francis Perrin recalled: "We lacked nothing. We learned easily, and not until I was in my teens did I take any formal exams – and then only to enable me to get the necessary paper qualifications."[13] Inside the security of this privileged, élite group, where Irène's teachers and schoolmates were all family friends, the classes seemed more like play than work. However, without occasion to meet new people, Irène would never acquire the art of casual conversation or feel comfortable in wider social circles.

Relatively free from rules or reprimands at home, Irène was shocked to discover that her mother could be strict and demanding at school. Marie insisted that her students keep their lab benches free of clutter and taught them to be meticulous in their observations and reports. Once, when Irène was caught daydreaming during a mathematics problem, Marie seized her daughter's notebook and threw it out the window. Quietly descending the two flights of stairs, Irène retrieved the book and somehow produced the right answer before returning to her seat. Thus, towards the end of these two valuable years, Irène had been carefully imprinted with important qualities: accuracy, tidiness, and the ability to pay complete attention to the task at hand.

Before the advent of antibiotics and diagnostic x-rays, illness and death were part of every household. During the severe winter of 1910, Irène was devastated when Grandpé succumbed to pulmonary pneumonia in his eighty-third year. The following summer passed slowly. Irène desperately missed her grandfather and longed for her mother's attention. Unfortunately, a combination of depression, poor health, and intense scientific activity kept Marie and her daughters apart. Although Irène and Eve spent the month of August at the seashore with their Polish aunts, Irène's plea for her mother's presence is filled with sadness. "When are you coming back … I shall be so happy when you come back because I badly need someone to caress … I have made a fine paper envelope to hold your letter. There is only one in it."[14]

Although Irène and her mother were very close, it is unlikely that Marie ever confided in her daughter about her intimate friendship with Paul Langevin. When the scandal broke, Irène was at gym class. Before André Debierne could retrieve her, schoolmate Isabelle Chavannes had unknowingly passed her a copy of *L'Œuvre,* believing it to be a story about Marie Curie's latest scientific achievement. "The Sorbonne Scandals," as the lead story was called, depicted Marie Curie as "an ambitious Pole who had ridden to glory on Curie's coat-tails and was now trying to latch onto Langevin's."[15] When Irène reached her mother's side, she stubbornly refused to leave until Henriette Perrin persuaded her to join Aline in another room.

When the Langevin affair exploded in the media at the beginning of November, many harsh anti-feminist allegations were rehashed, this time with vindictiveness. At the height of the excitement, news that Marie had been awarded her second Nobel Prize barely made an impact. Two weeks later, Marie accepted the Nobel Prize in Stockholm and spoke on the isolation of pure radium, which both stabilized her position in the scientific world and honoured the memory of Pierre. Accompanied by Irène and Bronia, she survived the official ordeal only to be rushed to a private clinic upon her return. Separated from her mother, who lay close to death, Irène agonized over another *L'Œuvre* headline that read, "Foreigners at the Sorbonne. Laboratories invaded by a mob mostly made up of foreign individuals. The numbers of women are constantly increasing, the most commendable of them are there because they are looking for husbands."[16]

Although Irène witnessed the effects of her mother's crisis, she could not fully understand its dimensions. Weakened by serious health complications, Marie appointed André Debierne as temporary head of the institute. Bronia spirited Marie away during her long convalescence to a small house in Brunoy, rented under the name of Dluska. Irène was essentially removed from the situation and sent to Montpellier to stay with her uncle. Worried about the future, Irène constantly inquired about her mother's health and well-being. Adding to the intrigue, Marie insisted that Irène use double envelopes, addressed to Mme Sklodowska and channelled through André Debierne, so her whereabouts could not be traced.

This sudden separation raised many questions for Irène. Why was her mother using her Polish name? Why must she, Irène, hide her treasured "Curie" heritage? Without a surname, or a grandfather to explain the logic of things, Irène's identity must have been shaken. In a letter that found its way to the sanitorium in the Savoie region where her mother was hospitalized after a relapse, Irène tells of visiting Langevin while *en route* to the Perrins at their Quai de Bethune apartment.[17] Perhaps she was still seeking an explanation for the terrible events that had thrown them into turmoil. In her future life, any rumours that would circulate about her own husband's passion for other women would be met with complete indifference.[18] As for Paul Langevin, whatever anger Marie and Irène felt towards him would not hinder their future scientific affiliations.

Almost six months later, Irène and Eve were invited to join their mother at the seaside cottage of Hertha Ayrton, an eccentric British physicist.[19] Of the two girls, Hertha found Eve easier to tame. The solemn and unsociable Irène could not be drawn out until Hertha discovered they could discuss mathematics together.

In July, the girls accompanied their mother on an outing in the Bregaglia Alps and the Engadine. It was there that Irène first met Albert Einstein and his son, Hans. Naturally, much of the conversation centred on physics, spoken in a curious mixture of French and German. Einstein was duly impressed by Irène's aptitude for science and pledged to her his friendship and support, promises he kept throughout his lifetime.

In between her travels, Irène prepared for the second half of her baccalaureate exam, required for entry into the University of Paris. On her sixteenth birthday, she received a letter from her mother that included a new trigonometry problem and the news she had waited anxiously to hear: she could now freely use their Curie name in all correspondence. Enormously relieved, Irène studied more diligently than ever. The mathematical salutation at the bottom of nearly every letter became an amusing game between them, ending with notes like "The derivatives are coming along all right; the inverse functions are adorable. On the other hand, I can feel my hair stand on end when I think of the theorem of Rolle, and Taylor's formula,"[20] a formula Irène had earlier referred to as "the ugliest thing I know."[21]

Many of Irène's letters to her mother are lively, interesting accounts of her life, sprinkled with sensitivity and good humour. She talks about reading German and English literature and the trouble she has separating the two languages. One particular piece comments on the political situation in England and highlights one of her maturing concerns: "I have ... seen that an English Minister is almost killed every day ... by the English suffragettes, but it seems to me that the[y] ... have not found a brilliant way of proving they are capable of voting."[22] From this time forward until her marriage to Joliot, Irène's relationship with Marie deepened, so much so that their stories seem inseparable at times.[23]

The family was reunited in the fall of 1913 in a large seventeenth-century building on Quai de Bethune, near the heart of Paris' famous Left Bank. The bohemian quarter full of artists, poets, and sidewalk cafes without doubt attracted young Eve, who was charming and fashionable. Irène, on the other hand, worked hard at her studies and longed for her quiet garden at Sceaux. She passed her first baccalaureate, the French equivalent of a high school degree, at the Collège Sévigné within the year.

Marie lived at the Quai for the next twenty-two years with only a few simple furnishings inherited from Eugène. The summer after Irène's baccalaureate she rented a small villa in L'Arcouest, a tiny fishing village in northern Brittany.[24] Anxious to be in the country, Irène and Eve left Paris early along with their governess. Marie intended to join them, but the German army invaded France on 2 August 1914, and she was unable to leave.

Irène wrote from Brittany, pleading to be allowed to return. Not only did she see a role for herself at her mother's side but she felt increasingly uncomfortable with the country people: "It means more because you yourself were accused of being a foreigner ... they say I'm a German spy ... I'm not very frightened about all this but I'm very upset. It makes me sad to think people take me for a foreigner when I'm so profoundly French and I love France more than anything else. I can't help crying every time I think about it."[25]

Marie Curie's answer was direct. She implored patience and appealed to Irène to protect the rights of those who suffered discrimination. She also suggested that Irène work hard on her physics and mathematics because during the war a lot of scientists would unfortunately be lost. "If you can't work for France right now, work for its future." And: "I'm well aware how much you have already become a companion and friend."[26]

Finally, on Irène's birthday, Marie wrote giving her daughter permission to return home as soon as Paris appeared calm and functional. Another letter, dated 20 September 1914, authorized her trip back alone: "If you can bring luggage, take the leather-covered hamper. There isn't much time for taking care of one's clothes here, so if you can, bring them."[27] When Irène received the news, she was unable to travel because she had cut her foot while rock-climbing with some of her l'Arcouest friends. Trying to hide her extreme disappointment, she wrote: "I'm beginning solid coordinates because I can't understand differential equations at all."[28]

By the time Irène joined her mother, Marie had established a fleet of donated vehicles outfitted with dynamo-powered x-ray units. One car carried Marie, Irène, and a military staff consisting of a doctor, an assistant, and a driver. The apparatus itself was fairly straightforward, but the task was daunting. In many cases the men were very badly wounded. Three years earlier, the Ministry of War proposed that soldiers wear helmets, but out of fear of looking too German, the Chamber dismissed the idea until 1915, when the situation could no longer be ignored.[29] By the time the first radiology unit had set off towards the front on 1 November 1914, 310,000 French soldiers had already died and 300,000 were wounded.[30]

For a young person of seventeen who had lived within a protected circle of friends and relatives, the scenes of horror and misery were a great shock. A short nursing course given by Les Dames de France left Irène ill prepared for what she would see and experience. The tragedies she witnessed marked her with a lifetime horror of war.

Despite the hardships, Irène found the risky and unpredictable life on the road exhilarating. They never knew whether they would reach an intended destination or where they would sleep. Food was scarce. They bartered their way through army checkpoints, changed flat tires, hunted down replacement parts for the car, and installed equipment against a backdrop of thundering cannons. During this time Marie and Irène became earnest collaborators and new expectations emerged in their special relationship. Irène assumed an almost "male" role as companion, in a sense replacing her father at her mother's side. She attributed their success at the front to the fact that her personality complemented her mother's. "[I was] more like my father and, perhaps, this is one of the reasons we understood each other so well."[31]

At eighteen, Irène went alone to set up x-ray facilities in the military hospitals of Amiens and Ypres. Teaching surgeons the finer points of locating embedded shell fragments was not easy, especially when one Belgian military doctor was "the enemy of the most elementary notions of geometry."[32] After several unsuccessful attempts to remove a piece of shrapnel from a man's leg, the stubborn doctor finally heeded Irène's technical advice. In an age when science and medicine were not symbiotic and few women pursued either profession, Irène's goals were severely challenged. Canadians and other allies are indebted to her skill and bravery, for without doubt she helped to save many of their lives near the trenches in Flanders.[33] She later wrote, "My mother had just as much confidence in me as she had in herself."[34]

After the Treaty of Versailles, Irène was awarded a military medal for her civilian services while her mother's contribution was overlooked. Marie's granddaughter, Hélène, attributes this rude government oversight to the fact that, nine years later, Marie's name was still privately associated with scandal in certain political circles.[35]

In 1918, at the age of twenty-one, Irène was named *préparateur* to her mother and was teaching classes in radiology to future medical technicians at the Institut du Radium. And during her absences Marie trusted Irène to keep the new institute in order: "Do as much as you can to keep people from making a hash of things while I'm gone."[36] At twenty-three, Irène obtained her second baccalaureate in mathematics and physics at the Sorbonne, as she moved towards the prerequisites for her PhD. Marie's suggestion that Irène work on the atomic weight of chlorine, which demanded great chemical precision, led to her first publications in the Academy of Science's *Comptes Rendus*.[37] One outstanding requirement remained, the *agrégation*, a special certification needed for teaching in the French academic system. Because of her war service, Irène was missing one-third of the classes in natural science. This meant postponing her doctoral work for another year in order to complete the missing courses. Attempting to bypass the rigid requirements, Marie appealed (unsuccessfully) to the minister of Education and the vice-rector of the Sorbonne. These actions would later reinforce the idea that Marie not only sought to protect her daughter's scientific education but discriminated in Irène's favour at the institute.[38]

The Curie family went on a grand tour of the United States in 1921. The trip was organized and financed by Marie (Missy) Mattingly Meloney, an empathetic journalist who wrote for a well-known women's magazine in New York. The entire venture was organized for the purpose of obtaining sufficient radium for Marie to continue her scientific work. The demands of the American trip were extravagant; university tours, speeches, excited crowds – many people waited for hours to glimpse the famous radium woman who had discovered the cure for cancer. Unable to continue at such a hectic pace, Marie's fragile health soon failed. Irène was then approached by the trip's organizers to receive honorary degrees as her mother's proxy. As a bonus, Irène was personally invited to give three lectures on radium, which she delivered in her best English.[39] Irène would return to America twice more, while Missy remained a loyal friend and confidante.

There were few individuals outside her academic circle whose friendship Irène accepted. One such person was Angèle Pompëi, whom Irène met while touring the Auvergne with a group of Sorbonne graduates. Drawn together by their common interests, Angèle was able to sustain the long silences required by Irène. Both women were avid hikers and often camped out or slept in mountaineers' shelters when it rained. Irène, the more adventurous of the pair, was fascinated by lightweight mountain equipment. She was known to modify her clothing by sewing on extra pockets to hold useful tools and gadgets.[40] During these holidays, she and Angèle often spoke of the need to fight for social progress and the importance of obtaining the vote for women. Irène's strong convictions eventually led her into the political arena where she represented both women and science.

Early in her career, Irène's dual status as daughter and personal assistant of *la patronne* became a problem. While some residents of the institute failed to recognize her scientific ability, others privately referred to her as a "lump on the log."[41] They grumbled about unfair privileges, and nasty rumours circulated about Marie's bias towards her daughter. Many people felt that Irène had been promoted too rapidly for her talents and Irène, with her blunt indifference, was disinclined to smooth ruffled feelings. Part of the jealousy focused on her work with polonium, the first element that Marie had isolated in the shed on Rue Lhomond. Polonium is a powerful emitter of α-rays, and this property became the focus of Irène's doctoral thesis in March 1925.[42] Polonium was not plentiful and Marie controlled the largest quantity in France, to which Irène laid the greatest claim.

To a large extent Irène felt safe within the confines of the laboratory. She worked under her mother's tutelege, insulated from petty rivalries, and moved within a small but affirming social milieu. Students at the institute would long remember Irène answering technical questions or participating in a quick discussion, while lifting her acid-stained smock to warm her rear in front of the radiator on cold mornings.[43] She was also seen in the corridor "shaking the radioactivity out of her hair and clothing" and was banned from the instrument room lest she set off their finely calibrated instruments.[44]

In public, the slight young woman with the green eyes and cinder hair gave the impression of cool severity. Privately, her L'Arcouest friends, who had known her since the days of the cooperative school, enjoyed her spontaneity and love of dancing. When interviewed by a female reporter from *Le Quotidien* in March 1925 after earning her doctorate, Irène confidently stated that family obligations "are possible on condition that they are accepted as additional burdens. For my part, I consider science to be the paramount interest of my life."[45] She might have continued fleshing out the subtleties of polonium had Frédéric Joliot not followed the advice of his old teacher, Paul Langevin.

After leaving the École de Physique et de Chimie, Frédéric had trained in an industrial lab, then served in the military. His educational background did not include the École Normale, which made him ineligible for the Sorbonne and a coveted scientific career.[46] Langevin, convinced that Joliot deserved a second chance, urged him to apply for a Rothschild grant and recommended him to Marie, who accepted him almost sight unseen. Once at the institute, Joliot was placed under Irène's supervision. He knew very little about radioactivity, and she helped him acquire a good deal of expertise. Irène's speed and dexterity when handling polonium was unsurpassed by her co-workers. Frédéric was equally impressed by her numerous articles on α-rays. In their future collaborations he would go to great lengths to prove his scientific ability, but Irène was always the better chemist.

Born in 1900, Frédéric Joliot was a tall, clean-shaven young man with an athletic build and cheerful disposition; his dark eyes and fine features reminded his friends of a youthful Maurice Chevalier. Since childhood he had idolized the achievements of Marie and Pierre Curie, clipping their combined portrait out of a popular magazine at the age of six. His sister had the photograph framed and it remained one of his treasured possessions.[47] Like Irène, Frédéric had never been a follower of religious doctrine but held strong opinions on social justice, patriotism, and the futility of war. Excited and gratified to be living out a cherished dream, he was a keen and enthusiastic student. His talkative nature and flamboyant hand gestures invoked hoots of laughter in the usually staid halls of the institute. Dispensing with formalities, he invited people to call him "Fred."

When Irène was too busy to answer his scientific questions, he would wait for her at the end of the day. Gradually, she allowed him to escort her home. Their evening walks evolved into quiet detours along the Seine, where the couple discovered a genuine liking for each other's company. As Fred grew to know Irène better, his opinion of her changed: "I began to notice her. With her cold exterior, forgetting sometimes to say good morning, she did not arouse a feeling of sympathy in the lab. But I discovered in this young woman, whom the others saw somewhat as an unpolished block, an extraordinarily sensitive and poetic person who in many ways was the embodiment of what her father had been. I had read much about Pierre Curie, I had heard from teachers who knew him, and I found in his daughter the same simplicity, commonsense, and ease."[48] Joliot later confessed to a Japanese colleague, "If I hadn't been able to marry her I decided to remain a bachelor."[49]

When colleagues learned that the unsociable Irène was to marry the dashing, energetic Joliot, three years her junior, gossip spread throughout the small scientific community. Irène remained unperturbed, caring little about what others thought might be the real reason behind their union. She had quietly chosen the man best suited to share her life, a decision based on love. In a letter to Angèle, Irène wrote: "[Fred and I] have many opinions in common on essential questions."[50] While Angèle was not surprised by the news, Marie was astonished. It seems odd that the changes in Irène's attitude would have escaped her mother's notice, given how interwoven their existence had become. Earlier in the year Marie had accepted an invitation to visit Rio de Janeiro. Perhaps her insistence that Irène still accompany her veiled a hope that her daughter would reconsider her relationship with Joliot.

News of the engagement also surprised Eve, who felt that Irène and Fred were very different: "She was as calm and serene as he was impulsive. By nature very reserved, she found it difficult to make friends, while he was able to make human contact with everyone. She took little interest in her appearance and dress, while he was good-looking, elegant, and always a great success with the opposite sex. In argument, Irène was incapable of the least deceit or artifice, or of making the smallest concession. With a hard obstinacy she would present her case, meeting her opponent head on, even if he occupied a high social position. Frédéric, on the contrary, without yielding on anything basic, knew magnificently how to use his intuitive understanding to put his opponent in a condition to accept his arguments."[51]

The marriage took place privately in Paris on 9 October 1926, after a short engagement. Following the ceremony, the couple returned to the laboratory for a few hours, then curiously spent their wedding night apart – Fred at his mother's place in Montparnasse, Irène at Quai de Bethune. Marie Curie reluctantly accepted her son-in-law, whose only vice seemed to be that he smoked too much. This pleased Irène but sometimes she, too, was sad over the loss of prior intimacies: "My mother and my husband often discussed things with such warmth, replying to each other with such speed, that I could not get a word in, and I had to insist on being allowed to speak."[52]

Fred finally completed his license in 1927, which enabled him to augment their family income with a teaching job. From 1928 onwards, the couple's names appeared frequently on papers submitted to the Academy of Sciences and they became known as Joliot-

Curie, though they still signed individually on scientific articles. Fred remained sensitive to public criticism about his modified name. On 3 May 1934, he wrote to Irène about an invitation sent to her by the Physical Society in London to attend its October Conference. "I am somewhat vexed by the phrasing of the letter, which gives the impression that I am at your orders. It is that wretched question of the signature which people interpret badly and which pains me."[53]

Pierre Savel, a colleague who joined the lab in 1931, was introduced at the end of his first morning to a young, pregnant woman whose lab occupied the ground floor. "She [Irène] would appear every morning and evening, not only to find out how the installation [of the cloud chamber] was going, but to take Joliot away. It needed all her persuasiveness to get him to agree to leave the lab, either for lunch or to go home at night. Each day, he would start work the moment he got in, only putting [off] his white overall after endless pleas from his wife."[54]

Motherhood proved an enjoyable experience to Irène, and despite her earlier comments on the priority of science, she devoted much time to her children. After the birth of Hélène she told Angèle, "I realized that if I did not bring children into the world, I would never be able to console myself that I had not made that astonishing experiment while I was still capable of it."[55] As with her first child, Irène was at her workbench only hours before their son, Pierre, was born in 1932. Already suffering from anemia, she contracted a severe case of pleurisy that lasted for months. This was the first real indication of a disease, probably leukemia, that began to take the lives of so many people she knew.[56] Fortunately, Fred was willing to share the burdens of children and household while she rested periodically in a sanatorium.

In 1930 André Debierne had said to Joliot: "You've come too late to study radioactivity. The families and the decay series of these bodies are known, and there is hardly anything left to do than to work out the third and fourth decimal place in their characteristic different qualities."[57] Fred and Irène were to prove him wrong. They were both interested in polonium's α-rays, but original research demanded new techniques and instruments. It was a dangerous undertaking as polonium is extremely toxic to the body, tending to concentrate in organs such as the lungs, liver, and spleen. The first task was to prepare batches of highly radioactive polonium, then develop methods of tracing what happens when α-particles collide with atomic nuclei. A more sophisticated means of detecting these particles was also needed, and the Joliot-Curies perfected the art of "particle" counters. At the end of this work, the French team was probably the best technically and certainly the best equipped in terms of polonium. They were now ready, in 1932, to make their key contributions to nuclear physics. It was a collaboration *par excellence*.

In the fast-growing field of atomic physics, a few hundred players published weekly accounts of their recent findings. All were perplexed when, in 1930, German physicists Walter Bothe and Hans Becker reported an experiment whose interpretations were unclear. They discovered that when the elements boron and beryllium were bombarded by α-particles, a low-intensity radiation resulted that could pass through ten centimetres of lead. As no other explanation could be found, Bothe and Becker believed they were seeing shortwave electromagnetic radiation of hard x-rays.

Having worked behind protective screens of lead and wood, Irène and Fred were fascinated by this mysterious radiation and used their most powerful tool, polonium, to study it. With an improved Geiger counter to measure intensity and a cloud chamber to photograph the tracks left by alpha particles, they succeeded in demonstrating two new properties of the Bothe-Becker phenomena but didn't know how to interpret them. The British team of J. Chadwick and H.C. Webster were hard at work on the same problem in Cambridge. Inspired by a paper published by the Joliot-Curies in the January 1932 issue of *Comptes Rendus*,[58] Chadwick and Webster hurried to test and proclaim the correct interpretation.[59] The neutron, predicted by Rutherford as early as 1920, had finally been identified and the Joliot-Curies suffered their first major miss.

In the 1930s, physicists had some knowledge of the behaviour of electrons, protons, and neutrons, but the subatomic picture was far from complete. From April to May 1932, the Joliot-Curies worked at the Jungfraujoch scientific station, 3,500 metres high in the Swiss Alps, doing experiments on cosmic rays and their effects on atomic nuclei. American physicist Carl Anderson had sometimes detected positrons in the upper atmosphere, thought to have resulted from complex "atomic" collisions. Reaching the conclusion that their research was leading nowhere, Fred wrote to his mother, "We are working hard here, and when the weather is good we ski on the glaciers."[60]

At roughly the same time, Anderson was studying thousands of "tracks" in his Wilson Chamber when he happened upon the remarkable phenomenon that confirmed the existence of the positron, or anti-electron, as predicted in 1930 by British scientist P.A.M. Dirac. Once again, the Joliot-Curies had missed a major scientific event, even though they had been the first to photograph a pair of electron/positron particles, originating from a single ray of energy.[61] In a competitive spirit, they again modified their instruments and, in 1933, were the first to measure the mass of the neutron accurately.[62] While doing these experiments, another "atomic" event occurred. When bombarding aluminum with alpha rays, electron/positron pairs were curiously absent, and instead, protons were detected. The Joliot-Curies suggested that these protons "transmuted" into a neutron and a positron, a daring new hypothesis that signalled their most important work so far in nuclear science.[63]

In October 1933 at the Seventh Solvay Conference in Brussels, Irène and Fred anxiously presented their results to an élite physics community. Out of a distinguished group of forty-six theorists and experimentalists, only Albert Einstein was missing, having already emigrated to the United States. Set against a backdrop of bearded men and two women (Marie Curie and Lise Meitner), whose black dresses blended into the audience, Irène's short-sleeved blouse and Fred's clean-shaven face gave the impression of youth and inexperience. As usual, Irène asked Fred to deliver their joint report, something she always asked of him because she felt awkward in front of an audience. Both were excited about their research but shocked by its reception. Instead of praise and support, a heated controversy erupted. When Langevin called upon Lise Meitner for her opinion, she said, "My colleagues and I have done similar experiments. We have been unable to uncover a single neutron."[64] Fred and Irène were especially dismayed to hear Meitner disagree. For the remainder of the conference most of the delegates avoided the couple, swayed by Meitner's influence. Others thought the Joliot-Curies had rushed their interpretation because they had narrowly missed identifying the neutron and misread the evidence of the positron. Only Neils Bohr and Wolfgang Pauli offered encouragement.

Back in Paris, Irène and Fred devised a series of new experiments. One evening, Fred was conducting a polonium experiment when something unusual happened. After bombarding an aluminum sheet with alpha particles, he noticed that positrons continued to appear even though all the neutrons had been absorbed. The emission did not lessen when the polonium source was removed, as the Geiger counter clicked for several more minutes. Having worked intensively with counters, it was second nature to associate the clicking with a diminishing half-life. Excited, Fred rushed to find Irène, who immediately recognized the phenomenon as a new radioactive element in the process of nuclear decay. Four days after the initial observation, their report was presented to the Academy of Sciences and published in *Nature* on 19 January 1934.[65]

Physical and chemical proof of their new radioactive element was now essential in order to convince the scientific community of its existence. With a half-life of three minutes and fifteen seconds, isolation had to be done quickly or the "artificial" element would degrade into stable silicon. Irène, the skilful and patient radiochemist, brilliantly isolated the new substance while Fred, the inventor, assisted by fine tuning the proper instruments. Together, they proudly presented a vial containing their new creation, phosphorus-30, to Marie. A second paper was presented to the academy entitled "Chemical Separation of the New Elements that Emit Positive Electrons."[66] Their discovery, artificial radioactivity,[67] paved the way to nuclear energy. Medicine would also benefit, as radioactive isotopes could now be produced in large enough quantities to satisfy the needs of doctors who specialized in treating cancer patients.

Although Marie heard rumours that the next Nobel Prize for chemistry would go to Irène and Fred, she did not live long enough to congratulate them. On 4 July 1934, she died of pernicious anemia at the age of sixty-seven. Eve and a doctor sat by Marie's bedside for sixteen hours, but Irène was too distraught to remain in the room. In a recent popularized version of Irène's life, the notion is raised that finally Irène was free to pursue her own life.[68] In fairness to Marie Curie, Irène's search for herself began when she chose to marry Fred.

The Joliot-Curies were pleased by the international recognition of their work, but their trip to Stockholm held mixed memories for Irène. She received her leather-bound certificate and medal from Gustave V, the same king who had awarded them twice to her mother and once to her father. At the official address Irène spoke first, recounting the physics of nuclear decay. Fred, considered to be the physicist, dealt with the chemical aspects of radioisotopes. Irène insisted that their lecture be divided this way because she felt that her father had never been given enough credit for his part in the isolation of radium. She was therefore determined to protect her husband from similar omissions. Fred went on to envision chemical chain reactions of an explosive character, warning that the release of atomic energy could be misused. Like Marie and Pierre before them, they chose not to take out patents on artificial radioactivity. Unlike Marie and Pierre, Fred and Irène keenly felt their responsibilities as scientists in a world of change.

The Nobel award brought instant fame and prosperity to the Joliot-Curies. Fred was quickly promoted to director of research at the Caisse nationale de la Recherche Scientific, a position he held for the rest of his life. On his recommendation, the Caisse bought the aging Ampère plant at Ivry and transformed it into the Atomic Synthesis Laboratory where artificial radioelements could be mass produced. Because they found themselves working in different institutions, Fred and Irène's intimate collaborations took a different form. When Fred accepted the chair in nuclear physics and chemistry at the Collège de France, Irène had already replaced him at the University of Paris. With their leadership secure at the Sorbonne, Collège de France, Caisse Nationale, and Ivry, they controlled every piece of serious nuclear work in France.[69]

Prior to 1940, atomic discoveries could have been made by any gifted team of physicists given enough chance, opportunity, and perseverance. The Joliot-Curies were on the verge of finding Chadwick's neutron and Anderson's positron. The neutron would later bring fame to Enrico Fermi, who read with interest Irène and Fred's concluding sentence in *Comptes Rendus*: "Sustained radioactivities, analogous to those we have observed, can without doubt be brought about by bombardment with other particles."[70] Assembling his own apparatus, Fermi and two other colleagues began to bombard different elements with neutrons, which appeared to transform them into other radioactive isotopes. For example, uranium produced transuranium isotopes with half-lives of ten seconds, forty seconds, thirteen minutes and ninety minutes, setting off a flurry of scientific investigation elsewhere. Lacking a good explanation of why it happened, Otto Hahn and Lise Meitner confirmed the existence of Fermi's thirteen-minute isotope and, by 1937, added nine others to the list. The suggestion made by chemist Ida Noddack about uranium simply splitting and not decaying into other radioactive elements was not taken seriously.

Irène found it difficult to believe in the findings of Hahn, Meitner, and other physicists. In collaboration with Yugoslav physicist Paul Savitch she attempted to characterize one of the newly proposed elements. Instead, she reported seeing nuclei resembling the substance actinium, an element lighter than uranium that she labelled "R-3.5."[71] After these findings were published, a letter from Hahn and Meitner suggested that if Irène didn't retract her paper they would print their criticism of it at the next opportunity. When Joliot met Hahn at an international congress in Rome that spring, Hahn said that while he respected Irène's expertise, he was convinced she was in error. Later, Hahn was overheard to say, "This damned woman. Now I will have to go home and waste six months proving that she was wrong."[72]

Irène and Savitch countered with another report in May that further distressed Hahn and Meitner. Hahn wrote another letter, this time pressuring Fred to exert his influence to stop Irène from contradicting the German results. When Irène showed Hahn's letter to Savitch, he replied, "I feel we should continue our work and let Hahn do what he wants."[73] Meitner, commenting from her forced exile in Scandinavia, voiced the opinion that Irène was still relying on her mother's chemical knowledge, "and that knowledge is just a bit out of date today."[74] Subsequently, Hahn refused at first to read the French team's latest articles, calling their R-3.5 "actinium-like" element "curiosium."[75]

Then, in an effort to settle the French-German dispute, he returned to his bench and uncovered a host of radioelements he had never seen before. Rapidly dispatching a note to Meitner, Hahn challenged her for a plausible explanation. Discussing the matter with her nephew, Otto Frisch, over the holidays, Meitner proposed the idea of nuclear fission[76] and Hahn published their theory several weeks later.[77] When Irène read the German article, she was appalled to discover that she and Savitch had missed the greatest scientific event of the century. Undaunted, Fred presented a new paper to the Academy of Sciences on "experimental proof of explosive splitting of uranium and thorium nuclei under bombardment from neutrons,"[78] along with the first ever photographs of fission fragments.

Even though Irène and Fred were apart during the day, their collaborations continued well into the night. Indeed, Fred felt it necessary to verify every essential calculation with Irène, who was always willing to listen no matter how ill she felt. In 1939, when she asked him where their studies were headed, Fred's answer was swift: neutrons and a chain reaction. Shortly afterwards, he produced the first experimental proof of a controlled chain reaction at the Collège de France and devised a simple atomic generator to house it. Along with Hans Halban, Lew Kowarski, and Francis Perrin, Frédéric Joliot applied for five nuclear energy patents with France's future needs in mind.

Strengthened by Fred's influence, Irène was determined to speak out on non-scientific matters that she felt strongly about. In a long interview in *Journal de la Femme* on 23 November 1935, she said: "I am not one of those ... who thinks that a woman [scientist] ... can disinterest herself from her role as a woman, either in private or public life ... If [the Nobel award] had thrust my name, the name of a woman, a little more in the limelight than on other days, I feel it is my duty to affirm certain ideas that I believe useful for all French women. Therefore, I have accepted the presidency of several meetings where the rights of women are discussed."[79]

In the five years that preceded the Second World War, the Joliot-Curies became very much tied to the political turmoil of their day. Even though internal problems in France were primarily economic, violence, which had not been seen since the suppression of the Commune in

1871, erupted in Paris on 6 February 1934.[80] This civil unrest was broken for a short time by the Popular Front government, which rose to power in 1936. Under the direction of Leon Blum, the Popular Front also provided the Communist party with an opportunity for growth. Armed with the slogans "Defence of Liberty, Defence of Peace, and Economic Demands,"[81] a new comprehensive program was proposed for a forty-hour work week, an unemployment plan, and improvement in public works. Blum was anxious to bring women into the government for the first time.[82] He selected three women whom he considered progressive in outlook and universally known: Irène Joliot-Curie, Suzanne Lacore (a well-known worker in child welfare), and Cecile Brunschweig (a champion of women's issues). Irène's appointment was the first in France to link science with national development.

To the astonishment of her friends, Irène agreed (albeit reluctantly) to serve in order "to make it easier for other women to also enter the government."[83] To Missy she wrote, "Fred and I thought I must accept it as a sacrifice for the feminist cause in France."[84] A limit was self-imposed on her term of office, so that neither her research nor her health would suffer. She reported for work in June 1936, her only mandate to consolidate scientific research.

Predictably, Irène's blunt mannerisms were instantly misinterpreted. One day when a new secretary asked her to sign a letter of refusal for a government function, Irène refused because it contained the phrase "I regret I cannot attend." Irène was not sorry she couldn't attend and insisted that the line be removed.[85] Every day she ploughed through enormous stacks of mail consisting of government business, solicitations for money to fund special projects, and appeals from people explaining health problems. When she felt she could do nothing to help, she was honest and straightforward, an attitude most people found refreshing. After six months she handed the portfolio over to Jean Perrin, her agreed-upon successor, but not before awarding a grant to her old friend Eugènie (Feytis) Cotton for improvements to her school of physics for girls in Sèvres.[86]

After war was declared, Irène wrote a comforting letter to Hélène, who was staying in L'Arcouest with her brother Pierre. This letter is similar to the one she had received from her mother as a young girl: "You have heard us speak about these things often enough to know that its coming [the war] does not surprise us ... We are busy organizing the laboratories to work for national defence. If all this lasts long, and it is quite likely, I shall send you the books you will need to do your studies for the year."[87] When Fred was offered refuge in Britain for himself and his family, he declined. Both he and Irène felt it their duty to remain behind in order to keep French science alive.[88]

Periodically confined to bed, much of the news surrounding the occupation reached Irène through her immediate family. Food was scarce, and items such as meat, butter, milk, and eggs had all but disappeared from the marketplace. She continued her scientific work at home and made arrangments for Hélène to attend school and board with a family in Besançon, near the Swiss border. Young Pierre would winter with them in Antony, not a pleasant prospect as most of the coal needed by Paris and the suburbs for heating had been shipped to munitions plants in Germany. Irène's biggest worry was Fred's increasing involvement with the resistance, something she knew about but never discussed. Clearly, it was safer not to know Fred's whereabouts, or with whom, exactly, he was associating.

Even though his lab had been taken over by the Nazis, Joliot was fortunate in that the physicist whom the Germans unwittingly sent was a colleague and anti-Nazi sympathizer, Wolfgang Gentner. Gentner had previously worked with the Joliot-Curies during the discovery of artificial radioactivity. Among other things, he was able to keep Joliot's cyclotron, the first in western Europe, from being dismantled and shipped across the German border. Gentner was also the first person Irène turned to when Fred was arrested a second time for encouraging resistance among university students.[89] The second person she called was Angèle Pompëi, who helped her wait out what must have surely been the longest day of her life.

Thanks to Gentner's swift intervention, Fred was only briefly interrogated before being allowed to leave of his own accord. However, following the arrest and execution of a number of prominent scientists, including physicist Jacques Solomon, Langevin's son-in-law,[90] Fred officially joined the *Front National,* serving as president until the end of the occupation. The French communists, who gained new momentum after Hitler violated the Nazi-Soviet Non-aggression Pact, were regarded by both the Germans and the Vichy government as more troublesome than any other single group. Although Irène was sympathetic to communist ideals, she never became a Party member. One suggestion offered by a colleague is that she was far too independent to submit to Party discipline, which had already exerted enormous pressure on Fred.[91]

In June 1941 Irène undertook a risky mission to retrieve the historic one gram of "American" radium, along with some other precious metals belonging to their laboratory, that had been hidden at Clairvivre during the Paris evacuation. As she was often ill and resting in the mountains, it was believed that her movements would not be questioned (Fred was not permitted to leave the occupied zone.) Accompanied by two technicians with a series of travel documents to prove their religious affiliation, her journey was successful, if exhausting.

From November 1941 to May 1943 Irène returned periodically to the mountains, but her health failed to improve. While visiting a special tuberculosis sanatorium in Leysin, Switzerland, she was able to correspond with Fred through Gentner's wife, who was a Swiss citizen.[92] Trying to keep busy by writing a history of radioelements, her letters to Fred and the children are a *mélange* of sadness and affection. Again, offers were made for her to leave France, this time by Missy in New York and her sister Eve in London. Irène wrote Missy: "Til now nothing is changed in our intentions. We think we have a mission to fill: to prevent the dispersal of the scientific workers from our laboratories and the loss of the radioelements needed for our work that were collected by my mother and ourselves. If we find that, for some reason or other, we cannot be useful anymore, then we will try [to leave] ... If there are no other reasons than the hardship of life, next winter, it is not so very dangerous for [my health]."[93]

By 1944 Fred's activities in the resistance were beginning to threaten his family. He was dismissed from his university post, just as Langevin had been shortly before his arrest, and detained a third time for questioning. Rumours circulated that he would soon be taken by the Nazis. The time had come for Irène, Hélène, and Pierre to seek asylum in Switzerland, their safe passage facilitated by Wolfgang Gentner, who was now also suspect because of his dual role. Fred assumed a false identity and spent his days preparing for the inevitable "war of the streets." He collected and crafted materials that could later be transformed into weapons and stored them in a secret place at the college. On August 23 Fred joined in the fighting at Saint-Germain-des-Prés, hurling his handmade Molotov cocktails at German tanks.

Irène's journey to Switzerland with her children was carefully orchestrated. Collecting Hélène at Besançon, they set out in the company of two guides for the strenuous hike across the Jura mountains on 6 June 1944. Each carried a few light provisions except for Irène, who insisted on bringing, among other things, a large, heavy tome on physics. At the picturesque village of Porrentruy they settled into a house overflowing with refugees. Their presence was soon discovered by the local civil administrator, who invited them into his own home for several days. From Porrentruy they continued on to Lausanne. During this time Irène received a visit from Paul Langevin, who also made his escape aided by the same group that had helped the Joliot-Curies. Irène, Hélène, and Pierre would wait three long months for any news of Fred.

With the liberation of Paris, the family was reunited by September 1944. Irène had made a satisfactory recovery during her stay in Lausanne thanks to rest and a package of new American antibiotics sent by Missy. She was able to return to her laboratory and her research, a study of the gamma rays of ionium (thorium-230).[94]

Although Paris did not suffer to the same extent as various other cities during the war, when peace arrived food and other necessities remained scarce. In a letter to Ellen Gleditch, Irène wrote: "I have received the bouillon [beef cubes] and a few small things from Mme [Eva] Ramstedt. I can't thank her because I don't know her address."[95]

Fred was decorated with the croix de guerre, promoted to commander of the Légion d'honneur, and made head of the Centre National de la Recherche Scientific, where he began to reconstruct much of France's besieged scientific community.[96] In October 1945 he was appointed high commissioner of the Atomic Energy Commission (AEC) and Irène was included among the other AEC appointments. Irène's task was to locate uranium for the reactor's fuel source. Having always been interested in both minerology and geology, Irène trained over three hundred uranium prospectors in a laboratory at the Museum of Natural History. The prospectors then scoured the outcroppings within France's borders, successfully locating a deposit of pitchblende in sufficient quantity. In the meantime, Fred worked on construction of the reactor called "Zoë," or Zéro énergie Oxyde et Eau lourde.[97] On 15 December 1948 he announced to the press that France's first atomic pile had come online. Technically, the reactor could only deliver five kilowatts of energy, but the psychological rewards were great. To the people of France, Zoë symbolized the nation's rise from the ruins of defeat.

Irène and Fred visited Moscow for the two hundred and twentieth anniversary of the Russian Academy of Sciences. It was her first trip to Moscow and his third. They returned to the college with stories of hard-working Russians who were entitled to higher education on the basis of their abilities, not their station in life. These and other communist ideals began to look more and more attractive to French intellectuals as their country restructured.

Fred's continued involvement with the Communist party caused Irène to be detained on Ellis Island at the start of her third and final trip to the US. She had accepted an invitation from the Joint Antifascist Refugee Committee to speak in support of Spanish refugees, but her visa was refused until vigorous protests were made on her behalf. When the dust had cleared, she visited Einstein at his home in Princeton and met with Robert Oppenheimer and Neils Bohr at the Institute for Advanced Study.[98] Speaking at a Refugee Committee function in Boston, Irène responded to questions on the danger of nuclear weapons. Several scientists at the meeting scoffed at her belief that "radioactive fallout can ruin crops, injure plants, and destroy human beings."[99] They were certain that she was exaggerating.

As the cold war gained momentum, anti-communism became the most prominent American political issue. When anti-communist sentiments barred Marguerite Perey[100] from obtaining a travel visa, Irène voiced her outrage in a letter to Gleditch: "It was the American Consul in Strasbourg that refused her [Perey] saying that everyone at the University of Strasbourg is a Communist!!! ... I assure you that I never had this impression when I was there."[101] In reply to Gleditch's concerns about the Washington meeting of the Commission on Radioactivity in September 1951, Irène writes; "Fred and I, we are unable to take part and prefer to wait for the next one in 1952. We have visas but there is no guarantee that they will be honored."[102]

In March of that year, Fred was attending a meeting of the World Committee of Partisans for Peace in Stockholm when he found himself going from one hotel to another, trying to find one that was willing to lodge a "Red."[103] This attitude extended to the scientific community as colleagues who were afraid of being labelled communist by association avoided him. At the college, "many preferred to walk downstairs five or six flights rather than take the same elevator."[104] Likewise, the traditional Sunday open house, inherited from Marie and Pierre Curie, became a rather lonely affair. In 1954 Irène's application to the American Chemical Society was rejected. The worst snub came in 1955 at the first United Nations Atoms for Peace Conference (which included a Russian delegation); Irène and Fred were not invited to participate because the French government felt their presence was unnecessary. When Fred lost his important ties with the government, the Communist party let him slip, more or less, into oblivion. Scholars are undecided whether Fred chose to remain in the Party of his own will (even though he knew he was being manipulated), or because he was unable to leave for some undisclosed reason.[105]

Struggling against her disease, Irène continued to pursue her many personal interests. She succeeded André Debierne as director of the Curie Institute and, with a team of twelve senior scientists and sixty researchers, was finally promoted to full professor at the Sorbonne.[106] Still unsuccessful at gaining membership to the prestigious Academy of Sciences, to which she presented herself several times as a candidate, Irène took the rejection lightly: "Well, at least they are consistent in their thinking!"[107] In fact, she applied to draw attention to the omission of women within the academy's hallowed ranks.

Irène's commitment to feminism and the peace movement was evident in her public as well as private actions. She was the principal speaker at several functions, including the International Women's Day conferences held in London and Paris. She headed a large French contingent at the World Congress of Intellectuals for Peace in Wroclaw, Poland, which later gave birth to the World Peace Movement. She also answered a plea in the Paris newspapers asking families temporarily to adopt coal miners' children for the duration of a long strike in 1948.[108] Irène and Fred adopted two girls, aged nine and five, who stayed with the Joliot-Curies long enough to attend the wedding of their daughter Hélène and Michel Langevin, Paul Langevin's grandson. Hélène had also become a physicist, while their son, Pierre, had become a promising biologist with a deep interest in music.

Determined to overcome her fatigue, Irène planned a second trip to Oslo in 1953 to visit Ellen Gleditch and do a little hiking in the mountains. Her first trip had been in 1946 when she received an honorary doctorate from the university. She told Gleditch, "I love to walk but I climb extremely slowly because of my lungs ... What are you planning to wear? Heavy boots, strong shoes, sandals, espadrilles?"[109] Back in Paris, she told her friend Eugènie "to breathe, to eat, the most elemental functions are painful for me."[110] To Angèle, who took Irène for drives in the country when she could no longer go for long walks, she confessed that she was becoming lazy.[111]

In October 1954 Irène attended the twentieth anniversary of Marie's death in Warsaw and returned to France to be fêted with Fred for their own discovery of artificial radioactivity. A peculiar change of heart on the part of the academy resulted in their being awarded the prestigious Lavoisier Gold Medal. Her last publication was on radium[112] – a fitting topic as it brought her work full circle to her mother's discovery of the element fifty-four years earlier. In 1955 she undertook her last scientific project, the design of a new Institute for Nuclear Physics and Radioactivity to replace the dated and overflowing Curie laboratory. The chosen site at Orsay, thirty-five minutes from Paris on the métro, would contain a large particle accelerator and room for physics to grow.

Irène died at the Curie hospital of leukemia on 17 March 1956, at the age of fifty-nine. In her last public message she had called for a World Congress of Mothers for "the defense of their children against the danger of a new war ... that would be atomic."[113] The French cabinet reluctantly decided to give her a state funeral, but the Joliot-Curies had become so unpopular that she was little mourned. It was believed that her illness was caused initially by the large doses of x-rays to which she was exposed during World War I. Ellen Gleditch, however, believed that it was due to a serious polonium spill that occured one day at the laboratory. In order to protect her co-workers, Irène had sent them out while she herself cleaned up.[114] Fred was devastated by her death but vowed to complete the work at Orsay that had been so close to her heart. As he was certain at one point that he would predecease her, he now appeared to be racing against the clock. At his request, he assumed the directorship of the Curie laboratory and taught Irène's classes at the Sorbonne. He died two years later of liver cirrhosis brought on by overexposure to polonium. With his death a unique collaboration of thirty years came to an end. His final plea to the world was to use atomic energy for peaceful means.

Irène wanted scientific progress to benefit humanity in a world that included women as equals. While the direction of her research may have been influenced by Fred's ambitions, Irène was not necessarily drawn towards those experiments that would bring the greatest recognition. Her most important individual work, on the R-3.5 "actinium"-like isotope, was propelled by her own curiosity, and her basic approach to science remained philosophical: "If an explorer tries to satisfy a taste for adventure in Research, it would seem that the tranquil life of the laboratory has little to offer … Yet we may find ourselves in the presence of singular facts … A task once begun develops in an unexpected fashion, opening new paths for future work. And thus we satisfy our spirit of adventure."[115]

10 … And Some Other Women of the French Group

MARELENE F. RAYNER-CANHAM
and
GEOFFREY W. RAYNER-CANHAM

The preceding chapters have looked at some of the women who worked with Marie Curie in the early years. There were many more. In addition to the seven women mentioned so far, the Paris group always contained a significant proportion of women researchers. Most of these women left little record of their lives and vanished into obscurity once they'd left the Institut Curie. One particular example was Lucie Blanquies, who worked in the Curie laboratory in the early years from 1908 to 1910,[1] yet we have been unable to trace what became of her. We know from her publications during her brief time at the Curie Institute that Blanquies compared the α-rays produced by different radioactive elements[2] and also identified another decay product in the actinium series.[3] Her presence was mentioned by May Sybil Leslie in a letter to her former supervisor, Arthur Smithalls[4] (see chapter 5). Yet we found no other record of her except that she "resurfaced" forty-seven years later as a guest at the celebrations of the fiftieth anniversary of Marie Curie's first lecture at the Sorbonne.[5]

J. Samuel Lattès was among the women researchers in Paris in the 1920s. She worked at the Institut Curie from 1921 to 1928, when she obtained a doctorate. Lattès published a number of papers on the biological effects of radiation.[6] Elaine Montel, who worked briefly at the institute from 1927 to 1928, published only one paper on her research but it is interesting that all the references in it are to work by other women: Maracineanu, Roná, Joliot-Curie, and Chamié.[7] While details about these two women are sparse, we had greater success with two others: Eva Ramstedt and Irén Götz.

EVA RAMSTEDT

Just as Gleditsch was known only in her native Norway, so Eva Julia Augusta Ramstedt remains unknown outside of Sweden.[8] Ramstedt was born in Stockholm on 15 September 1879, the daughter of Johan Ramstedt, mayor of Stockholm, and Henrika Torén. She graduated from the Ahlinska School in Stockholm in 1898 and spent the following year in Paris. She entered Uppsala University in 1900, from which she obtained the equivalent of a PhD in 1910. The 1910–11 year was spent with Marie Curie in Paris, her research resulting in two publications, the first on the solubility of radon in organic liquids[9] and the second on the solubility of the decay products of radium in water.[10] It was during that year that Ramstedt met both Gleditsch and Leslie, and the three remained friends all their lives.

She returned to Sweden and spent 1913–14 as a research assistant at the Nobel Institute for Physical Chemistry and as a part-time teacher at the Ahlinska School. From 1916 to 1920, she continued to teach at the Ahlinska School while holding a research position at Stockholm's Hogskola (subsequently renamed the University of Stockholm). At about this time she and Gleditsch jointly wrote a book on radioactivity.[11] From 1919 to 1932, Ramstedt held various faculty positions at the University of Stockholm and also taught mathematics and physics at the teacher's college. Life was not easy for her. In a letter to Frederick Soddy, whom she had met in 1920 while travelling with May Sybil Leslie and Ellen Gleditsch, she wrote: "We are working under rather difficult conditions here. The building is hopelessly too small, every room is crowded. I work in cellar rooms not fit for working at all. This year I am happy though because I have an assistant. I am myself much taken with teaching and committee work for the University and research work cannot be done as I should like it. Now my assistant can work even if I cannot."[12]

Even this minor effort at research must have collapsed since Ramstedt, in her periodic correspondence with Marie Curie, commented: "As for me, I have been unfaithful to science, never finding time for research. Nevertheless, my present work interests me a lot and as well I hope to be able to give a course on the atom at the University this winter."[13] Ramstedt later expressed much the same feelings in a letter to Lise Meitner: "When reading about the scientific work, I sometimes find it very upsetting to have left it, but at the same time I like my present work very much."[14]

Ramstedt was chair of the International Federation of University Women from 1920 until 1946 and vice-chair of the corresponding Swedish organization from 1920 to 1939. In recognition of her contributions to the study of radioactivity, the University of Stockholm awarded her a DSC in 1942. She continued teaching until her retirement in 1945, though she lived another twenty-nine years, dying on 11 September 1974 at the age of ninety-five. Ramstedt seems to have received little recognition even in her native land, where she is mentioned only in biographical dictionaries[15] and in a brief survey of Swedish women scientists.[16] No complete biography of her has been published.

IRÉN GÖTZ

Women flocked from all over Europe to work at the Curie Institute, among them Irén Júlia Götz. Götz was born in Magyarovar, Hungary, on 3 April 1889. She obtained an honours doctorate from Budapest University in 1911,[17] writing her thesis on the quantitative determination of radium emanation.[18] Upon graduation, she arranged with André Debierne to work in the Curie Institute.[19] She spent two years in Paris and published some of her work,[20] then returned to Hungary to marry Laszlo Dienes, who was well known for his left-wing sympathies. Unable to find work in her field, she commenced research at an agricultural station from which she published a number of papers on the feeding and health of animals.

In 1919 Götz was appointed by the revolutionary government to teach theoretical chemistry at the University of Budapest, thus becoming the first woman lecturer at a Hungarian university.[21] She was at the university at the same time as Elizabeth Róna,[22] and it is probable that they knew each other. Unfortunately, Götz did not hold her position for long. Though the communist dictatorship, the "Red Terror," had been oppressive, the virulently anti-communist government that assumed power on 6 August 1919 was far worse.[23] The "White Terror" is believed to have been responsible for about five thousand deaths and about seventy thousand were jailed or placed in internment camps. Left-wing intellectuals were favourite targets of the new regime, but Götz was able to flee the country, crossing the border into Austria, then travelling in Rumania where she taught at the Pharmacological Institute of Cluj University and performed research in electrochemistry between 1922 and 1928. Always interested in radioactivity, she wrote a text on the transformation of the elements and the modern concept of matter in 1926.[24] From Rumania she moved to Germany, working as a chemical advisor in Berlin from 1928 to 1931. Finally, she moved to the Soviet Union where she became the head of a department at the Nitrogen Research Institute in Moscow, a post that she held until 1939. She died of typhus in 1941.

PART THREE
The British Group

Ernest Rutherford was the key figure in the British school and it is with his name, rather than with a geographic location, that we associate this group of individuals. We have uncovered considerable information about four of the women who worked directly with Rutherford. The first of these is the Canadian physicist Harriet Brooks (chapter 11), Rutherford's first graduate student. Brooks worked with Rutherford in Montreal from 1898 to 1901 and again from 1903 to 1904. She spent 1902–03 with J.J. Thomson at Cambridge and part of 1906–07 with Marie Curie in Paris. Following in Brooks's footsteps, the American Fanny Cook Gates (chapter 12) worked with Rutherford at McGill University from 1902 to 1903 and then with Thomson in 1905. In 1903–05 Thomson had taken on Jesse Mable Wilkins Slater (chapter 15) to help with his researches on thorium. May Sybil Leslie (chapter 5) joined Rutherford's group at Manchester University in 1911–12 after her two years with Curie and she, in turn, was followed by Jadwiga Szmidt (chapter 13) in 1913–14. As noted elsewhere, Szmidt had met Leslie and Gleditsch when all three were working in Paris during 1911. Among the women researchers in later years was Elizaveta Karamihailova (chapter 19) of the Austro-German group, who joined Rutherford in 1935. Other prominent members of the British group, including Charles Barkla and William Ramsay, also took on women researchers (see chapter 15), though we focus only on the three women associated with Frederick Soddy, the longtime collaborator of Rutherford. The first of these was Ruth Pirret (chapter 15), one of Soddy's researchers between 1909 to 1910. Pirret probably overlapped with Winifred Beilby (chapter 15), Soddy's wife, who helped her husband from about 1909 onwards. However, it was Ada Hitchins (chapter 14) who was the stalwart researcher with Soddy, working with him from 1912 until about 1927, except for a brief period during the First World War.

11 Harriet Brooks: From Research Pioneer to Wife and Mother

MARELENE F. RAYNER-CANHAM
and
GEOFFREY W. RAYNER-CANHAM

Next to Marie Curie, Harriet Brooks performed some of the most important work in the very early days of radioactivity. She was a talented scientist who worked with the three greatest names of the time, Ernest Rutherford, Marie Curie, and J.J. Thomson. Typical of the women researchers, however, Brooks had low self-esteem, and despite Rutherford's encouragement she eventually abandoned her research work in favour of a life as wife and mother.[1] In this Brooks was an exception since most of the pioneer women never married. It is easy to forget the pressure they must have been under to conform to a more conventional view of womanhood.

Born on 2 July 1876 in Exeter, Ontario, Brooks was the third of nine children of George Brooks and Elizabeth Worden. Although most of the academic women of the time came from fairly affluent middle-class families,[2] Brooks was again an exception. Her father, George Brooks, was a commercial traveller for a flour company, and supporting a large family on his salary involved considerable hardship. The family moved frequently around Ontario and Quebec because of his job. Harriet completed her grade school education at the Seaforth Collegiate Institute in Seaforth, Ontario, and then enrolled at McGill University when the family moved to Montreal in 1894. Only Harriet and her next youngest sibling, Elizabeth, continued their studies beyond high school.

Brooks was an outstanding student, winning a scholarship or award in each year of her studies, and it was the monetary scholarships that enabled her to cope with the costs of a university education.

Her first two years at McGill were mainly devoted to the study of mathematics and languages (Greek, Latin, French, and German), while the last two years were spent almost exclusively on physics. A popular student, she was elected class president in her third year. She graduated with an honours BA in mathematics and natural philosophy in 1898 and was awarded the highly coveted Anne Molson Gold Medal for outstanding performance in mathematics.[3]

Simultaneously with the BA, she received a teaching diploma. This was not uncommon for women students as it opened the option of a teaching career, one of the few acceptable professions for an educated woman at the time. The diploma was awarded by the McGill Normal School, a teacher training school associated with the university. To fulfil the requirements of the diploma, the McGill students took pedagogy courses in the evenings. Tuition at the normal school was free, but students had to sign a contract committing themselves to three years of teaching after graduation.[4]

Upon completion of her BA, Brooks was hired by Ernest Rutherford as his first graduate student researcher. Rutherford, who had just arrived from England, had already been hailed as an outstanding scientist, and his offer of a position must have indicated his high regard for Harriet's talents. Her initial work was on electricity, the subject of her first publication[5] and of her master's thesis, which she completed in 1901.[6] Rutherford always believed that it was important to start young researchers with straightforward topics to build their confidence,[7] and electricity was his own specialty prior to his foray into the poorly understood world of radioactivity.

In the fall of 1899 Brooks took up a concurrent appointment as nonresident tutor in mathematics at the newly formed Royal Victoria College, a women's college at McGill University. The teaching duties presumably enabled her to work off her commitment to the normal school while also supplementing her income. Characteristically Brooks devoted herself to the task and later received a glowing reference from the then principal of the university for her teaching performance.[8]

Rutherford had reported in 1900 that thorium gave out some radioactive substance apart from the common radioactive rays.[9] He gave the name "emanation" to this puzzling material that could be carried away by air currents. At the time it was unclear whether emanation was a gas, a vapour, or a finely divided solid. After Brooks had completed her master's thesis, Rutherford gave her the challenging task of unravelling the mystery. Brooks's work showed that the emanation was a gas (which we now know to be radon), with quite different properties, such as a lower molecular weight, from those of radium.[10]

This discovery was crucial to the progress of radioactive research. At that time, the radioactive elements were believed to maintain their own identity while the radiation was being released. Identifying the gas as having a lower molecular weight indicated that it could not simply be a gaseous form of thorium.[11] It was this result that led Rutherford, together with Frederick Soddy, to the later realization that a transmutation of one element to another had occurred.[12] Yet Brooks's critical pioneering step in the discovery process has been long overlooked.

Brooks then turned her attention to a comparison of the radiations from radioactive substances.[13] This, too, was important work, for it showed that the β-radiation consisted of fast-moving negative particles and that the radiation was the same, regardless of its source.

In 1901 Brooks obtained leave from the Royal Victoria College to take up a position at Bryn Mawr College as fellow in physics to pursue studies towards a PhD.[14] Her correspondence with Rutherford over the next six years provides a glimpse of her travels and an insight into her personality. Many of her letters indicate low self-esteem and her reliance upon Rutherford for encouragement and support. From time to time she protested that Rutherford praised her too highly. In her first letter from Bryn Mawr, she commented: "Please do not give me any more credit than I deserve in that comparison of radiations. You are quite too generous in that respect."[15]

During her year at Bryn Mawr (1901–02), she was awarded the prestigious Bryn Mawr President's Fellowship for graduate study in Europe. This prestigious award originated with the first president of Bryn Mawr, M. Carey Thomas, who had been insistent on the need to encourage brilliant women students. She argued that "if the graduate schools of women's colleges could develop one single woman of Galton's 'x' type – say a Madame Curie, or a Madame Kovalewsky[16] born under a happier star – they would have done more for human advancement than if they had turned out thousands of ordinary college graduates."[17] It was to encourage the most outstanding students that Bryn Mawr provided these scholarships for travel to Europe, where the recipients could study with the greatest minds of the time – providing that the host universities would accept women.

Brooks informed Rutherford of the good news: "I'm afraid that your generosity in placing me as a collaborator where I am really nothing more than a humble assistant has rather imposed on the faculty of Byrn Mawr, for last night, they awarded me the European Fellowship and much to my satisfaction would prefer that I should make use of it at Cambridge ... There are of course other difficulties to be met besides the ones I have mentioned money and the objections of my family for instance who will think me wholly out of my mind but I think I can overcome them."[18] Financial considerations were obviously still of concern to Brooks. And some of her family must have been less than encouraging about a young woman who intended to travel the world rather than return home as the vast majority of women graduates did at the time.[19]

Rutherford arranged for Brooks to work in the Cavendish Laboratory, Cambridge, with his former supervisor, the famous J.J. Thomson.[20] In one of her letters to Rutherford from Cambridge, she again displayed her lack of confidence: "I am afraid I am a terrible bungler in research work, this is so extremely interesting and I am getting along so slowly and so blunderingly with it. I think I shall have to give it up after this year, there are so many other people who can do so much better and in so much less time than I that I do not think my small efforts will ever be missed."[21] Yet comments by Thomson, Rutherford, and others indicated that she was extremely skilled and very gifted in her field. Writing in 1906 on physics at McGill University, A.S. Eve remarked: "Miss Brooks has published several papers on various radio-active phenomena, and this lady was one of the most successful and industrious workers in the early days of the investigation of the subject."[22]

After her year in England, Brooks did not return to Bryn Mawr to complete a doctorate but instead resumed her duties at the Royal Victoria College as non-resident tutor in mathematics and physics, at the same time rejoining Rutherford's research group. She might have thought that further absence would result in the loss of her position at McGill, but her return might also have been prompted by the unsatisfactory nature of her stay at the Cavendish. The atmosphere in the laboratory was quite aggressive[23] – certainly not an environment with which Brooks would have been comfortable. Thomson himself was fairly tolerant of women students but he made only brief visits to the research laboratory each day, and Brooks's discoveries ran contrary to his views on the nature of radioactivity. It was probably for these reasons that Brooks wrote regularly to Rutherford for his advice on her research results.

During this second period at McGill (1903–04) Brooks made a major discovery: she observed that a non-radioactive plate placed inside a radioactive container itself became radioactive.[24] This she ascribed to a volatility of the radioactive material, enabling atoms of the material to transfer from one surface to another. We now explain the observation as the recoil of the radioactive atom; when a particle is released from an atom, the atom itself will be thrust in the opposite direction, often with enough momentum to eject the atom from the radioactive material and deposit it on another surface. By this means, the product from the decay can be removed and examined separately from the parent material.

Rutherford recounted Brooks's findings at the Bakerian Lecture he delivered in London.[25] Initially agreeing with Brooks that the experiment showed some form of volatility, a letter from one J. Larmor positing the momentum argument changed his view.[26] Brooks's observation was overlooked and the effect was rediscovered four years later by the teams of Hahn/Meitner and Russ/Markower.[27] Rutherford wrote to Hahn pointing out that Brooks had observed the phenomenon years earlier,[28] and in his later autobiography Hahn noted that "Brooks may have been the first researcher to have observed the phenomenon of radioactive recoil."[29]

In the same year she also published some of the work performed in 1902–03 at the Cavendish Laboratory.[30] This massive study on radioactive decay indicated that the rise and fall of radioactivity from the heavy elements could best be interpreted in terms of two successive radioactive changes. This discovery laid the groundwork for the crucial concept of radioactive decay sequences and formed the backbone of Rutherford's Bakerian Lecture. Throughout the presentation, Rutherford gave credit to the work of "Miss Brooks."[31]

After only one year at McGill, Brooks accepted a position as tutor in physics at Barnard College, the women's college of Columbia University in New York.[32] Although the annual salary of US $1,000 was an improvement over the $750 (Canadian) she received from McGill, her main reason for moving was probably a romantic attachment to Columbia physics professor Bergen Davis,[33] whom she had met at the Cavendish.

At Barnard, Brooks spent most of her time teaching advanced physics and her life was relatively uneventful until the summer of 1906, when she became engaged to Davis. At the time it was commonly expected that women had to choose between a career and marriage. Brooks wrote Dean Laura Gill at Barnard to inform her of the impending marriage. Gill replied: "I feel very strongly that whenever your marriage does take place it ought to end your official relationship with the college."[34] Brooks, however, refused to submit to the norms of the time:

> I am quite sure that my duties will be, if anything, better performed under the new conditions but if I find that my new relationship, at any time, interferes

in the slightest with my professional work, I shall, of course, at once tender my resignation. But failing such an outcome I should not be justified in resigning.

I think also it is a duty I owe to my profession and to my sex to show that a woman has a right to the practice of her profession and cannot be condemned to abandon it merely because she marries. I cannot conceive how women's colleges, inviting and encouraging women to enter professions can be justly founded or maintained denying such a principle.

I am sorry to have thus to appeal from your decision but I cannot acquiesce without violating my deepest convictions of my rights.[35]

Brooks had a strong supporter in Margaret Maltby, head of physics at Barnard, who intervened on her behalf: "I most sincerely hope that you will not object to her [Brooks] continuing her work in the department. She is greatly interested in her research and teaching. She enjoys both and she is a very good teacher. I think I can sympathize with her thoroughly. Neither you nor I would like to give up our active professional lives suddenly for domestic life or even for research alone. I know of no woman to take her place – no one available who has the preparation and the personality and ability to teach, and the skill in physical manipulation, that she has."[36]

But Gill, quoting the view of the college trustees, replied: "The College cannot afford to have women on the staff to whom the college work is secondary; the college is not willing to stamp with approval a woman to whom self-elected home duties can be secondary."[37] In any event, the engagement was broken off. Rutherford wrote to Hahn: "You may be interested to hear that Miss Brooks has broken off her engagement, got tired of the young man!"[38] Brooks initially agreed to continue as tutor but the following month resigned her position at Barnard College.

Many of the letters from Brooks to Gill were written in the Adirondacks from Summerbrook, the summer residence of John and Prestonia Martin, two prominent Fabian socialists. Among their other guests that summer were Maxim Gorky, the Russian writer and revolutionary, his companion, Maria Andreyeva, and his secretary, Nikolai Burenin.[39] It is not clear how Brooks met the Martins. Prestonia Martin advertised Summerbrook as a retreat for "tired" teachers and it may be that Brooks went there to get away from her problems at Barnard. An alternative possibility is that Brooks met Andreyeva at a lecture that the Russian woman gave in New York City. The writer John Dewey recalled: "A group of women had asked my wife if she would give the use of our apartment to Mme Andreyeva to speak on the condition of women in Russia; a sensational newspaper reported that this was a reception to her, and the Barnard girls were the guests."[40]

Brooks stayed most of that summer with the Martins. Then in October 1906 she voyaged with the Gorky entourage across the Atlantic to Italy. From there she travelled to Paris, where she became an independent researcher in the Laboratoire Curie under the direction of André Debierne.[41] At Christmas and again in the spring, Brooks returned to visit Gorky, Andreyeva, and Burenin, who by then had moved to the island of Capri. Through later years, Brooks corresponded with Andreyeva and even gave her her passport at one point, so that Andreyeva, a *persona non grata* in pre-revolutionary Russia, could return periodically to visit her son in St Petersburg. One assumes that Andreyeva, who acted as English translator for Gorky, could give a passable imitation of Canadian-accented English.[42]

Although Brooks never published any of her work at the Curie Institute, we know that she began to study the emanation from the element actinium and continued her experiments on the recoil of the radioactive atom; for three subsequent articles from the institute, two by Debierne and one by Curie's research associate, Mlle Lucie Blanquies, cited her work.[43] Curie invited Brooks to stay for the 1907–08 year, but Brooks decided instead to apply for a position at the University of Manchester in England.

The position at Manchester was the John Harling Fellowship. This most prestigious post required the holder to devote the whole of one year to research. Rutherford was about to move to Manchester, and it seems that Brooks was keen on rejoining her former mentor. Rutherford regarded Brooks as a great asset to his research group, commenting in his reference for her: "Miss Brooks has a most excellent knowledge of theoretical and experimental Physics and is unusually well qualified to undertake research. Her work on 'Radioactivity' has been of great importance in the analysis of radioactive transformations and next to Mme Curie she is the most prominent women physicist in the department of radioactivity. Miss Brooks is an original and careful worker with good experimental powers and I am confident that if appointed she would do most excellent research work in Physics."[44] Rutherford was a blunt and forthright man, and his glowing recommendation contrasts sharply with Brooks's sense of inadequacy as expressed in her letters.

Brooks was still waiting to hear about the Harling fellowship when she suddenly decided to end her physics research and marry Frank Pitcher, who had been a laboratory demonstrator at McGill University while she was a student. To see why Brooks gave up science for marriage, we should look first at the career choices that were open to her as a woman physicist. She could have continued as a research assistant, either with Rutherford at Manchester or with Debierne in Paris. Or she could have looked for an academic teaching position at another women's college in Canada or the United States; but without a PhD, she could only have hoped for a position of lower rank, such as tutor or lecturer.[45]

Moreover, as a single woman in her thirties she faced societal pressure towards marriage. This was aptly summed up by M. Cary Thomas, first president of Bryn Mawr College: "Women scholars have another and still more cruel handicap. They have spent half a lifetime in fitting themselves for their chosen work and then may be asked to choose between it and marriage."[46] Numerous courtship letters from Pitcher emphasizing the need for her to "settle down"[47] indicate that Brooks was initially very reluctant to give up her peripatetic research life. But Brooks's two close women friends counselled marriage. Prestonia Martin, her Fabian friend, wrote in her book on feminism that "child rearing is the noblest work an intellectual woman can do."[48] And Mary Rutherford, Ernest's traditionalist spouse, clearly thought her pressure had been effective, for she later wrote to Brooks: "Kind regards to Mr Pitcher. Mrs Gordon [Harriet's sister, Edith] evidently thinks I am responsible for him!"[49]

After her marriage in London on 13 July 1907, Brooks returned to Montreal with Pitcher. In the following years they raised three children, two of whom died as teenagers.[50] Brooks gave up any active role in science. Her acquiescence to a purely domestic role may have been due to the traditionalist values of her husband and of the upper-class milieu in which she now found herself.[51] Rutherford, in any case, had departed for England, and the radioactivity research program at McGill had declined considerably.[52] Without his presence and support, it would have been hard for her to re-establish a scientific career. Instead, Brooks became active in various organizations, such as the Women's Canadian Club and the Canadian Federation of University Women.

Harriet Brooks died on 17 April 1933. From accounts of her lingering illness, it would seem possible that she suffered from a radiation-related disease. Rutherford wrote to Arthur Eve when he heard the news of her death: "She was a woman of great personal charm as well as of marked intellectual interests. I am afraid her domestic life was not without serious trials which she bore with astonishing fortitude. My wife and I held her in great affection and her premature death is a grievous blow to us."[53]

Why has Brooks been overlooked by historians of science? She was repeatedly mentioned by Rutherford in his text and in his research papers, more so than his other students of the period. It seems clear that much of Rutherford's early work on radioactive decay and particularly on radon was performed by Brooks, and her role is recognized in the three major biographies of Rutherford.[54] Perhaps one reason why she has been overlooked in the history of radioactivity is that it was years before the significance of her experimental work became apparent; the phenomenon of premature discovery is well known in the history of science.[55] Alternatively, the historical silence surrounding this fascinating scientist may simply reflect our tendency to focus on the more prestigious figures (mostly male) in the advancement of science. Obviously, justice needs to be done, for Brooks accomplished much in her brief scientific career. Moreover, the career-versus-family dilemma that she faced is still a problem today.

12 Fanny Cook Gates: A Promise Unfulfilled

MARELENE F. RAYNER-CANHAM
and
GEOFFREY W. RAYNER-CANHAM

Gates was an adventurous spirit, and her life typifies the career challenges of the gifted academic woman at the beginning of the twentieth century. From her birthplace in the United States she voyaged to Germany, Switzerland, Canada, and Britain. Most of her life was spent teaching physics at Goucher College but she passed many summers at other institutions pursuing research. Gates's work in the field of radioactivity was accomplished through research positions with Ernest Rutherford and J.J. Thomson. In her later years, she took a powerful but scientifically marginalized position as dean of women at the University of Illinois.

Born in Waterloo, Iowa, on 26 April 1872,[1] Gates was the daughter of John Cook Gates, a lawyer, and Adelia St John. After completing her secondary education at East Waterloo High School, she attended Northwestern University, Illinois, which had become predominantly a women's college at that time.[2] She graduated with a Bachelor of Letters in 1894 and a Master of Letters in 1895, both in mathematics. During the 1894–95 year, she was a fellow in mathematics at Northwestern University and an instructor in mathematics at Northwestern Academy.

The next two years were spent at Bryn Mawr College, first as a scholar and then as a fellow in mathematics. Gates may have become interested in pursuing an academic career while at Bryn Mawr, for as Wein commented: "Bryn Mawr women would not be prepared for marriage but for independent careers, particularly in academia, in which they were expected to excel."[3] Like Harriet Brooks, she was awarded a President's European Fellowship in her second year. Gates chose to spend the fall of 1897 at the University of Göttingen as a student of mathematics and physics and winter 1898 as a student of physics at the Polytechnik Institut at Zurich. Zurich was well known as a liberal institution that had been welcoming women students for many years,[4] while Göttingen had just become the first German university to award doctorates to women.[5] Interestingly, the only two North American pioneer women in radioactivity, Brooks and Gates, both attended Bryn Mawr and were granted these fellowships.

Upon her return to the United States in 1898, Gates accepted a position in physics at the Women's College of Baltimore (later Goucher College). She was a forthright character, insisting that a number of pieces of physics equipment be purchased prior to her arrival.[6] Once installed in her position at the college, she sent President Goucher an extensive list of additional physics equipment that she required immediately. The total cost came to $730, an enormous figure for that period. She also cautioned the president to expect a request the following year for equipment for the study of spectra and x-rays. At the end of the letter she boldly addressed the question of salary: "You doubtless recall that when last we spoke of the work, I said I should not feel justified in continuing to work for another year for less than $1200. Should the Committee feel unable to make the advance in salary, may I ask to be informed if possible before the twenty-sixth of April?"[7]

One assumes that her request for a raise was settled to her satisfaction, for she remained at Goucher for thirteen years, rising through the ranks from instructor to associate professor and head of the department. The volume of her correspondence with instrument manufacturers indicates considerable success in badgering the president over the years for more equipment.

Although the physics facilities at Goucher were adequate for teaching, Gates's research necessitated spending summers at larger institutions. She developed particular ties with the University of Chicago, spending the whole summer of 1899 there. Gates was torn between research in mathematics or physics. She sought advice from E.H. Moore of the Department of Mathematics at the University of Chicago, who recommended in a letter that she pursue research in physics as it would complement her area of teaching. She also asked him where she might spend her year of leave in 1902–03, to which he replied: "Don't worry about 1902–3 now. Try during the summer to make yourself known to Chicago people and catch ideas and next year think more definitely of 1902–3."[8]

Gates must have decided to spend 1902–03 at McGill University in Montreal as a graduate student with Ernest Rutherford. There is no indication why Gates chose to work with Rutherford, though in his biography of Rutherford Wilson[9] suggests that she was perhaps attracted by the success of Harriet Brooks.[10] Rutherford wrote to J.J. Thomson about her work: "Miss Gates a professor from a women's college Baltimore has been working here at the effect of high temperature on the mobilization of excited radioactivity on wires. She finds that this activity is transferred from the wire to the outer cylinders at a white heat but it is not lost. It is difficult to decide whether it is an actual volatilization of excited radioactivity or an effect of platinum dust at a white heat."[11]

Gates's work in Montreal[12] was important in determining the nature of radioactivity. One of Rutherford's earlier experiments had shown that heating a platinum wire coated with a radioactive material lost the radioactivity if the wire was heated to incandescence.[13] This observation appeared to support the traditionalist view that radioactivity was simply a physical or chemical phenomenon.[14] Gates showed that the radioactivity had not been destroyed but had been deposited on surrounding surfaces. Her findings suggested that the substance had simply volatilized from the hot surface and condensed on the cooler surroundings. As Heilbron has commented, the results of her work showed that "the rule that radioactivity could not be affected by physical or chemical treatment survived without exception."[15]

Her other piece of work, "On the nature of certain radiations from the sulphate of quinine," was also of significance.[16] Researchers at the time were claiming to find rays everywhere they looked. Many believed in the N-rays of René Blondlot[17] and the "black light" of Gustave LeBon.[18] LeBon also claimed that the phosphorescence of quinine sulfate and other compounds, when they were heated or cooled, was a transient form of radioactivity. If this were true, the finding would again be proof that radioactivity was a simple physical phenomenon. Gates's work showed that the emissions from the quinine sulfate were certainly not radioactivity: they were simply electromagnetic radiation. Once again, Gates had provided the evidence to show that the radioactivity phenomenon was different from a simple chemical or physical process.

Upon her return to the United States, Gates wrote four letters to Rutherford, the first three of which were addressed from Goucher College. The initial letter mentioned that she was sending her research results back to Rutherford. She asked Rutherford to comment on her conclusions: "On the whole I am very little pleased with the paper and I feel that you will be also. Please criticize it as severely as you feel inclined and I will try and improve its final form."[19]

The second letter[20] described the research work that she was doing at Goucher. She mentioned attending a meeting of the American Philosophical Society and she expressed pleasure and surprise that other scientists had shown interest in her work. She was in fact highly regarded, as evidenced by her election as a fellow of the American Physical Society and as a member of the Société Française de Physique.[21] The letter comments upon the difficulties that she was facing, such as the conversion of the building from direct current to alternating current, and the large number of students that she had. That she had so many women students suggests that she was an excellent teacher and also, perhaps, that female students at the time were less averse to the physical sciences. Gates showed her dedication by helping to set up a chapter of Phi Beta Kappa in April 1904.[22] In the official acceptance of the chapter, the secretary commented: "I congratulate you most heartily upon the success of your students; no higher or sincerer compliment could be given to your own success than just such an outcome for your work. But then, just such an outcome was predicted for you by all your friends here, who are, I assure you, none the less happy that in this instance they can say truly 'I told you so.'"[23]

In the same month Gates asked Rutherford for a letter of reference.[24] She described in her letter how a major fire at Goucher had ended her hopes for a new science building, more time for research, and a better salary. She believed that she had to look elsewhere if she was to grow professionally. Her final letter to Rutherford, addressed from Waterloo, Iowa, thanked him for his letter of reference and expressed hope that things would improve at Goucher College: "President Goucher has promised me a good assistant for next year a Miss Vaughan who has spent two or three years in graduate work in Physics at the University of Chicago, so that with the hope of relief from much of the laboratory routine, I promised to return to Baltimore next year. I realize however that opportunities there are apt to be at a stand still for some time to come and I shall therefore be on the look out for a better position and I am sure the good words you were so kind to write will be of good service to me."[25]

During the summer of 1905, Gates worked at the Cavendish Laboratory with J.J. Thomson. Thomson had a lesser opinion of her experimental work than Rutherford, to whom he wrote: "We had Miss Gates over for some months she got on very well though she is hardly a great experimenter."[26] At the time Thomson was actively researching the theory of universal decay – that all matter was radioactive.[27] His negative view of Gates's abilities might have been simply a reflection of her inability to find radiation in all elements, as he was anticipating.

Gates's stature in the scientific community was indicated by a paragraph in the "Scientific Notes and News" section of *Science*: "Miss Fanny Cook Gates who has been engaged in research work in the Cavendish laboratory since last April, has resumed her duties as head of the physics department in the Women's College of Baltimore."[28] Gates had not been back in Baltimore for long before she obtained leave again, this time for 1907 and 1908. She spent most of those two years at the University of Pennsylvania, though for part of 1907 she worked in the chemistry department of the University of Chicago. Her research work during this time on "the conductivity of air caused by certain chemical changes" won her a doctorate from the University of Pennsylvania in 1909. Her thesis work was never published. Gates was also active outside the world of science, becoming chair of the Educational Committee of the Maryland Federation of Women's Clubs, a post that she held from 1909 to 1911.

By 1911 she must have felt that the time had indeed come to search for a better appointment, for she left her permanent position at Goucher to work as a researcher at the University of Chicago for two years while taking courses in education and natural science. She then accepted the position of professor of physics, professor of mental physics, and dean of women at Grinnell College, Iowa, a move home that may have been prompted by parental illness. She stayed only three years at Grinnell College before accepting a position as associate professor of physics and dean of women at the University of Illinois, Urbana-Champaign, in 1916.

The position of dean of women is among those jobs identified by Rossiter as "women's work" in science.[29] With the expansion of coeducation in the US there was a need for a position that involved the supervision of women students and the administration of services for them. The duties of the position were somewhat vague, but in general it was considered suitable for a faculty woman. Gates was an extremely active dean.[30] She planned and organized the beginning of the residence hall program and established the first cooperative houses for women at the university. Mary Lou Filbey, chronicler of the deans of women at Illinois, commented that in her first year, Gates became the "most powerful and respected Dean of Women that the University had had."[31] She also served on the policy and planning committee of the university.

When appointed, Gates had been promised an assistant dean and a secretary and President James had indicated that her request for an assistant would be approved. However, in the budget for 1917 she was specifically denied an assistant. Showing an impulsive streak, Gates went ahead and hired an assistant dean without authorization from the president. She told James that the assistant dean would be arriving on campus shortly and that if he did not reconsider, she (Gates) would resign. To Gates's surprise, James accepted her resignation (unlike Goucher, who seemed to have acquiesced to her every ultimatum). She arranged a meeting with the president at which she agreed to return to the university under his terms.

This was at least her second attempt at hiring without approval. In 1901 during her time at Goucher, Gates received a letter from Robert Millikan of the physics department at the University of Chicago.[32] Millikan was enquiring whether or not there was to be a vacancy at Goucher for the 1902–03 year. Apparently, Gates had contacted him about such a vacancy and he had recommended a Miss Chase. However, Millikan noted, "apparent contradictions between your letter and one from the President have occasioned some uncertainty." This suggests that, once Gates had arranged her year at McGill, she had forged ahead to find a replacement without President Goucher's approval. Since Gates obtained her year in Montreal, Goucher must have acceded to her wishes.

With the death of her father in December 1917, Gates requested a leave of absence for the remainder of the 1917–18 academic year to recover from extreme fatigue. She returned to the university and prepared the 1918–19 budget, in which she once again threatened to resign if the position of assistant dean was not approved. President James accepted her resignation without delay. Gates then applied for an administrative position with the YWCA in New York. However, rumours of her impetuous and somewhat erratic behaviour had become general knowledge and the YWCA asked James whether it was true that she had a drug-addiction problem. James replied that no one at Illinois had any reason to believe that she has. Later, James sent a positive comment on Gates to the YWCA: "She did not like the limited scope for her activity which the office here presented and therefore was looking all the time for a wider field. She needs a wider opportunity for her abilities and energy."[33] James's responses must have satisfied the YWCA, for they appointed Gates as general secretary.

Regrettably, she did not find a "wider opportunity," and the remaining years of her life were spent moving back and forth between positions. She stayed at the YWCA from 1918 to 1919 and then became the special teacher of physics at the Lincoln and Brealey School from 1920 to 1921, returning to the YWCA as general secretary for 1921–22. After this, she took a one-year position as head mistress of the Phoebe Ann Thorne School of Bryn Mawr College before returning to the Lincoln and Brealey School for 1923–28. In poor health, she accepted a position in 1928 as part-time teacher of physics at the Roycemore School in Evanston, Illinois, where she continued working until her death on 24 February 1931 in Chicago.[34] The cause of death was not reported, but years of work with radioactive substances might well have caused a radiation-related disease as they did for so many of the early researchers in the field.[35]

To have achieved a doctorate in her day, Gates must have had considerable talent and determination. Her correspondence indicates a strong will and her European travels must have been a tremendous challenge for a single woman from a rural American background. Gates obviously felt that she could accomplish more than a small women's college gave her scope for, and perhaps this was her mistake; for at that time there were few directions open to women academics.[36] President Goucher accepted the need to accomodate Gates's wishes in order to keep one of his most talented faculty members. He provided funds for equipment, an assistant, and frequent leaves. President James, however, looked upon Gates as an easily replaceable dean. Thus, after he reneged on his original promise of an assistant, Gates's life at Illinois became a battle of wills that overshadowed her accomplishments as dean. Her peripatetic existence after leaving Illinois proved to be a considerable descent from the promise of the early years of her career. Yet she deserves to be remembered as one of the early women in the field of radioactivity who, but for the marginalization of women in science, might have progressed much further.

13 Jadwiga Szmidt: A Passion for Science

MARELENE F. RAYNER-CANHAM
and
GEOFFREY W. RAYNER-CANHAM

Yet again we tell the story of a woman who devoted her life to physics, and, like so many others, she had to voyage far from her native country to become involved in the forefront of atomic research. Jadwiga Szmidt was one of three women who worked with both Curie and Rutherford, the others being Harriet Brooks and May Sybil Leslie. After returning to Russia, she joined the famous physics research group of Ioffe. As with most of the women in this compilation, details of her life are sparse, but we do have various items of correspondence from which we can reconstruct her character and experiences.

Szmidt[1] was born in Lodz (then in Russia, now in Poland) on 8 September 1889.[2] She is noted as being a member of the evangelical Lutheran faith, which would have been very unusual in that Catholic city. McGrayne has commented that almost all of the famous women scientists have come from minority religous backgrounds, particularly those that emphasize education for women.[3] Szmidt trained as a teacher at the Women's Pedagogical Institute in St Petersburg[4] at a time when teaching was one of the few options open to women interested in science. As Koblitz explained, "in 1881 the government closed the St Petersburg higher women's courses. When the courses reopened in 1889, it was discovered that decisive steps had been taken to prevent women from entering the sciences. The physical-mathematical and natural science facilities had been greatly reduced, their laboratories had been closed, and most natural science courses had been cancelled. Study languages or art, the government in effect was saying – that ought to be harmless enough. Only after many years were the natural science and physical-mathematical courses permitted to grow naturally again."[5]

After graduating in 1909, Szmidt accepted a position at Tagantseva's Girls Secondary School in St Petersburg. She taught physics, both the classes and the laboratory sessions, for the next two years. In 1911 she travelled to Paris for a teachers' training session at the Sorbonne. While there, she obtained permission to spend the summer semester working in the Curie laboratory[6] (no doubt after paying the required fee), even though she did not have a formal undergraduate degree. This was a particularly fortunate summer for her to be there, as May Sybil Leslie, Ellen Gleditsch, and Eva Ramstedt were also working with Curie at the time. She became particularly friendly with Leslie and Gleditsch and maintained contact with them over the years. Unfortunately, there are no accounts of her research for that summer. In the Curie Archives only her name, "J. Schmidt," is recorded in the 1910–11 list of researchers.[7]

Szmidt returned to St Petersburg, where she resumed her teaching post. Her sojourn at the Curie Institute must have fired her interest in research, for she started to spend part of her time working on optotechnics in the laboratory of the Russian physicist A.L. Grishun, who had been one of her professors at the Women's Pedagogical Institute.

In 1913 Szmidt made the journey across Europe again to work with Rutherford at the University of Manchester. Her year in Britain was remarkably productive for someone who had no formal degree in physics or chemistry. This was probably due in large part to the way Rutherford supervised the laboratory. Robinson, a member of the research group at the time, described it thus: "In the middle years [the Manchester period] he generally had in the laboratory between fifteen and twenty men [and women] engaged in full-time work in radioactivity, and it must be remembered that as a rule a large proportion of these consisted of young workers for whom he had to provide fairly constant supervision as well as the original idea and scheme of attack. The supervision he provided in full measure ... His young assistants could therefore have the unusual satisfaction of doing productive work during their apprenticeship."[8]

Szmidt's first study was on a comparison of the gamma rays emitted by different radioactive elements and their study by means of absorption by different gases.[9] Her other work showed that the radiation emitted by bombarding metal surfaces with β-rays was identical to that produced when x-rays were directed upon the same metal surfaces.[10]

During her work on the absorption of gamma rays by gases, Szmidt had a chemical accident. The circumstances were described by E. Andrade, one of her fellow researchers. Giving Szmidt "an ardent feminist" the pseudonym "Natasha Bauer," he said that the accident had occurred while, alone in the laboratory, she had tried to open a jammed bottle of toxic sulfur dioxide gas. The escaping gas had asphyxiated her and she had been lucky to survive. Andrade was in Rutherford's office when she appeared later, and he recounted the conversation between them: "'What's this I hear, Miss Bauer?' he said as soon as she had shut the door, 'What's this I hear? You might have killed yourself!' to which she replied 'Well, if I had, nobody would have cared.' 'No, I daresay not, I daresay not' said Rutherford, 'but I've no time to attend inquests.' I believe that he was more than half in earnest, thinking of a morning wasted away from the laboratory, but from the lady's face this was not the reply she expected."[11] In fact Rutherford's gruff response hid a real concern for his students that, in his relations with Szmidt, surfaced after her departure.

Ellen Gleditsch arranged to stay with Szmidt in Manchester on her way back from the United States to Norway. Rutherford commented to Boltwood on the planned visit: "I understand from Miss Schmidt that Miss Gleditsch is coming to stay with her some time next week."[12] Gleditsch found that Szmidt had not settled into the way of life at Manchester. She persuaded Szmidt to accompany her to the laboratory for the traditional afternoon tea. Gleditsch mentioned this incident in a letter to Boltwood: "Miss Szmidt does not seem to have found it easy to adjust to English life, and I was tempted to laugh at her stories. My presence gave her courage to go to the afternoon tea, something she very seldom did."[13] It was unfortunate that Szmidt had been missing the afternoon tea sessions. Robinson described how important they were: "I am sure that the laboratory tea-table, situated in the radioactivity training laboratory, was far from being the least important bench in the laboratory. Rutherford provided tea and biscuits every day, and nearly always attended himself, sitting at the table, with the rest of us perched on stools and at the neighbouring benches. It was a period of relaxation and general gossip, but the meeting often resolved itself into an informal colloquium, with Rutherford taking rather more than a chairman's part. It was here, too ... that Rutherford's essential friendliness was most apparent."[14]

Szmidt must have become more contented with her life in England, for her later correspondence from Russia expressed a longing to return and resume research with Rutherford. Towards the end of her year in Manchester Szmidt visited Leslie, who was working at the time at His Majesty's Factory in Litherland, Liverpool.[15] On her journey back to St Petersburg, Szmidt also stayed with Gleditsch in Oslo for a while.

Over the next few years she corresponded with Rutherford. Some of the letters are lost, but in the first that remains she commented on the insignificance of her contribution and her appreciation of Rutherford's assistance: "I just saw the Phil. Mag. for August, where my note is published. I am ashamed you had a lot to do with the proofs etc., the note was not worth it, and I thank you so much ... I always hope to return to research some day or other."[16]

The next letter (undated, but about October 1915) thanked Rutherford for sending her an article about the death of the brilliant scientist Henry Moseley, who had been killed on the battlefield at Gallipoli.[17] Szmidt mentioned that her health had improved and that her work with war refugees took all of her time and prevented her from resuming research; as part of her refugee work she was organizing Polish schools. Rutherford became quite anxious about her health. He wrote to Ellen Gleditsch: "I have heard at intervals from Miss Szmidt, and feel a little concerned, from stray remarks of hers, about her health. I think she has little idea how to take care of herself physically."[18] Gleditsch must have written to Szmidt, perhaps to encourage her to look after herself. Szmidt mentioned the letter from Gleditsch in her next communication to Rutherford: "Mlle Gleditsch had communicated that she found the half-life of Radium to be near 1660; She intended to let you know this, so probably has written." Szmidt also noted that she had commenced studies towards a formal academic qualification: "I have finished my examinations (they were twelve) at the University for Men and was accorded 1st class in Mathematics. But to receive the diploma I have to write up an original research work in pure Mathematics, Physics, or Astronomy. Prof. Rozhdestvenskii was kind enough to accept my first paper on the energy of alpha-rays on the condition to translate it into Russian and to supply a historical note."[19]

Having acquired a formal qualification in physics, Szmidt returned to active research in 1916, this time with the Russian physicist A.F. Ioffe[20] at the Polytechnic Institute in St Petersburg. Ioffe commented in a letter to his spouse: "I am quite pleased with Yadviga Richardovna,[21] she has a wide understanding and has a good comprehension when pursuing her problems and shows particular laboratory skills."[22] In 1918 Ioffe became the leading figure in the creation of the Leningrad Physico-Technical Institute (LFTI).[23] This institute was the cradle of Soviet physics, dominating research work in the Soviet Union between 1919 and 1939. Ioffe's views dominated the direction of the work: the institute had to serve the Soviet state but it also had to form an integral part of European physics. Szmidt was to be one of its links with physics in the West.

The early days of the institute were not promising. With the political uncertainties of civil war, international isolation, and a lack of funding, organizing a research institute was very difficult. Furniture had to be acquired from the Winter Palace while chemicals and other equipment were obtained from the former Agricultural Institute. This was on top of the lack of food and heat. In 1920 Szmidt wrote a long letter to Rutherford expressing her feeling of isolation and despair. Like most of the pioneer women, she belittled her own abilities:

I am sending this letter in duplicate – if I get the chance I shall send several copies of it by different ways – as I am very anxious that you should get one of it ... It is only directly that we got to know of the splendid work you have done, as already in 1918 we had a stop regards literature, so that we know very little about what has been done since. Moscow, by some chance or other, systematically gets hold of some German magazines, but we have to go there to do our reading. Then I wish to tell you about myself, though there is nothing interesting to tell. Since I left England I did some pedagogical work, usually, to earn my living (just now I am lecturer at the Polytechnical Institute) and at the same time I always was trying to do some research. I work under Prof. Ioffe whom you will know as one of Röntgen's pupils and one of Russia's best physics-men; I have started several experiments concerning Röntgen rays and vacuum apparatus. The frustrating thing is, that I do not believe in the possibility of getting any positive results under the conditions that are existing for work here, and I could not call my work a successful one. It is my wish to work in England once more, of course supposed your guidance and advice. I could manage to get away, but in this case I would have to leave in Petrograd all I own and would come to England with practically no capital to live on. Now it is the object of my letter to ask you if there is any chance of my staying in England so that I could earn my living, modest as it be, and have spare time enough to do some work at the laboratory under you. If you think it possible, let me know, and I would be only too thankful and ready to do at the Laboratory any work you require. I wish to point out what there is against my coming:

1. You will remember I am not too good a worker.
2. The life we lead here is not too favourable for intellectual progress.
3. I would have to go through the whole literature 1918–1920.

And there is not more than one thing I can say in my favour and that is that I wish so much to do some work.[24]

Nothing came of Szmidt's plea and she remained at LFTI, though she was able to visit Rutherford in 1921.[25] It was also in 1921 that she did a critical favour for a young Soviet physicist, Peter Kapitza.[26] Kapitza, who was working at the institute, heard "exciting things about Rutherford and his school" from Szmidt. Having already planned to visit England, Kapitza asked Szmidt to send a letter of recommendation to Rutherford. Presumably she did, as Kapitza sent the next thirteen years working at Cambridge with Rutherford. Szmidt's recommendation may very well have marked a turning point in the life of this future Nobel laureate.

In January 1923 a new research building had been completed, and later in the year Szmidt married A.A. Tshernyshev, an outstanding Russian electrophysicist and director of the Technical Department at LFTI. From this point on she used the name Szmidt-Tshernyshev. Szmidt was one of the few women in atomic science to marry and continue her research. At the time, continuing to work generally meant staying single or marrying a supportive fellow scientist; the normal expectation was that a woman would leave academia upon marriage.[27]

Szmidt wrote to Curie in 1923 seeking a reference for a French visa so that she and her new husband could attend the International Congress of Electrotechnical Engineers in Paris. Noting that Curie might not remember her, she listed all the people who had worked at the Curie laboratory when she was there. Szmidt continued: "Would you kindly permit me to give your name [as a reference]? I would be extremely grateful to you as I have no acquaintances in France and a name as famous in the World of Science as yours would undoubtedly accelerate the formalities. At the same time, I am asking you to allow me to hand over to you in Paris [copies of] my modest contributions of which two were written under Professor Rutherford."[28] It was Curie's secretary who responded, giving permission for Szmidt to use Curie's name in the visa application.[29]

By 1924 Szmidt had been appointed director of the Electrovacuum laboratory[30] where she worked on cathode-ray oscilloscopes and the principles of television, an indication of the move towards applied research at LFTI. Much of this work was accomplished with her husband, and it resulted in a publication on the principles of television[31] and a patent on the oscilloscope design.[32]

Life was again deteriorating at LFTI by the late 1930s, this time because of ideological pressures.[33] The Stalinist ideologues and Party officials pressured the research physicists to conduct their research at the dictate of the Party program and to break off contacts with Western researchers. This attack was followed by the Great Terror of 1936–38 during which the physics community was decimated in the Stalinist purges. There are no records of Szmidt in these years, but with her known links to the West, it is quite possible that she suffered in the purges, unless Ioffe managed to save her.

As well as being active in research, Szmidt translated scientific works. Her linguistic skills were exceptional; in addition to her native languages of Polish and Russian, she was fluent in English, French, German, and Italian. In 1935, with Professor V.Ya. Frenkel, she translated Semenoff's classic, *Chemical Kinetics and Chain Reactions*, into English.[34] She also translated Michael Faraday's text *Experimental Researches in Electricity* from English to Russian but never lived to see its publication in 1947. Szmidt died in April 1940 at the comparatively young age of fifty. The exact date and cause of death are not recorded, but her husband died on the eighteenth day of the same month – more than a coincidence, one would suspect. Frenkel mentioned a tragedy in the Tshernyshev family but gave no further details,[35] and no other information could be found. Certainly, the two deaths within the same month suggest that natural causes were unlikely.

Szmidt, then, was yet another woman drawn to the study of radioactivity by the excitement of this new field that promised to unlock the mysteries of the atom. Though her time with Curie and Rutherford was not particularly fruitful, her later life in St Petersburg enabled her to flourish in the dynamic environment of Ioffe's institute. Unfortunately, we will never know how she felt about her isolation in the Soviet Union, away from the great discoveries that were taking place in Britain, France, and Germany.

14 Ada Hitchins: Research Assistant to Frederick Soddy

MARELENE F. RAYNER-CANHAM
and
GEOFFREY W. RAYNER-CANHAM

Although many of these pioneer women played significant roles in their supervisors' success, few received credit for their contributions. In the several accounts of Frederick Soddy's career, it is rare to find any acknowledgment of the work of Ada Hitchins, though she worked with him for fifteen years, latterly as his private research assistant, and was instrumental in the discovery of element 91, protactinium.

Frederick Soddy had very few assistants throughout his career and Hitchins was the only one to stay with him for an extended period.[1] Born at Tavistock in Devon, England, in 1891, Hitchins was the daughter of William Hedley Hitchins, a supervisor of Customs and Excise.[2] The Hitchins family moved to Campbelltown, Scotland, where Ada graduated from high school in 1909 with English, Latin, French, dynamics, and higher mathematics. From 1909 to 1913 she studied at the University of Glasgow. Originally enrolled for a MA, she changed to a BSc in 1913, graduating with an honours degree in geology, botany, and, with special distinction, chemistry.

In her last year at Glasgow she began to research under Frederick Soddy,[3] one of the later figures in the story of radioactivity. A British-born chemist, he had emigrated to Canada, where he joined Ernest Rutherford at McGill University in Montreal.[4] After returning to Britain in 1904, he obtained an appointment as lecturer at Glasgow University.[5] When Hitchins started to work with him, he assigned her to investigate the half-life of "ionium," which had first been isolated as a new "element" by Bertram Boltwood in 1907.[6] It had been conjectured since 1904 that radium was produced when uranium-238 decayed. However, the rate of production was so slow that it was believed some intermediate species might exist. With the discovery of ionium, Boltwood had filled in this missing "element," though we now realize that ionium was not a new element but simply an isotope of thorium, thorium-230. In 1915 Hitchins's work on this isotope was published.[7] She had selectively extracted uranium from ores and established a half-life for ionium. Moreover, the uranium solutions showed a continuous increase in radium content, proving experimentally that radium was definitely formed from the decay of uranium.

In October 1914 Soddy took a position at Aberdeen University as professor of chemistry. With him he brought Hitchins, his other research student, John A. Cranston, and the sealed flasks containing all of the uranium/radium solutions. At Aberdeen Hitchins obtained a position as Carnegie Research Scholar,[8] which indicates the regard in which she was held at Aberdeen. The scholarships were part of a trust fund set up by American philanthropist Andrew Carnegie; the one-year appointments were to be given, along with one hundred pounds sterling, to graduates of Scottish universities for research and study.[9]

When Cranston was drafted for military service in March 1915, Hitchins took over his line of research. The work involved identifying precursor of actinium and resulted in the discovery of a new element, protactinium – element 91. Although she was not listed as a co-author on the publication, the acknowledgment notes that "...the experiments were continued for a time by Miss Ada Hitchins, BSc, Carnegie Research Scholar, until she also left to engage in war duties. Her valuable assistance has contributed very materially to the definiteness of the conclusions that it has been possible to arrive at."[10] In his reminiscences of the period, Cranston acknowledged Hitchins's role in completing the work.[11]

Another part of her work during this period consisted of the preparation of new radium standards for the calibration of the gold-leaf electroscopes that were used in radioactivity measurements. Soddy later wrote that "Miss Hitchins made a number of new radium standards and determined the quantity of radium in the radium-barium chloride preparation."[12] Hitchins was also involved in the determination of the atomic weight of lead from radioactive ores; Soddy mentioned that the analyses had been performed by Hitchins and himself.[13] Hitchins's results were later refined by Stefanie Horovitz.[14] It was the difference in atomic weight of lead from lead ores and lead from thorium and uranium ores that led to the concept of isotopes.[15]

With the drafting of male scientists, the First World War expanded employment opportunities for qualified women, particularly in government and industrial laboratories.[16] In September 1916 Hitchins left Aberdeen to undertake war service for the British government, working in the Admiralty steel analysis laboratories. Soddy was left without research students for the rest of the war, except for a brief period in 1917 when Beatrice Simpson acted as his personal research assistant.[17] After the war, with the return of the surviving male researchers, Hitchins lost her position in the analytical laboratories – a common experience among women chemists, like other women scientists, at the time.[18] However, she was successful in gaining employment as a chemist with a Sheffield steel works.

Hitchins rejoined Soddy in 1921, the year he was awarded the Nobel Prize in chemistry. By this time he had moved to Oxford University. Soddy had great difficulty in attracting graduate students there,[19] so Hitchins's role became particularly important. Initially she was made technical assistant and set to work on radium extraction in a project involving the Radium Corporation of Czecho-Slovakia.[20] In 1922 she became Soddy's private research assistant.[21] Hitchins refined the half-life measurements of ionium (thorium-230) and determined the ionium/thorium (thorium-230/thorium-234) ratio in a number of mineral samples.[22]

With the discovery of isotopes and the group displacement law, the fundamental problems of radiochemistry seemed to have been solved.[23] Soddy focused increasingly on the commercial applications of radioactive isotopes, a topic that had always fascinated him.[24] Much of Hitchins's time at Oxford was spent devising improved ways of extracting radioactive elements from pitchblende and thorium from Indian monazite sand.[25] Considering the levels of radiation to which she must have been exposed over the years, it is surprising that she did not rapidly succumb to a radiation-related illness. Soddy noted in a later reference for Hitchins that "she has also charge of my radioactive materials ... and has worked up considerable quantities of radioactive residues and other materials for general use." He also made clear his esteem for the woman he had employed as a research assistant through much of his active period as a scientist: "I regard Miss Hitchins as an exceedingly accomplished chemist with a wide knowledge and experience of difficult chemical and mineral analysis. It is much to her credit that in her work with extremely rare and difficultly obtainable materials she has never had any accident or lost any of the material."[26]

The Hitchins family had emigrated to Kenya, and Ada proposed to join them in the summer of 1927. Soddy wrote to the British Colonial

Office bringing her qualifications to their attention: "Miss Hitchins is an exceedingly accomplished chemist, and I feel that her special technical knowledge and experience are likely to be thrown away in a new country, unless she can obtain a position in connection with the Government services."[27] Hitchins did indeed obtain a position in Kenya with the colonial adminstration as Government Assayer and Chemist in the Mining and Geological Department, thus ending her connection with the world of radioactivity.[28] She worked in the department from 1932 until her retirement in 1946. According to the department's annual report of 1946, our last record of Hitchins, "she filled the post of Chemist and Assayer gaining an outstanding reputation for accuracy and complete reliability, and her loss was keenly felt by the mining industry. It is recorded with appreciation that after a short holiday she returned from time to time [after retirement] to the Department to carry out urgent assay work, since by the end of the year Mr W.P. Horne appointed as Chemist and Assayer had not arrived in the colony to take up his duties."[29]

Given Soddy's regard for her, it is doubly regrettable that Hitchins, who was with him for much of his renowned career, has been overlooked in accounts of his life. Scientific research usually involved teamwork. Yet Soddy's biographers, like those of so many other famous scientists, convey the impression that his discoveries were made by the "great man" alone.

15 ... And Some Other Women of the British Group

MARELENE F. RAYNER-CANHAM
and
GEOFFREY W. RAYNER-CANHAM

Though Rutherford was the central figure of the British group, the research workers were more diffuse than those of the Paris-centred French group. In this chapter we will briefly look at women who worked with some of the other major figures of the British school, scientists whose fame has now dwindled but who were Nobel laureates around the beginning of the twentieth century: Charles Barkla, William Ramsay, Frederick Soddy, and J.J. Thomson.

Charles Glover Barkla, who received the Nobel Prize for physics in 1917,[1] had an all-male research team until the First World War. When his team was drafted into military service or into war-related work, Barkla was able to continue his researches on x-ray scattering "by utilizing the services of the most able women graduates."[2] The first of these recruits was Janette Gilchrist Dunlop, who had graduated from Edinburgh University with an honours degree in mathematics and natural philosophy in 1914. Awarded a Carnegie Research Scholarship, she worked with Barkla on the relation between x-ray scattering and atomic structure.[3] Dunlop only stayed with Barkla until 1916 when, according to her obituary, she was "compelled by war-time teacher shortage" to enrol in a teacher-training program.[4] This marked the end of her research activity. Margaret Pirie White followed in Dunlop's footsteps, graduating one year later. She obtained a position as assistant in natural philosophy in 1916, which was upgraded to the rank of lecturer in 1920.[5] White had one publication with Barkla[6] before resigning in 1921 to marry one of Barkla's other students, Robert Taylor Dunbar.[7] In later years,

Barkla had two other women researchers working with him: A.E.M.M. Dallas and Gladys I. Mackenzie.

William Ramsay, Nobel laureate in chemistry for 1904, employed two women assistants in his early work on radioactivity. The sole record of these women is Otto Hahn's memoirs. Hahn, who was working with Ramsey at the time, was involved in the study of materials so radioactive that they glowed in a darkroom. He remarked: "It was not only Ramsay who took an interest in my work. So did two young ladies who were also working in the laboratory. One of them, a Miss O'Donoghue, was especially well disposed to me. I had once shown her my preparations in the darkroom, where they were fluorescent even in the dim light, and since then she always liked coming into the darkroom with me. But I never dared to kiss her. I had been warned that in England a kiss was tantamount to an engagement, so that it might lead to one's being sued for breach of promise."[8]

Of the three additional women about whom we have a significant amount of knowledge, Jesse Mable Wilkins Slater worked with J.J. Thomson, while the other two, Winifred Moller Beilby and Ruth Pirret, worked with Frederick Soddy

JESSE MABLE WILKINS SLATER

Slater was a student with J.J. Thomson at Cambridge University in the very early days of radioactivity. She was born on 24 February 1879, the daughter of John Slater, an architect, and Mary Emily Wilkins.[9] Educated at South Hampstead High School and Bedford College, London, she was admitted to Cambridge University in 1899 as a Gilchrist Scholar at Newnham College. There she obtained first-class honours in both parts of the examinations in physics and chemistry. Until 1948 Cambridge did not formally offer degrees to women,[10] thus Slater's BSc, awarded in 1902, was granted by the University of London. She was then invited to work with J.J. Thomson at the Cavendish, where she studied the decay products of thorium from 1903 to 1905 as a member of Bathurst College.

One of her papers,[11] based on work in which Thomson claimed the existence of a new type of ray, was well done but not of great significance. Her other line of research[12] proved to be of real importance and it was particularly noteworthy for its thoroughness. When the "emanation" (radon-220) from the thorium decayed, it produced a solid deposit. Most radioactive decays were by that time known to follow an exponential law. However, Slater showed that this particular decay was more complex in nature. She correctly ascribed it to the presence of two substances, each having its own decay rate. Further, she managed to separate the components and show that the first product had a half-life of eleven hours and the second a half-life of fifty-five minutes. This is remarkably close to the contemporary value for these isotopes: lead-212 at 10.6 hours and bismuth-212 at 60.6 minutes. In fact, this was the first evidence for the existence of bismuth-212, and Slater is actually credited with its discovery.

After completing her research work in 1905, Slater took a teaching position at King Edward VI High School for Girls in Birmingham, which held the foremost position among girls' schools in England for science teaching.[13] From 1909 to 1913 she occupied a position at Cheltenham Ladies College, after which she was appointed assistant lecturer (approximately equivalent to a North American assistant professor) in physics and chemistry at Newnham College, her Alma Mater. In 1914 she was promoted to lecturer (equivalent to an associate professor), a position that she held until 1926. For her time, this must have been among the highest academic posts held by a woman in science in Britain.

Slater obtained leave from her Newnham position to undertake war-related duties. For the first three years of World War I, she was a part-time nurse. Then, on the basis of her research work, she was called for full-time duty as a radiographer at British military hospitals in France. She was also associated with a French hospital, holding the rank of Officier de l'Instruction Publique. This work would probably have brought her into contact with either Marie Curie or Irène Joliot-Curie, the leading radiographers of the time. In 1926 Slater married and, as was to be expected at the time, gave up her career. She died on 25 December 1961.

WINIFRED MOLLER BEILBY

Beilby was a rare example of a woman who commenced her research work only after marriage. Born on 1 March 1885 in Edinburgh, Scotland, she was the only daughter of Sir George Thomas Beilby (1850–1924), an industrial chemist. When she was fourteen years old the family moved to Glasgow, where she was educated at home by a governess. Beilby had wide-ranging interests, she acquired a dedication to the cause of women's suffrage from her mother, and she loved outdoor activities, such as skating.[14] At the age of seventeen she was sent to Paris to study art, another of her passions.

It was through her father that she met Frederick Soddy. Sir George, who had helped finance Soddy's research, mounted a fund-raising campaign in 1904 to raise money for equipment needed for radium research. Soddy and Beilby became engaged in 1906[15] and married in 1908. Beilby wanted a quiet, early-morning wedding so that they could go off quickly on their mountain-climbing honeymoon. In keeping with Beilby's love of the outdoors, the couple spent most of their vacations climbing some of Europe's highest and most hazardous peaks.

Beilby became interested in Soddy's work and helped him considerably, publishing her work on the gamma rays emitted by radioactive atoms in 1910.[16] Soddy himself commented: "She did quite a bit of work for me at Glasgow. You can read it all in my communications. It was a tedious investigation and she stuck at it, like Marie Curie, I used to say."[17]

When Soddy moved to Oxford University, Beilby apparently gave up research to devote herself to the necessary social life, spending her free moments painting and gardening. At Oxford Soddy developed a fascination with economics that ultimately became an obsession, and it was remarked that "she [Beilby] had a strong sense of fun and humour which enabled her to withstand her husband's moods when he was battling with the delinquencies of the banking world or the mistakes of the government."[18]

Towards the end of 1935 Beilby fell ill; she died suddenly in August 1936, quite possibly as a result of lengthy exposure to gamma rays during her research work. Her passing came as a grievous shock to Frederick Soddy and it was the cause of his resignation from Oxford. Soddy commented to his biographer, Muriel Howorth: "My marriage was my greatest achievement. My wife and I worked splendidly together."[19]

RUTH PIRRET

Pirret, who also worked with Soddy, graduated from the University of Glasgow in 1898 with a BSC honours degree in pure science (mainly chemistry).[20] An outstanding student, she became a school teacher in Arbroath after graduation. Starting in October 1909, she worked for at least six months with Soddy. Lord Fleck, another researcher with Soddy at the time, commented: "T.D. MacKenzie was the research student who was the most important worker in those early years. Miss Ruth Pirret was also a contributor, but on a less prominent scale."[21]

Pirret was studying the ratio of uranium to radium in minerals, an extension of the work of Ellen Gleditsch. It had been argued at the time that the older the mineral, the higher the proportion of radium, and Pirret and Soddy confirmed some of Gleditsch's findings. However, they made a much more important discovery: that uranium existed in nature as a mixture of two isotopes. As they wrote about their finding: "We are therefore faced with the possibility that uranium may be a mixture of two elements of atomic weights 238.5 and 234.5, which, like ionium, thorium, and radio-thorium [thorium isotopes 230, 232, and 228], are chemically so alike that they cannot be separated."[22] A subsequent report based on a continuation of their studies[23] proved to be more pedestrian. Nothing further is known of Pirret. She may have returned to school teaching or she may have got married, which would almost certainly have ended her research career.

PART FOUR
The Austro-German Group

As we mentioned in chapter 1, Stefan Meyer can best be considered as the "father" of the Vienna group, Vienna having been the central European research Mecca in the early days of radioactivity. Lise Meitner (chapter 16) was introduced to the study of radioactivity by Meyer in about 1906. Even after Meitner moved to Berlin to work with Otto Hahn, she kept in touch with the Vienna women and also developed links with Gleditsch and Ramstedt of the French group. Stefanie Horovitz (chapter 17) followed in Meitner's footsteps at Vienna, though Horovitz was directly associated with Otto Hönigschmid. Marietta Blau (chapter 18) spent fifteen years in Vienna, punctuated by a short stay in Paris with Curie. She then fled the Axis countries to work with Gleditsch in Norway before proceeding first to Mexico and then to the United States. Elizaveta Karamihailova (chapter 19) travelled from Bulgaria to attend the University of Vienna. She stayed in the city for thirteen years, then spent four years with Rutherford at Cambridge before returning to her native country. Elizabeth Róna (chapter 20) from Hungary worked with Hahn and Meitner in Berlin, then spent fifteen years at Vienna at Meyer's invitation. She spent a year with Curie and visited Rutherford, living with Karamihailova during her stay at Cambridge. She, too, found refuge with Gleditsch after fleeing the Nazis and settled finally in the United States. Chapter 22 looks at Berta Karlik, who made no major discoveries herself but was a major link between Blau, Gleditsch, Karamihailova, Meitner, and Róna. Karlik was the one woman who stayed in Vienna throughout the Second World War, though she worked briefly in England (with Bragg and Rutherford), France (with Curie), and Sweden (with Róna) between the wars. We have included the German chemist Ida Tacke Noddack (chapter 21) in this part, though she had no actual connection with the Vienna-centred, physics-oriented group.

16 Lise Meitner: The Foiled Nobelist

SALLIE A. WATKINS

Next to Marie Curie, Lise Meitner is the best known of the early women atomic scientists.[1] Her doggedness enabled her to survive the anti-women bias of many of the scientists in Berlin. Her years of work in the field of radioactivity culminated in a flash of inspiration: what had been thought to be a nuclear fusion process involving the heavy elements was, in fact, nuclear fission. Unfortunately, it was her longtime colleague, Otto Hahn, who ultimately received the Nobel Prize for this discovery.

The twelve-month span from November 1878 through October 1879 was a halcyon year for physics, marking the births of four great contributors to the field. Lise Meitner, born 7 November 1878, led the way, followed by Otto Hahn on 8 March 1879 and Albert Einstein on 14 March 1879, while Max von Laue was born 9 October. A century later, in September 1979, there was a grand celebration at the Max Planck Institute in Munich commemorating the lives of these four great scientists. But of this nothing could be known when Elise (always called Lise) joined her older sisters, Gisela and Auguste, as the child of Hedwig Skovran and Philipp Meitner in Vienna.

The city into which Lise Meitner was born near the turn of the century sparkled with brilliant theatre, immortal music, and a dazzling social life for those fortunate enough to belong to the middle or upper classes, as the Meitners did. "Social relationships tended to be quite close between families in the academic and professional world ... Doctors, lawyers, government officials, and professors belonged to the comfortable upper-middle class, with spacious but not elegant housing, at least one but usually two servants ... and time for leisurely vacations in the Tirolean mountains."[2]

As a respected and prosperous attorney, Philipp Meitner would have moved easily in this urbane culture, though he and his family were not native to the city. Lise Meitner told a cousin that her family "came from Moravia from a small place near Mahrisch-Weisskirchen where my grandparents had a kind of estate."[3]

As the years passed two more girls, Carola and Frida, and three boys were born to Hedwig and Philipp Meitner. The Meitners lived in "a very cultured middle-class environment ... It was a prolific family ... and an economical way of life was encouraged despite the distinguished position of the father. Lise Meitner therefore saw it as her duty early on, to contribute to the income of the family through private tutoring; she coached young girls who had to study on their own for the university entrance examinations, since girls were not admitted to the gymnasium at that time ... Music was very cherished in her parents' house."[4] In terms of religion, Otto Robert Frisch (Auguste's son) later observed that, "although her [Lise's] parents came from Jewish stock, her father was a freethinker, and the Jewish religion played no role in her education."[5] In fact, at the age of thirty Lise Meitner converted to Christianity and was baptized as a Protestant, though there is no evidence that she was ever a devoted practitioner.[6]

Motivated by their parents, the Meitner children were drawn to the professions.[7] It was said in later years that to enjoy bona fide membership in the family, one needed either to become a doctor or to marry one. Leading the way, Gisela became a medical doctor; Auguste was a precocious pianist who at the age of twelve played at one of the grand soirées thrown in the homes of wealthy Viennese;[8] Lise became a physicist; Carola did not go to university but married Rudolf Allers, a psychologist who achieved world renown in his profession; Frida acquired a doctorate and, after settling in the United States, became a college professor of mathematics. Of the boys, Moriz (Fritz) became an engineer and Walther became a chemist, after earning a doctorate in the field. Historical records are unclear on the third son, Max. One report is that he was delicate from childhood, dying of a heart attack at an early age; another, that he took his own life in a dual suicide.

As a child, Lise Meitner followed the usual school patterns of Viennese girls of the day. Since Austria had only recently become part of the Austro-Hungarian Empire, its educational system paralleled that of the other German states. Girls attended state-supported primary schools; then, if their parents could afford it, they went on to private schooling. Lise Meitner attended elementary school for five years, then spent three years at a school for middle-class girls.[9] She then went to a typical private high school for girls where the standard curriculum included such subjects as religion, natural history, diction, German, French, English, Italian, geography, arithmetic, physical science, penmanship, drawing, needlework, singing, and gymnastics. Many of these schools provided two years of advanced education and a third year of practice teaching.[10] Finally, she attended the Höhere Töchterschule in Vienna's District Six, finishing there in 1899.[11]

Earlier, Meitner had shown exceptional curiosity in the areas of natural science and mathematics. Frisch says she told him that her interest in physics was kindled by her fascination with the interference patterns formed when oil floats on a puddle of water.[12] As for mathematics, Frida quotes their mother as saying that when Lise was eight, she was caught covering the crack under the door of her room with a towel so that she could read her mathematics book far into the night. When her eyes became too heavy, she slept with the book under her pillow.[13]

By conventional standards, Lise was not pretty in comparison with her sisters. She seems to have accepted this with grace. As an adult, "she was not the least bit interested in being glamorous. [It was] said that when she wore a hat, there was a special machine that would drop the hat on her [head] as she was walking through the front door."[14] We can picture Meitner smiling[15] at this story – a light-hearted person making the best of the situation in which she found herself. At the same time, her plain features and her place as a middle child in a large family[16] may have fuelled her drive for academic excellence: "Although I had a very marked bent for mathematics and physics from my early years, I did not begin a life of study immediately. This was partly due to the ideas which were then generally held with regard to the education of women and partly to the special circumstances in my native city, Vienna."[17]

As a woman, Meitner was unable to pursue the traditional course of study in preparation for admission to the university. This would have meant attending a gymnasium (academic high school), and at that time such schools existed only for boys. The university did admit women, though they had to study privately to prepare for the gymnasium leaving examination, the Matura. Gisela blazed the trail for Lise, passing the Matura in 1899.[18] Lise began a private course of study in 1899 with Dr Arthur Szarvasy as tutor. Meitner recalled: "Dr Szarvasy had a real gift for presenting the subject matter of mathematics and physics in an extraordinarily stimulating manner.

Sometimes he was able to show us apparatus in the Vienna University Institute, a rarity in private coaching – usually, all one was given were figures and diagrams of the apparatus. I must confess that I did not always get correct ideas from these, and today it amuses me to think of the astonishment with which I saw certain [types of] apparatus for the first time."[19] Meitner applied herself so diligently to her studies that she completed the usual four-year course of preparation for the examination in two years. Of the fourteen young women who took the test, only Meitner and three others passed.

The University of Vienna had been open to women for several years – in theory. But as usual, there was a major time delay between the enactment of legislation and a change in human behaviour and practice. When Lise Meitner entered the university, "a female student was regarded as a freak."[20] Here she found a strange culture, that of a homogeneous subset of Viennese society composed of groups of young men who had formed fast friendships during their gymnasium days. Meitner, nearly twenty-four when she entered the university, began an uneasy coexistence with these men.

The Institute for Theoretical Physics was housed in a "temporary" building, a primitive converted apartment house at Türkenstrasse 3. "The entrance looked like ... a hen house, so that I often thought, 'If a fire breaks out here, very few of us will get out alive.'"[21] Since there were no proper benches for lectures, students had to sit on chairs, precariously balancing notebooks on their laps. "The floor was an ancient inlaid one, through which gaping crevasses ran, in which even today untold amounts of mercury might remain. Each step made the entire room shake ... and even the outer walls trembled when a strong wind blew outside."[22] The situation was so bad that before a meeting in Vienna in 1894, Franz Serafin Exner, then head of the Physical Chemistry Institute, left the city because he was so ashamed of the miserable quarters of the physics institute.[23]

If Meitner found her fellow students and the physical environment less than hospitable, she found warm encouragement among her professors. As her friend Berta Karlik commented: "In particular she often spoke of the contagiously enthusiastic lectures of Ludwig Boltzmann; it was probably he who gave her the vision of physics as a battle for ultimate truth, a vision she never lost."[24]

From 1901 until the end of 1905 Meitner studied mathematics, physics, and philosophy. Initially, she was uncertain whether she wanted to become a mathematician or a physicist: "My first term, I studied differential and integral calculus with Professor Gregenbaur. In my second term he asked me to detect an error in the work of an Italian mathematician. However, I needed his considerable assistance before I found the error, and when he kindly suggested to me that I might like to publish this work on my own, I felt it would be wrong to do so and ... unfortunately annoyed him forever. This incident did make it clear to me, however, that I wanted to become a physicist, not a mathematician."[25]

By Meitner's own admission, she had begun by attending too many lectures,[26] becoming what was known as a "finch" in European universities; professors lectured to their own students, but if the windows were open no one could prevent the small birds from listening in. Finches were usually found at the very back of the lecture hall, looking apologetic and taking more notes than the registered students.

In her first year at the university, Meitner also had to do her practice teaching at a girls' school. This sort of "safety net" was traditional for a university student of the time, to allow for the possibility of future employment as a teacher should the need arise. But teaching the French language convinced her that her career lay elsewhere.

Meitner showed her serious dedication to university studies when she announced at the family dinner table that she would have no time to read the daily newspaper for the next several years. She would rely, she said, on other family members to keep her abreast of current affairs in the world outside the university. Unable to resist this golden opportunity, her mischievous sisters and brothers invented a war; every evening they recounted stories of the day's military engagements and the numbers of casualties on each side, speculating about the long-term outcome.[27] Evidently Lise's conversations with professors and fellow students at the university focused on the business at hand rather than on such small talk as current events, else the mongers of the specious war would have been exposed.

Though her privacy was a guarded treasure, Lise Meitner was no recluse. One person with whom she studied for a time was Paul Ehrenfest; the record of their work together is revealing: "Ehrenfest looked up Lise Meitner shortly after his return to Vienna from Göttingen, when he heard that she had a detailed set of notes on all of Boltzmann's lectures. [No surprise, this; Meitner routinely worked in such a fashion.] He proposed that they study Boltzmann's ideas together, as well as some of the background work by C.G. Jacobi and others in analytical dynamics, a suggestion that she quickly accepted since his gift for clarifying the ideas of theoretical physics was already quite evident. Their joint study was a great stimulus to her scientific development, but she found Ehrenfest's need to probe into the very soul of anyone with whom he came into contact was too much for her, and it prevented a very close acquaintance."[28]

The years passed quickly, and on 20 November 1905 Lise Meitner submitted her dissertation to the physics faculty as a part of the requirement for the Doctor of Philosophy degree. Using mercury ointments, she had studied Maxwell's formula for the conduction of electricity in inhomogeneous materials, to determine whether this relationship would hold for heat conduction as well. She was able to show that it did. The dissertation was approved a month later.[29]

On 11 December Meitner had passed the *Rigorosa,* or final state examination, summa cum laude, and on 1 February 1906 she was awarded the DrPhil degree in physics from the University of Vienna (this degree signified something lower than a modern PhD but higher than a master's). Further original research reported in a second dissertation, the *Habilitationsschrift,* was required for the *Venia Legendi* (right to teach). For Meitner, this was to come sixteen years later.

Meitner stayed on in Vienna for another year and a half. Ehrenfest had drawn her attention to the scientific papers of Lord Rayleigh, whose studies on optics now prompted her own first independent work. "I was able not only to explain an experiment which Lord Rayleigh had carried out ... [with] unexpected results, but also I could predict some other consequences and give them experimental proof."[30]

During the summer of 1906, Boltzmann, Meitner's brilliant and beloved mentor, committed suicide; it was a severe shock to all the members of the physics department. The physics institute was taken over temporarily by Stefan Meyer – a decisive turn of events for Meitner's later work,[31] since Meyer had become involved in the new field of radioactivity. Meitner, having lost a mentor, sought a new direction and studies in radioactivity were a natural step for her. Moreover, "as an aspiring scientist she ... [had been] fascinated by the newspaper accounts of the discovery of radium by Pierre and Marie Curie in 1902."[32]

In her new field Meitner chose to investigate the deflection of alpha rays as they passed through matter; were they in fact deflected at all? She designed and performed experiments to detect the deflection, if any. Using ThB (lead-212) and AcB (lead-211) preparations, she studied the absorption of alpha and beta radiation in platinum, gold, silver, copper, tin, and aluminum and succeeded in detecting alpha-ray deflection.[33]

Although the work was challenging, Meitner felt a need to deepen her knowledge of physics. Boltzmann's lectures had not treated such new developments in theoretical physics as the breakthroughs made by Max Planck, Albert Einstein, and others. Further, Anton Lampa, who directed the Department for Elementary Practical Work, was rather sceptical of the modern developments in physics.[34] Meitner might perhaps have concurred with Erwin Schrödinger's observation about "the tendency of the golden Vienna heart to place amiable nincompoops in key positions ... where they blocked progress ... Thus atmospheric electricity and radioactivity, which really had their beginnings in Vienna, were taken out of our hands, and anyone who felt inspired to work seriously in these fields had to go abroad, as for example Lise Meitner from Vienna to Berlin."[35]

Whether or not Meitner's analysis paralleled Schrödinger's, her conclusion was the same: she must leave Vienna. Following his death, Boltzmann's vacant chair had been offered to Max Planck. Planck declined the post, "primarily ... because of the unexpected interest the Berlin faculty showed in retaining him."[36] But he had visited the physics institute in Vienna and Meitner had had the chance to observe him in action. She decided that she could profit by spending several semesters in Berlin as his student. Her parents agreed with her and pledged financial support.[37]

Uprooting herself from Vienna and settling in a new city at a university to which women were not yet admitted must have been traumatic for Meitner. She had to find living quarters and transportation; painfully shy, she needed to take the initiative in meeting people, knowing that even her speech betrayed her as a foreigner. Would her acquaintance with Planck smoothe the path? "I registered with Planck at the University in Berlin so as to attend his lectures, he received me very kindly and soon afterwards invited me to his home. The first time I visited him there, he said to me, 'But you are a Doctor already! What more do you want?' When I replied that I would like to gain some real understanding of physics, he just said a few friendly words and did not pursue the matter any further. Naturally, I concluded that he could have no very high opinion of women students, and possibly that was true enough at the time."[38] Nonetheless, Meitner became one of Planck's closest friends.[39]

What was the general situation for women in Prussia when Lise Meitner arrived there in 1907? During the late 1880s and early 1890s, the German Women's Movement had begun to pressure the government to reform secondary education for girls, provide access for women to higher education, and replace male teachers with females in girls' schools. In Prussia the petitions were systematically denied both by Parliament and the universities, until finally, in 1894, the Ministry of Education issued regulations for secondary girls' schools, as well as for the education and examination of female teachers. According to the historian Stock: "The program the government set up provided for nine years, instead of the ten years in the better existing schools. The Ministry of Education gave the excuse that nine years of study was the maximum a girl's health could take."[40] As for admission to the universities: "The matter came up in the Prussian Parliament in 1898. The arguments used against the presence of women in male universities were that it was shameful for the women and unwholesome for the men, and that knowledge would suffer from women's inability to study seriously."[41]

In 1897 Arthur Kirchhoff had reported responses to a questionnaire on the admission of women to university studies. Planck wrote that, "if it is at all compatible with academic order, I shall readily consent to their admission – on approval and always revocably – to my lectures and my practical courses, and in that respect up to this point I have had nothing but favorable experience. [But on] the other hand I must keep to the fact that such a case must always be regarded just as an exception ... Generally it cannot be emphasized enough that nature herself prescribed to the woman her function as mother and housewife, and that laws of nature cannot be ignored under any circumstances without grave damage, which in the case under discussion would especially manifest itself in the following generation."[42]

In some regions of Germany, admission of women to the universities was made legal as early as 1900. Eventually even Prussia capitulated; in 1908 women were granted permission to matriculate as regular students in universities. Meitner arrived in Berlin, the centre of Prussian thought, with a doctorate in physics one year before women were admitted to university studies there. Paid employment was out of the question in academe, so she relied on her parents' support to eke out a modest living.

Her plan was not only to attend theoretical lectures but also to do experimental work, just as she had at the University of Vienna. With the goal of becoming an experimentalist in mind, she applied to the head of the Institute of Experimental Physics at Berlin, Heinrich Rubens:

> He told me the only space he had was in his own laboratory, where I could work under his direction, that is, to a certain extent with him. Now it was quite clear to me then, as a beginner, how important it would be for me to be able to ask about anything I did not understand, and it was no less clear to me that I should not have the courage to ask Professor Rubens. [Meitner had coolly appraised her career needs, at the same time factoring in a personal liability – shyness.] While I was still considering how I could answer without giving offense, Rubens added that Dr Otto Hahn had indicated that

he would be interested in collaborating with me, and Hahn himself came in a few minutes later ... The only difficulty was that Hahn had been given a place in the institute directed by Emil Fischer, and Fischer did not allow any women students into his institute ... He finally agreed to my working with Hahn, if I promised not to go into the chemistry department where the male students worked and where Hahn conducted his chemical experiments. Our work was to be confined to a small room originally planned as a carpenter's workshop; Hahn had it fitted out as a room for measuring radiation. For the first few years, I was naturally restricted to this work and could not learn any radiochemistry.[43]

Meitner's recollections of a time almost sixty years earlier suggest that she was passively accepting of Fischer's mandate. Yet this acceptance was punctuated by daring escapades in which she sneaked up from the basement and hid beneath classroom benches to eavesdrop on chemistry lectures.[44]

If becoming Hahn's partner seemed like a victory, it was a hollow one. Education for women in Prussia was one thing; employment in academe was another. Meitner was a "guest" at the institute, receiving no salary. Her parents supported her until December 1910, when her father died. She later confided to Frisch that she had lived for long periods on black bread and coffee.[45]

The following year, her good friend Planck hired Meitner as an assistant with a small salary. Hahn wrote that she was one of the first female scientific assistants in all of Prussia.[46] She added: "The chief work of ... [Planck's] assistant was to correct beginners' weekly problem sets and discuss the results with Planck, who would ask students who had done well to demonstrate for the others during the exercise hour."[47] This arrangement was ideal for Meitner, since it allowed her ample time to pursue her radioactivity research with Hahn.

Meitner was ingenious at finding ways and means to supplement her small salary as an assistant. For example, on 23 September 1911 E.M. Wellisch of Yale University sent her a money order for ten dollars in payment for her translation of a manuscript from English to German so that it could be published in a German journal.[48] She also wrote for a living. Otto Hahn later recalled that, "in addition to her work in the wood shop and her classes with Planck, Lise Meitner wrote many articles on problems in physics for the *Naturwissenschaftliche Rundschau*, then edited by Professor Sklarek, whom she had known in Vienna."[49]

Just before the First World War, Hahn received a commission on the commercial sales of mesothorium. He commented: "It was an extraordinarily large sum in those days (66,000 Marks). I gave a tenth of it to my colleague Lise Meitner, whose work on fractional crystallization was of very great help to me."[50] While Hahn's contribution eased her acute financial distress, Meitner was to know periods of economic uncertainty until the very end of her long life.

Meitner and Hahn had to develop the structure of their scientific partnership. Hahn had had a professional headstart. Though Meitner was a few months older, her studies had been delayed, as we have seen, because she was barred from attending the gymnasium. Coincidentally, the very month that Hahn passed his oral examination for the doctorate (July 1901) found Meitner sitting for the gymnasium completion examination. Hahn went on to military service in October as Meitner began university studies. A year later he began an assistant lectureship at his Alma Mater, Marburg. After two years there, he was offered a job as organic chemist at Kalle and Company Chemical Works.[51]

Since his position would involve occasional trips abroad, it called for a speaking knowledge of English and French. With the help of his mentor at Marburg, Theodor Zincke, Hahn secured a place in the London laboratory of Sir William Ramsay in the autumn of 1904. Ramsay gave him the task of extracting radium from barium salt by Marie Curie's method of fractional crystallization. Work on this problem led to Hahn's discovery of radiothorium.

From England, Hahn went to McGill University in Montreal for six months, studying with Ernest Rutherford. It was there that he discovered radioactinium. Meanwhile, Ramsay had urged Hahn not to enter the world of industry but to continue with radium research and try to get a lectureship in Berlin instead. Ramsay wrote to his fellow Nobel Prize winner, Emil Fischer, director of the Chemical Institute, on Hahn's behalf. Receiving a positive response, Hahn joined Fischer in Berlin at the beginning of October 1906. A year later, he and Meitner took up work together.

Initially, Meitner regarded Hahn as a mentor and teacher: "Hahn was of the same age as myself and very informal in manner, and I had the feeling that I would have no hesitation in asking him all I needed to know. Moreover, he had a very good reputation in radioactivity, so I was convinced that he could teach me a great deal."[52] But as the years went by Meitner quickly came into her own, assuming a leadership role in their work together. "Hähnchen, leave that to me, you haven't the faintest notion about physics!"[53] In an attempt to sort out the Hahn-Meitner roles, Werner Heisenberg later observed that she "not only asked 'What,' but also 'Why.' She wanted to understand … she wanted to trace the laws of nature that were at work in that new field [radioactivity] ... We may suppose that also in later joint work, Lise Meitner exercised a strong influence on the asking of questions and the interpretation of experiments and that Hahn mainly felt responsible for the thoroughness and accuracy of the experiments."[54] While Meitner was obviously at a disadvantage in terms of prior experience in the laboratories of acclaimed scientists – and in terms of major discoveries – her scientific mettle was without question.

For his part, Hahn, though a clever chemist, had never taken first place during his school days. Moreover, at university, after "some initial effort I almost entirely gave up going to ... physics lectures, and in spite of many attempts later on in life I could never make up for my lack of solid grounding in physics."[55] At the interface of physics and chemistry, the new field of radioactivity called for just such a partnership as that of Hahn and Meitner. But Hahn recalled: "There was no question of any closer relationship between us outside the laboratory. Lise Meitner had had a strict, lady-like upbringing and was very reserved, even shy ... for many years I never had a meal with [her] except on official occasions. Nor did we ever go for a walk together. [Was a young woman of her upbringing perhaps awaiting an invitation?] Apart from the physics colloquia that we attended, we met only in the carpenter's shop. There we generally worked until nearly eight in the evening, so that one or the other of us would have to go out to buy salami or cheese before the shops shut at that hour. We never ate our cold supper together there. And yet we were really very close friends."[56]

Even "after ten years of friendship Meitner addressed Hahn with the formal 'Sie' and began every letter 'Dear Herr Hahn!'"[57]

Nonetheless, *joie de vivre* was a feature of the Hahn-Meitner laboratory. Hahn remembered that "Lise Meitner ... took a hand in my musical education."[58] Meitner elaborated: "When our work went well we sang together in two-part harmony, mostly songs by Brahms ... Our personal and professional relations with our young colleagues at the neighboring Physical Institute were excellent. They often visited us and would occasionally climb in through the window of the carpentry shop instead of taking the usual way. In short, we were young, happy, and carefree [but sadly, in retrospect] – perhaps politically too carefree."[59]

Evidence of Meitner's growing stature as a physicist was accumulating. Scientific papers reporting the radioactivity work she had done in Vienna appeared in 1906 and 1907.[60] As a result, she was earning increasing recognition among her colleagues and superiors in the Berlin circle. A high level of intellectual comradeship was generated by the Wednesday colloquium, led first by Rubens and then by von Laue.[61] Meitner commented: "This group of young physicists made up an unusual circle. Not only were they brilliant scientists ... they were also exceptionally nice people to know. Each was ready to help the other, each welcomed the other's success. You can understand what it meant to me to be received in such a friendly manner into this circle."[62]

Besides participating in the Berlin colloquium, Meitner travelled abroad to present research results. At the annual meeting of the German Naturforscher in Salzburg, she reported on "minor pieces of work"[63] in which she and Hahn "had discovered and properly classified two new groups of beta emitters in the radium series"[64] in 1909. The occasion was memorable for Meitner for two reasons: it took her back home to Austria, and it was at the conference that she first met Albert Einstein.[65]

Meanwhile, the workspace allotted to Hahn had expanded to three rooms. When Meitner's confinement to the carpentry shop was lifted, she was free to move into the areas equipped for chemical separations, thus greatly extending her research opportunities. The first shared project undertaken by Hahn and Meitner in the three-room facility was that of verifying (or failing to do so, as it happened) a theory of Heinrich Willy Schmidt.[66] Schmidt had proposed that a given radioelement would emit beta rays of constant velocity, and that the absorption of these rays would be exponential in mathematical form. While Schmidt's assumption proved to be invalid, along the way Meitner and Hahn found several new beta emitters in the three naturally radioactive series, as confirmed by chemical separation techniques.

About the same time, Hahn correctly identified the process of radioactive recoil. He had been involved in a comparative study of Actinium-x (radium-223) and radioactinium (thorium-227) and had prepared a manuscript for publication. "But Lise Meitner, after reading my manuscript, said immediately, 'What you have observed there with actinium and with fairly thick layers of the preparation should be far easier to observe on alpha-emitting active deposits in infinitesimally thin layers.'"[67]

The physics underlying the recoil method is straightforward. An alpha particle is ejected with a large amount of forward momentum. When actinium-x is produced from radioactinium, the actinium-x, being positively charged, recoils (both by electrical repulsion and to conserve momentum) from the alpha particle of the radioactinium. The positively charged material can then be collected on a negative electrode. Infinitesimally thin layers of the parent preparation are helpful in increasing the surface-to-volume ratio.

Within a week after Meitner had suggested the broadened use of the recoil technique, she and Hahn had found they could produce radium-B (lead-214), actinium-C″ (thallium-207), and thorium-C″ (thallium-208) by the new method. A preliminary report was read by the two of them at a meeting of the German Physical Society on 22 January 1909.[68]

Returning to their beta ray work, Meitner and Hahn measured the velocity of beta rays by studying their deflection in a transverse magnetic field. "As there were no magnets in the Chemistry Institute, we carried out these experiments with Otto von Baeyer in the Physics Institute ... [If] our efforts [to lay down thin deposits] were successful, we raced out of the Chemistry Institute as if shot from a gun, up the road to the Physics Institute a kilometer away, to examine the specimens in von Baeyer's very simple beta spectrometer ... [These] investigations did enable us to discover the so-called line spectra of beta radiation, which, in fact, have no connection with primary beta radiation, although it took us – or rather me – until after the First World War to realize the fact."[69]

In 1912 Hahn became engaged to marry Edith Junghans. Meitner extended her friendship to Edith and later became godmother to the Hahns' only child, Hanno. In the same year the new Kaiser Wilhelm Institute (KWI) in Berlin-Dahlem was officially opened. Hahn was appointed head of the department of radioactivity research; Meitner joined the department as an unpaid guest worker. The move to the KWI held at least two advantages: Hahn himself, now a scientific member of the institute, had financial security; and Meitner and he had new, uncontaminated laboratories where they could study weakly radioactive materials. For some time the Hahn-Meitner department had been diverging gradually into two separate units, one for radiochemistry (Hahn), the other for nuclear physics (Meitner).[70] Meitner continued with her study of beta radiation, while Hahn turned his attention to the elements potassium and rubidium.

The year 1914 finally brought professional validation to Lise Meitner. The University of Prague, then still within the Austro-Hungarian Empire, offered her a permanent position. But Planck conspired with Fischer to defeat the effort to entice her to leave Berlin.[71] The outcome? "This recognition of her work brought it about that she was now taken on as a scientific member of the Kaiser Wilhelm Institute of Chemistry, and paid a salary."[72] The professional salary paid Meitner was, of course, subject to the vagaries of the German economy, which, both during and after the war, was disastrous. In the postwar period, anyone on a fixed or infrequently adjusted salary was in dire circumstances, what with rampant inflation. Hahn recalled: "Everyday at midday ... the value of money was reduced by half ... I remember that my wife used to ... meet me at the bus stop, to get the money and then bicycle straight off to the grocer's."[73]

The principal scientific problem pursued by Meitner and Hahn during the war derived from discoveries of 1913. By that year, several radioactive elements had been placed in their proper locations in the periodic table, leaving a noticeable gap between thorium and uranium. Otto Hahn and Lise Meitner had taken up the challenge of tracking down the missing element.

In the spring of 1914 they discovered a chemical separation technique that lent promise to their hope of identifying the mystery substance. Because of long half-lives and weak preparations, however, years of observation would be required. Fortuitously, this problem was well suited to the times, as war would remove both Hahn and Meitner from the laboratory for a while. When war was declared in August 1914, Hahn was immediately conscripted; Meitner continued to work in Berlin-Dahlem until the summer of 1915, when, after taking a course in x-ray technology and human anatomy at the City Hospital in Lichterfelde, outside of Berlin, she volunteered as an x-ray nurse in the Austrian army. Even then, "Hahn was ... assigned to Haber's group ... and often came to Dahlem, while I was able to get leave of absence from my voluntary position frequently."[74]

After about fifteen months in Austria, Meitner returned to Germany and full-time work in the laboratory in October 1916. From this point, the discovery of the new element is chronicled in a series of letters from Meitner to Hahn. This correspondence, translated and analyzed by Sime,[75] shows that Meitner "did nearly all the work; chemical separations, complex indicator experiments, exacting measurements with primitive electroscopes, procurement of material and equipment despite wartime shortages."[76]

By 17 January 1918, Meitner had reached a point where she could send Hahn a long and detailed progress report – and very good news.[77] Meitner had definitively identified the new element, measured the range of its alpha particles, and determined two possible half-life values. The Hahn-Meitner paper, "The Mother Substance of Actinium, a New Radioactive Element of Long Half-life," was submitted to the *Physikalische Zeitschrift* on 16 March and appeared later that year.[78]

If their work had not by now centred Hahn and Meitner in the scientific limelight, this discovery did. While Kasimir Fajans and

Oswald Gohring had identified a short-lived isotope of protactinium, and Frederic Soddy and John Cranston had prepared very small quantities of the element (too small to determine its decay characteristics),[79] the Meitner-Hahn achievement was clearly a triumph of skilled, painstaking, brilliant experimental work. The announcement of the discovery of protactinium by Meitner and Hahn marked the beginning of a series of papers by them on the properties and behaviour of that element.

The years after 1917 were especially busy ones for Meitner. A mature scientist now, having published ten solo papers, fourteen with Hahn, seven with von Baeyer and Hahn, and one with Franck (on radioactive ions),[80] she was long overdue for an independent professional appointment. Curiously enough, this came about through the instrumentality of Fischer, the very man who, ten years earlier, had denied her access to his institute. Now, "in later years, he was most kind in supporting me in every respect, and I have him to thank for the fact that in 1917, I was given the responsibility of setting up a department of radiation physics in the Kaiser Wilhelm Institute of Chemistry."[81] She and Hahn had parted company as he was interested, as always, in studying new radioactive materials and their chemical properties, while she found it more exciting to continue her investigation of radioactive emanations.

In 1922 Meitner finally received the *Venia Legendi* from the University of Berlin, where her only prior connections had been those of attending colloquia and serving as Planck's assistant. This allowed her to become a lecturer (*Privatdozentin*) at the university, but the only payment was the students' course fees. Her inaugural lecture was on the subject of cosmic physics, but, swayed by the prevailing views of women in that day, the press reported it as "Cosmetic Physics." Four years later, Meitner was named a titular professor at the University of Berlin.

After completing her protactinium studies Meitner returned in 1922 to the puzzling question of the beta spectrum, now in her own department with her own assistants. She had accepted a formidable challenge. The beta-ray question had long puzzled the scientific community. As far back as 1900, the Curies[82] and Becquerel[83] had shown, by observing the direction of their deflection in a magnetic field, that beta rays carry a negative charge.[84] Oddly enough, the photographic image of the rays was diffuse and extended, even though the entering beam was well collimated. The observers were left to conclude either that beta rays from the same source were not identical, or that they were emitted with a range of energies. The first of these choices was eliminated in 1909, when Bucherer[85] measured the limiting value of the ratio of charge to mass for beta radiation. Since his result came within about two percent of the accepted value for electrons, it was generally agreed that the beta radiation consists of a stream of fast-moving electrons.

Meitner firmly believed that primary beta rays must have well-defined energies. After all, the beta decay process represents a transition between two specific nuclei; hence it must involve the release of a definite amount of energy. Convinced that the photographic techniques in use at the time were at fault, she pursued a characteristic course of action. Throughout her scientific career, she was alert to experimental developments in other fields that might be modified for use in her own work. Now she adopted a method devised by Danysz that significantly improved the resolution of her photographic measurements.[86] The technique as she used it focused electrons by deflecting them through 180 degrees; the result was to emphasize spectral lines and to diminish the visual intensity of the continuous spectrum. With this evidence Meitner attributed the faint, continuous background to secondary effects.

This flew in the face of the beta-ray group at Cambridge. While working in Hans Geiger's laboratory in Berlin in 1914, James Chadwick had proved that the line and continuous spectra of beta radiation are distinct.[87] Now, at the Cavendish Laboratory, his co-workers C.D. Ellis and W.A. Wooster held that primary beta rays are responsible for the continuous spectrum, though they could not explain the spread of energies. Several researchers had proposed that the electrons with less than the maximum energy had radiated some in the form of gamma rays. But some beta emitters had been shown not to produce any gamma rays, such as bismuth-210. In a search for the missing energy, Ellis and Wooster set about measuring the average energy per beta disintegration. Using bismuth-210 and refined calorimetric techniques, they showed that beta rays are emitted from the nucleus with a continuous range of energies; further, the average value of 0.35 MeV per disintegration compared well with the average of 0.34 MeV as calculated from the continuous spectrum.[88]

In disbelief, Meitner and her assistant, Wilhelm Orthmann, repeated the Ellis-Wooster work using improved methods.[89] To their dismay, the results were in agreement with those reported by Ellis and Wooster. In fact, the question of continuous beta energies was not resolved until December 1930, when Wolfgang Pauli wrote his famous open letter to conference participants in Tübingen, in care of Meitner and Geiger.[90] Pauli proposed the existence of a neutral thief-particle that steals energy from the betas as they are emitted. Though Meitner's belief in the fundamental simplicity of nature had not served her well enough this time, her attempts to defend it had led to important new knowledge.

In 1925, by carefully measuring the beta spectral lines of actinium, she settled a long-standing argument with Ellis and H.W.B. Skinner regarding the sequence of events in fast nuclear processes. The question was this: when a gamma ray from the nucleus gives rise to a secondary electron, does the energy transfer take place in the electron shells of the original or of the new atom? Meitner showed conclusively that the secondary beta rays are associated with the new atom, so the gamma radiation is emitted after the disintegration of the original atom.[91] Thus, gamma radiation had to succeed beta decay, not initiate it, as Ellis had held.

Meitner must have derived a certain satisfaction from the letter Ellis wrote to her on 8 December 1925: "We have been at considerable trouble to settle the question of whether the gamma-ray sometimes precedes the disintegration as I deduced from Rutherford and Andrade's measurements, or whether the simpler standpoint advanced by you that the disintegration always happens first was correct. We have had three different ways of testing this and they all show that the gamma-rays come out afterwards, so you were right! I cannot help feeling a little annoyed that we were led astray by that old experiment."[92] There was a double triumph here: Not only had Meitner correctly solved an exceedingly difficult problem but her competition had conceded defeat.

But Meitner's hard work and clear scientific insights were not always duly recognized. During the early 1920s when she was concentrating on the beta-ray spectrum, Meitner had observed and correctly interpreted the radiationless transitions in which an electron from an outer shell jumps into a vacancy in an inner shell, transferring its energy so that another electron can be ejected.[93] Today this phenomenon is called the Auger effect, after Pierre Auger who published a series of papers on the subject in 1925 and 1926. Clearly, we have here a case of false attribution, and it has recently been examined as such. Why was Meitner not given, or did not herself claim, credit for the discovery? Sietmann remarked: "As a woman in the male dominated world of physics, she was an outsider … [Further, contemporaries] always described Lise Meitner as timid and restrained; showing ambition or struggling for recognition did not conform to her personality … If Lise Meitner were to be asked today why she did not claim priority for her research results, she would probably answer: 'At that time, it did not occur to me.'"[94] With another swing of life's pendulum, however, recognition and honour came to her once again.

On 7 February 1928 James Chadwick wrote to inform Meitner that a small informal conference on beta-ray and gamma-ray problems was to be held at the Cavendish Laboratory, Cambridge: "We should be very pleased if you could come to this Conference, and if you would read a paper to open the discussion on β-ray spectra for the first meeting ... We have a small fund at our disposal which will enable us to contribute ten pounds to the cost of your journey."[95] Persuasively, Chadwick's letter was followed by one written the next day by C.D. Ellis.[96] Meitner's response was quick and enthusiastic. On 14 February she wrote to Chadwick: "It is self-understood that I can make my speech in English, although my English is faulty; but you will handle the grammatical mistakes in good spirits and with patience ... I am very pleased also that this occasion will make it possible to greet Professor Rutherford, whom I once saw in Berlin 20 years ago."[97] Meitner wrote in a similar vein to Ellis the same day, adding a question about the expected length of her speech.[98]

Curiously, the central role Meitner was to play at the conference was revealed to her only gradually. She had known all along, of course, that her talk was to open the conference, but words such as "keynote" or "principal" had not been used. In July Chadwick dashed off a letter to inform Meitner of final conference arrangements: "In the [Friday] evening we invite you and the other guests to a dinner to be held in Trinity College. As you are the chief guest may I warn you that you may be expected to make a speech, if only a short one?"[99] Meitner was known to shy away from public recognition, and perhaps Chadwick had deemed it advisable to break the news of her status at the conference gradually. Meitner seems to have accepted the situation with grace. The letter of thanks that she wrote to Chadwick once back in Berlin was warm.[100]

Meitner had been vigorously pursuing other lines of investigation. When the Geiger-Müller counter was introduced in 1926, Meitner had immediately seen its usefulness in testing a new formula of Oskar Klein and Yoshio Nishina relating to the Compton effect, the collision of high-energy light quanta with loosely bound electrons in matter. As Meitner and Hupfeld worked through their experiments in connection with the Klein-Nishina formula, they observed an anomalous gamma-ray attenuation first referred to as the Meitner-Hupfeld effect, today understood as electron-positron pair production – the conversion of energy into matter.[101]

When we review Meitner's trail-blazing work of the 1920s, we understand why Otto Hahn would describe these years as a time when the work of Lise Meitner, rather than his own, brought to their institute recognition from the international community of scientists.[102] Her place as one of the handful of leaders in physics is charmingly brought out in connection with a Berlin colloquium on matrix mechanics given by Werner Heisenberg on 28 April 1926. Heisenberg "wrote to his parents ... the next day that all of the 'bosses of physics' – Einstein, Laue, Nernst, Ladenburg, Meitner – had assembled to hear him speak."[103]

One result of Meitner's acknowledged pre-eminence in physics was the presence of Franco Rasetti in her laboratory in 1932. In the late 1920s Enrico Fermi had begun to send members of his group at the University of Rome to other countries to learn new experimental methods. Rasetti had come to Meitner to learn techniques of handling radioactive substances, as well as the construction and use of the Wilson cloud chamber, which she had adapted for use in radioactive studies. When he returned home, Rasetti responded to Meitner's hospitality by adding her name to the mailing list for preprints of papers of the Rome group. Thus, she was among the first to learn of the neutron irradiation experiments of Fermi and his young collaborators, who had assumed the ambitious task of systematically bombarding all the known elements of the periodic table. Meitner commented later: "The most interesting results seemed to accrue from bombarding the then heaviest element, uranium: Fermi thought that this led to higher elements with atomic numbers 93 and 94, i.e. to transuranic elements."[104]

As Meitner followed the published results of the work of the Rome group, she began to realize how important chemistry would become to this type of investigation; she saw that neither the sketchy knowledge of chemistry common to a physicist nor occasional consultation with a chemist would suffice.[105] She therefore approached Otto Hahn with a proposal that they resume the partnership that had earlier been so productive for them. After persisting for several weeks, she "persuaded Otto Hahn to renew our direct collaboration ... with a view to investigating these problems. So it was that in 1934, after an interval of more than twelve years, we started working together again, with the especially valuable collaboration, after a short time, of Fritz Strassmann."[106] Meitner's investment in the neutron irradiation project heightened and intensified as the work progressed. Later, Strassmann was to declare: "I am convinced, L. Meitner was the intellectual leader of our team."[107]

The hard work of the team was pleasantly interrupted in September 1934, when Meitner and Hahn were invited to take part in the great celebration of the hundredth anniversary of the birth of Mendeléev.

The commemoration itself, held in Moscow, was followed by a conference on nuclear physics in Leningrad. Here Meitner gave a presentation that surveyed developments in nuclear physics.[108]

In retrospect, it is easy to identify another of the papers given at the Mendeléev conference as being of fundamental import. Ida Tacke Noddack, a German chemist, attacked the hypothesis Fermi had put forth to explain his neutron irradiation work (that elements of atomic number greater than uranium were being observed). Her argument, published soon after, pointed out that "it is conceivable, that the bombardment of heavy nuclei by neutrons might cause them to break up into larger fragments, which would be isotopes of known elements, but not close neighbors of the irradiated element."[109] Tennenbaum commented: "Lise Meitner and Otto Hahn attended Ida Noddack's 1934 lecture in Leningrad, and ... they received her paper, as did Fermi and the other leading researchers in the field."[110]

Why was Noddack's argument so little heeded? Frisch held the view that "her comments (published in a journal not much read by chemists and hardly at all by physicists) were regarded as mere pedantry,"[111] while Hahn argued: "Ida Noddack ... suggested that all the elements of the periodic system would have to be eliminated before one could claim to have found a transuranium element. Her suggestion was so out of line with the then-accepted ideas about the atomic nucleus that it was never seriously discussed."[112] Whatever the reason, Noddack's challenge to Fermi was not pursued by others working in the field.

Fermi's transuranic hypothesis was also called into question by a former co-worker of Otto Hahn, Aristide von Grosse, who had since emigrated to the United States. Von Grosse claimed that two of Fermi's "transuranics" were, in fact, isotopes of protactinium. This assertion threw down the gauntlet to the protactinium discoverers – Meitner and Hahn. They "soon demonstrated that Fermi was right, but ... also found that what happened when uranium was bombarded with neutrons was very complicated indeed."[113] Meitner and Hahn went on to bombard uranium for four more years.

Eventually, they proposed (with Strassmann) three isomeric series that they viewed as corresponding to the beta emitters observed by Fermi's group. The first of these involved five successive beta decays; beginning as it did with element 92, it thus ended with 97; the second series, also involving a chain of beta decays, went from 92 through 95, with unidentified decay products beyond 95; the third, with only one beta decay, would yield element number 93. The series hypothesis, as well as the work leading up to and following it, is chronicled in a sequence of publications between 1935 and 1938.[114]

Although several of the reaction products the Berlin group proposed had been definitively identified by them, Meitner "found it very disturbing to discover ... such a long chain of successive β-disintegrations, i.e. continually increasing nuclear charges with unchanged masses. One outcome of my concern was our precise examination of uranium under slow neutron bombardment."[115] Significantly, it was this very "examination of uranium under slow neutron bombardment" that led to the discovery of nuclear fission.

Yet Meitner was not on hand for that discovery. On 6 September 1933, the Law for the Restoration of the Career Civil Service[116] stripped Meitner of her professorship at the University of Berlin. In the eyes of the Nazi regime, her Jewish heritage rendered her unsuited to teach in an Aryan nation. On the one hand, this official governmental position was unthinkable, unspeakable, untenable. On the other, it interfered little with her own daily routine since she was primarily occupied with her research at the independent Kaiser Wilhelm Institute for Chemistry, though she did offer an occasional course of lectures at the university.[117] Hence the loss of her professorship had no substantive effect on her work, especially since she could still attend the Berlin Physics Colloquium. And Planck, her old friend and teacher, in an effort to preserve the character of German science, advised against emigration barring absolute necessity.[118]

Her work went on more or less calmly until 11 March 1938, when the Austrian leader Kurt Schuschnigg resigned and broadcast a plea to the army and the country's citizens to offer no resistance to a Nazi takeover. The *anschluss* was announced on 13 March. Now the protection of Meitner's Austrian citizenship was stripped away and her situation became one of crisis. Added to the political threat were murmurings from one of her colleagues that "the Jewess endangers the Institute."[119] When this complaint reached Hahn's ears, he chose to side with the institute. Without Meitner's knowledge he discussed her position with Heinrich Hörlein, treasurer of the Emil Fischer Society for Advancement of Chemical Research, which sponsored the institute. Hörlein proposed that Meitner resign.[120] When Hahn told Meitner what he had done, she was dumbfounded and angry: how could he have deliberately precipitated her downfall?[121]

Having been denied a German passport, Meitner left Berlin secretly on 13 July, accompanied by a Dutch friend, Dirk Coster, who had come to travel to Holland with her. She had two small suitcases, a purse with ten marks, and Hahn's mother's diamond ring (Hahn wanted her to be provided for in an emergency).[122] The train crossed the border without incident, ending months of anxiety for Meitner. But "[relief] turned to shock: She was ... uprooted from work, colleagues, income, and language, suspended between a past that was gone and a future that held nothing at all."[123]

In August Meitner travelled to Sweden, first to Göteborg on the west coast to visit Eva von Bahr Bergius, a friend from the early days in Berlin. From there she made her way to Stockholm, where, through the good offices of Bohr and others, a position had been arranged for her in the Nobel Institute for Physics under Manne Siegbahn. For her first nine years there, she was paid less than the starting salary of an assistant.[124] "My income is so meager that it is only by being very economical that I can pay for my room, meals, and small daily expenses (bus, postage, etc.)."[125]

Yet early in 1939 we find her sending money to relatives as they fled Austria one by one, having to abandon both professions and possessions. The 1939–42 correspondence has numerous references to her financial generosity to sisters, brother, nieces, and nephews.[126] How did she manage this? Once she had retired from the KWI in August 1938, she began negotiations to receive her retirement income.[127] She was informed that her pension had to remain in Germany as long as her residence was outside the German Reich without permission of the KWI.[128] She then arranged that her retirement salary be "paid to her sisters at Vienna [Gisela and Auguste] ... up to their flight."[129] It is likely that her sister Auguste was able to get this money to Sweden when she and her husband fled to Stockholm in February 1939.

This was not Meitner's first time in Sweden. As early as 1919 she had vacationed there with Frau Dr Emma Jacobsson.[130] Then, in April 1921, she had been a visiting professor in Lund – ironically enough, at the invitation of Siegbahn (and while Coster, then a student of Ehrenfest's, was working there).[131] Now, under less happy circumstances, Meitner took up residence at the Årsta Damhotell, where she lived until May 1939, awaiting the arrival of her possessions from Germany. Her existence was bleak, as she wrote to Hahn: "Probably you can't really imagine what it means for a person of my age to have been living in a small hotel room for nine months, without any comfort, without any scientific aids, and with the fear that nobody has the time required to advance my affairs [in Berlin]."[132] Her professional situation was no better: "I don't feel at all happy. I have here only a workplace, no position that would entitle me to anything. Try to imagine what it would be like if ... you had a room in an institute not your own, without any help, without any rights and with the attitude of Siegbahn who loves only big machines and who is very confident and self-assured – and there I am with my inner shyness and embarrassment!"[133]

Fortunately, Meitner's barren existence was brightened by occasional visits from relatives, former colleagues, and old friends. In September 1939 she wrote to Hahn: "O.R. [Frisch] is here for the return trip from Upsala ... He had professional business to do there."[134] Next she welcomed Bohr, who delivered a lecture in Stockholm on 12 October and spoke with Siegbahn on Meitner's behalf.[135] Then in November, Meitner was invited for a week to Bohr's Institute for Theoretical Physics in Copenhagen.[136] Hahn had also been invited, and Meitner met his train on 13 November.

Meitner and Hahn had much to discuss. Irène Joliot-Curie and Pavel Savitch had for some time been studying neutron-irradiated uranium, but by a chemical technique different from that used by the Berlin group. Hahn had written Meitner on 25 October: "Now, at the end of last week another paper of Curie and Savitch came out ... According to the instructions of Curie we have found [their strange decay product] ... (Perhaps it is even related to a radium isotope. But this I say only very cautiously and privately!)"[137] Meitner responded at once with many questions.[138] Following a daily rapid-fire exchange of mail, the Copenhagen meeting of Meitner and Hahn was perfectly timed.

Hahn returned to Berlin with instructions from Meitner to test for radium with all the weapons in his chemical arsenal. She did not believe that he and Strassmann would find it. "Fortunately, L. Meitner's opinion and judgment carried so much weight with us in Berlin that we immediately undertook the necessary control experiments," Strassmann later wrote.[139] These experiments led, finally, to Hahn's halting announcement to Meitner on Monday, 19 December: "We are coming closer to the frightful conclusion: our Ra isotopes do not act like Ra but like Ba."[140] On Wednesday Meitner responded: "Your ... results are very perplexing. A process initiated by slow neutrons leading to barium! ... It seems very preliminary to me to accept such a shocking result, but in atomic science we have experienced so many surprises, that one can say nothing without further evidence. It is impossible to do so."[141]

A few days later, Meitner was burning to discuss this bizarre finding with Frisch. Frisch had travelled from Copenhagen to Kungälv, near Göteborg on the west coast of Sweden, to spend a few days during the Christmas holiday with his aunt. This reconstructed version of Frisch's ruminations is based on the several personal accounts he has left us.[142] On his first morning there, he had found Lise Meitner at breakfast, brooding over a letter from Hahn. Frisch recalled that he had wanted to tell her of a new experiment he was planning with a large magnet,[143] but she wouldn't listen; he had to read Hahn's letter. His first reaction was that the findings were a mistake; Meitner insisted that she "knew the extraordinary chemical knowledge and ability of Hahn and Strassmann too well to doubt for one second the correctness of their unexpected results."[144] Intrigued, Frisch suggested a walk in the woods where they could do some clear thinking. He put on skis, but she claimed, and proved, that she could get along just as quickly on foot.

How could barium result from the neutron bombardment of uranium? The problem was a formidable one. So far as was known at the time, no fragments larger than protons or alpha particles had ever been chipped off nuclei. Hahn and Strassmann were bombarding with slow neutrons, so there was not nearly enough energy available to remove a sufficient number of such small particles to get down to barium. This would have meant going from atomic number 92 to 56, and from an atomic weight of about 238 to one of about 137. Impossible!

The chipping scenario dismissed, the discussion moved to a consideration of the nucleus itself. Both Meitner and Frisch recalled that some ten years earlier, George Gamow had proposed that the nucleus might behave like a drop of liquid. More recently, Niels Bohr had revived this theory, expanding it and fitting it to his compound nucleus formulation.[145] Meitner herself recalled the moment of insight: "The new process gradually became comprehensible in the light of Bohr's liquid-drop nuclear model, according to which the surface tension stabilizes the nucleus *vis-à-vis* small deformations. In the course of our discussions we evolved the following picture: if, in the highly-charged uranium nucleus – in which the surface tension is greatly reduced owing to the mutual repulsion of the protons – the collective motion of the nucleus is rendered violent enough by the captured neutron, the nucleus may become drawn out length-wise, forming a sort of 'waist,' and finally splitting into two more or less equal-sized, lighter nuclei which, because of their mutual repulsion, then fly apart with great force."[146]

The time for pencil and paper had come, and Meitner and Frisch sat down on a log and began to draw pictures and make calculations. Meitner made use of her detailed knowledge of the mass-defect curve to show that if a uranium nucleus were to split into two fragments, say, barium and another element near the middle of the periodic table, sufficient mass would be converted into energy to release a vast amount, about 200 MeV (million electron volts). For his part, Frisch had been calculating the opposing effects of surface tension and surface charge, finding that the two would cancel each other out around atomic number 100. Thus, the uranium nucleus would be stable but just barely so, and an impinging neutron could, perhaps, supply enough energy to split it.

Finally, classical physics made it clear that the two positive nuclei resulting from a split, at first in contact, would be quickly driven apart by Coulomb repulsion. A simple calculation showed that they would acquire such high speeds that their combined energies would be about 200 MeV. *Mirabile dictu*, the energy budget was balanced, and Meitner and Frisch saw that the division of a uranium nucleus could be explained on a purely classical basis.

So this was what had been happening all along as Meitner-Hahn-Strassmann, Enrico Fermi and his group in Rome, Irène Joliot-Curie and Frédéric Joliot in Paris, and others had bombarded uranium with slow neutrons. In the meantime, Frisch had to leave Kungälv to return to Copenhagen and his work at Niels Bohr's Institute of Theoretical Physics, while Meitner went back to her position at the Nobel Institute in Stockholm. Between them, over long-distance telephone lines between Stockholm and Copenhagen, they drafted a letter to the editor of *Nature* describing their discovery.

Frisch shared the whole story with Niels Bohr, who was preparing to depart for the United States (he left on 7 January 1939),[147] and with George Placzek. Placzek suggested an experimental check of the uranium-to-barium transformation using a cloud chamber. Not having one at his disposal, Frisch used an ionization chamber; the large pulses caused by ion fragments were easily detected. He did not get around to doing this experiment until nearly two weeks after his return to Denmark; moreover, he had held off on submitting the Meitner-Frisch letter to the editor of *Nature* long enough to prepare a second letter reporting his experimental confirmation of nuclear fission. Now, on 16 January, he was ready to post both communications.

But Meitner's triumph would soon turn to ashes in her mouth as circumstances conspired to deprive her of a leading role in the fission drama. In early January Hahn had announced the discovery in his name and Strassmann's alone.[148] The scientific community did not learn of the Meitner-Frisch theoretical contribution until their report appeared belatedly in *Nature* on 11 February.[149] In a further and unexpected turn, Bohr, in whom Frisch had confided before Bohr set sail for the United States, recounted the fission saga at a Washington meeting of the American Physical Society on 26 January. The news spread like wildfire, sending physicists scurrying to their laboratories to duplicate the experiment. Fission was old hat by the time the 11 February issue of *Nature* reached their desks. Thus, it was Hahn who was recognized as the discoverer of nuclear fission. In his award address for the Nobel Prize in chemistry he made no mention of

Meitner.[150] Others did recognize the value of her research; even before her contribution to the solution of the problem of nuclear fission, she was unsuccessfully nominated for the Nobel Prize no fewer than ten times between 1924 and 1937.[151]

But of more fundamental concern to Meitner than scientific credit was scientific truth. Since 1934 she and Hahn had been postulating the formation of transuranic elements, creating ever more exotic decay schemes to support laboratory results. What of all this now? New Year's Day 1939 found Meitner writing to Hahn: "You will completely understand that the question of the validity of the transuranium hypothesis also has a very personal side for me. If the entire work of the last three years was incorrect, the blame must not be one-sided. I was also responsible for that and must find some way to be a part of the retraction."[152] She went on to propose a double retraction – one by him, one by her. Writing to Hahn, Frisch interpreted Meitner's position: "She … meant that if the results achieved in joint work proved to be false, then the retraction of these results should not take place by you alone; otherwise people would say the three have blundered and now after the departure of one, the other two have brought things into order; this is naturally not a good recommendation for the third."[153] Frisch was sensitive to Meitner's shaky start in a wholly new professional life in a foreign country. Hahn, however, seemed not to be, and the dual retraction was never made.

When the United States pursued the fission process with a view to developing a superbomb, Meitner was invited to work on the project but refused.[154] She explained later: "I hoped that the newly-discovered source of energy would be used only for peaceful purposes. During the war, I used to say … 'I hope they will not succeed in making an atomic bomb, but I fear they will.'"[155]

Meitner's unhappy situation in Siegbahn's institute continued, but there were professional ties elsewhere to buoy her spirits. Early in 1939 she joined Frisch in Copenhagen to search for transuranic elements in the decay products of uranium fission.[156] They were unable to detect any by the experimental methods they used, though today we know that neptunium, plutonium, and others are indeed formed.

In August 1938, while concerned colleagues were working feverishly to secure a position for Meitner outside Germany, W.L. Bragg offered her a one-year situation at Cambridge.[157] When Meitner decided to accept Siegbahn's offer of five years at his institute, Bragg wrote in January 1939 inviting her to "come to Cambridge to give four lectures and stay for about a fortnight."[158] She was delighted to go to England, where in the end she gave three lectures.[159] This trip also gave Meitner the chance to see family members who had fled to

England from Austria. On 9 September she presented a paper on uranium disintegration to the British Association for the Advancement of Science in Dundee.[160] And so her work continued.

After the war, when travel became easier, Karl Herzfeld, head of the physics department at the Catholic University of America, invited Meitner to spend a semester there.[161] Meitner accepted Herzfeld's invitation and spent a busy several months in the United States. She gave a lecture course and a seminar on nuclear physics during the spring semester at the university[162] while also travelling elsewhere to give lectures and receive honorary degrees. Her off-campus lectures for March and April brought her to Princeton, Columbia, Brown, Harvard, and Duke Universities, and Wellesley and Sweet Briar Colleges.[163] Meitner's Washington semester also gave her a chance to become reacquainted with her sister Carola and her family, who had emigrated there.

In 1947, after her return to Stockholm, Meitner was promised a research professorship at the Royal Institute of Technology in Stockholm, contingent upon approval by the Swedish Parliament.[164] Her hopes remained high for several months until she learned that the motion had never been made in Parliament. Yet she did receive an improved salary from the Nobel Foundation. Eventually, a settlement on Meitner's retirement benefits from the Kaiser Wilhelm Institute was reached with its legal successor, the Max Planck Society. Still, her Swedish salary was to be subtracted from her German retirement pay.[165]

She remained at the institute until the end of 1953, when she moved to a laboratory of the Royal Academy for Engineering Sciences to take up "a sort of advisory post at the institute of [Sigvard] Eklund ... in order to have a room and a library at my disposal, that means a sort of Altersstübel [old person's room]."[166]

In the spring of 1959 Meitner again travelled to the United States, this time at the invitation of Walter Michels, head of the physics department at Bryn Mawr College.[167] Michels had obtained a grant from the Fund for the Advancement of Education to help provide an honorarium and expenses.[168] On her arrival in New York, Meitner spent a short time with her sister Frida, then went on to Philadelphia where she gave lectures at Bryn Mawr on the development of the status of professional women, the history of gamma-rays, and the early history of radioactivity. She also participated in the spring meeting of the American Physical Society and visited Duke University and Argonne National Laboratory.[169]

Her lectures and her friendly interaction with students were exceedingly well received; it was a pleasant time for her. As Michels wrote to Frisch, "she got plenty of rest and responded to delightful spring weather that held good all through her visit."[170] Yet even in this halcyon time, discord arose. Michels, then editor of the *American Journal of Physics*, had planned to publish Meitner's two physics lectures in the journal and possibly combine all three into a small volume. Meitner, for her part, intended to submit the manuscripts to her longtime publisher friend, Paul Rosbaud. Michels appealed to Frisch to mediate.[171] In the end, the first paper appeared as "The Status of Women in Physics" in the August 1960 issue of *Physics Today*, and the others went unpublished.

In 1960 Meitner retired to Cambridge, where Frisch was a fellow at Trinity College. She was suffering from an increasing hearing loss that forced her to curtail her activities. Still, she continued to travel. In 1963 she was invited to Vienna to address a large audience at the Urania Volksbildungsanstalt on "Fifty Years of Physics."[172]

Happily, in 1966, two years before her death, Meitner was named (along with Hahn and Strassmann) to receive the Enrico Fermi Award, which carried with it a monetary value of fifty thousand dollars. While this was indeed a pleasant surprise, the timing was poor. "Had it come earlier, she would have been in a position to do the travelling she had hoped to do, but by 1966 she was too infirm."[173] As time passed Meitner's strength gradually failed, until her life ended on 27 October 1968 at Hope Nursing Home, Cambridge.

Lise Meitner was remembered as a gentle, caring person, so modest and shy that she "always cried in embarrassment when she was presented with a new honor ... The intensity with which she carried on conversations ... was so great that she could forget everything else around her."[174] Following Meitner's visit to Bryn Mawr College, a student wrote: "Dr Meitner's smile is lit with hypnotic charm ... She talks with wit and grace on any subject the conversation turns to ... She is enchanted with the white mouse that two Bryn Mawr students keep as a dormitory pet. She delights in the warmth and fragrance of American spring. She tells, apologetically, of climbing only halfway up the Jungfrau when she was 75. She is sorry she cannot understand Schoenberg's music."[175]

Lise Meitner's long and fruitful life is well summarized in words she wrote herself. In her 1963 Vienna talk, she said: "I believe all young people think about how they would like their lives to develop. When I did so, I always arrived at the conclusion that life need not be easy; what is important is that it not be empty. And this wish I have been granted. Life has not always been easy – the first and second world wars and their consequences saw to that – but it has indeed been full, and for this I have to thank the wonderful development of physics during my lifetime, and the great and lovable personalities with whom my work in physics brought me into contact."[176]

ACKNOWLEDGEMENTS

It is with pleasure that I acknowledge the kindness of those who have granted me access to and facilitated my work on the papers of Lise Meitner: Ulla Frisch, wife of Otto Robert Frisch; Rolf Neuhaus, director of the Library and Archives for the History of the Max Planck Society; Marion M. Stewart and succeeding archivists at Churchill College; James G. Brennan at the Catholic University of America; and faculty members of the Department of Physics at Bryn Mawr College. I am deeply grateful to Ruth Lewin Sime for valuable discussion and for generous sharing of resources, as well as to Roger Stuewer for a meticulous critique of the first draft of this paper.

17 Stefanie Horovitz: A Crucial Role in the Discovery of Isotopes

MARELENE F. RAYNER-CANHAM
and
GEOFFREY W. RAYNER-CANHAM

Stefanie Horovitz is another individual who must have led a fascinating life but about whom very little information has survived. Her research concerned a study of the atomic weights of elements, a somewhat tedious field of work. However, it was the results of her studies that provided firm evidence for the existence of isotopes. Sadly, she was killed by the Nazis in Warsaw during the Second World War.

Horovitz was one of three women from Poland who made significant contributions to the study of radioactivity, the others being Curie and Dorabialska. Horovitz was born in Warsaw on 17 April 1887. Her father, Leopold Horovitz, was a well-known artist.[1] Around 1890 the family moved to Vienna, where Horovitz completed the university entrance requirements in 1907 (presumably, like Meitner, she had to take a private course of study to prepare for these examinations). She entered the faculty of philosophy at the University of Vienna that fall. In 1914 she received a PhD in chemistry under the supervision of Guido Goldschmiedt with a rating of "very good." Two publications resulted from her doctoral work on the rearrangement of quinone using sulphuric acid.[2] Horovitz then changed her field from organic chemistry to a study of the nature of the elements themselves, commencing work at the Radium Institute of Vienna with Otto Hönigschmid in 1913 or early 1914. It is probable that she became Hönigschmid's protégée through Goldschmiedt's recommendation, for Hönigschmid himself was a former student of Goldschmiedt.[3]

But Hönigschmid also had a positive attitude towards his women students.[4]

Hönigschmid was actually affiliated with the Technical University of Prague from 1911 to 1918 but he maintained research facilities in Vienna. It is unclear whether Horovitz stayed in Vienna, where she was listed as a research worker,[5] or whether she travelled regularly with Hönigschmid to Prague. It would seem that she visited Prague on at least one occasion, for Hönigschmid, writing from Prague to his friend Max Lembert, closed with the comment, "With best wishes also from Fraulein Doctor Horovitz, the beautiful graduate."[6]

Hönigschmid had a particular interest in the determination of precise atomic weights of the chemical elements. This was very important work, for the determination of the formulas of new chemical compounds relied upon knowledge of the atomic weights of the combining elements. The first table had been prepared by the American Chemical Society in 1893, after which the German Chemical Society published its own table in 1898. It became obvious that to obtain consistent and reliable data, chemists needed to collaborate in the development of an accurate, unified table of atomic weights. To accomplish this goal, the German Chemical Society invited chemical societies from around the world to participate in this endeavour.[7]

Accurate and precise determinations of the values were needed but much of the work at the time was of dubious quality. Hönigschmid had spent a year studying under Theodore William Richards at Harvard University. Richards had developed what were then the most precise and reliable values for atomic weights and he was later awarded the Nobel Prize in chemistry for his pioneering work in atomic weight determination. Hence the subsequent work of any of Richard's associates had instant credibility.

Frederick Soddy had proposed that the radioactive elements disintegrated to form lead. However, according to his proposed Group Displacement Law, the atomic weight of the lead produced from the decay of uranium should be 206.5 rather than the 207.1 found for common lead.[8] Soddy had proposed isotope as a name for forms of the same element that differed only in mass (the word isotope was actually devised by Soddy's friend, Dr Margaret Todd).[9] The strange concept of atoms being identical in everything but weight won few adherents and Soddy's own measurements had been regarded with scepticism. If it could be shown that lead from radioactive sources really had a different atomic weight than that of common lead, it would be a triumph for Soddy's law and for the concept of isotopes.

It was this task to which Horovitz was assigned: determining the atomic weight of the lead present in radioactive minerals. Horovitz and Hönigschmid were in the right place at the right time. The only European mine producing large quantities of pitchblende, the radioactive ore, was in Austria at St Joachimstal.[10] The majority of the ore was shipped to the Radium Institute at Vienna (most of the remainder going to Curie and Rutherford). At Vienna, the valuable radium was separated out, leaving lead in the residues. Thus, the first task for Horovitz was to separate out very pure samples of lead from these residues, not an easy task in itself. These careful extraction procedures were followed by vital mass measurements that could be used to determine the atomic weights. Each of the measurements was made to the nearest one hundred thousandth of a gram – a precision that is unusual even today.

Her results were startling to the chemists of the time. The atomic weight of the lead from the disintegration of uranium and radium was found to be 206.736 compared to 207.190 for "normal" lead.[11] The difference of 0.4 units, although seemingly small, was enough to demolish the belief that atomic weights were invariant. It was the first authoritative evidence for the concept of isotopes and it provided strong support for Soddy's Group Displacement Law. As a result, the paper produced on their work by Horovitz and Hönigschmid is regarded as one of the crucial publications in chemistry in the first half of the twentieth century.[12] The discovery of atoms of the same chemical identity but different masses caused astonishment, even consternation, among many chemists of the time.[13]

Following this first result, Horovitz analyzed new samples from St Joachimstal as well as samples from two other mines: pitchblende from German East Africa; and bröggerite from Norway, the latter almost certainly supplied by Ellen Gleditsch. Her results this time were even more convincing, giving values as low as 206.046.[14] A difference of over one mass unit could not be explained by experimental error. During this period, Hönigschmid mentioned her in a letter to Lise Meitner that "Miss Horovitz and I worked like coolies. On this beautiful Sunday we are still sitting in the laboratory at 6 o'clock … I am sending you greetings from Miss Horovitz, who does not believe that you still remember her. I have just argued with her about that."[15] Possibly Horovitz and Meitner had met in 1907, the year Horovitz started at the University of Vienna and Meitner departed from Vienna for Berlin.

Over the next two years, Horovitz determined the precise atomic weights of uranium, "ionium," and thorium.[16] The study of ionium was important since it showed that ionium had the same spectroscopic and chemical properties as thorium, the only difference being the atomic weight. Thus, ionium was not a separate element at all but, as we now know, simply an isotope of thorium, thorium-230. So in one experiment Horovitz disproved the existence of an element and found the second established case of isotopes.

These were her last publications before she disappeared from scientific research. Unfortunately, Hönigschmid's correspondence was destroyed in the Second World War, so we have no letters between Horovitz and Hönigschmid at all.[17]

Many years later, Horovitz was discussed in an exchange of letters between Kasimir Fajans and Elizabeth Róna. Fajans, who knew that Róna had also worked at the Radium Institute, asked whether she knew of the fate of Stefanie Horovitz.[18] Róna replied that Horovitz had been at the institute before her time but that she would try to find out from Berta Karlik, Marietta Blau, or Professor Przibram.[19] One assumes that she was unsuccessful, for it was Fajans who revealed the tragic news: "You probably have not received any information from Vienna about the fate of Dr Stephanie Horovitz. I learned about it from a mutual relative at Warzawa. Stephanie moved there after World War I and after her parents had died in Vienna to join her married sister. She was not active in chemistry and both were liquidated by the Nazis in 1940."[20]

So ended the life of a women who, in her brief career, had made a crucial discovery in the study of the atom. Like the other women in her field Horovitz's contributions have been forgotten, and although "Hönigschmid and Horovitz"[21] are credited with proving that radioactive lead has a lower atomic weight than normal lead, it was Hönigschmid who received the honours. Sadly, Hönigschmid's life, too, came to a sudden end. Near the end of the Second World War, he and his wife killed themselves during the Nazi administration when, "after destruction of the [Munich] Institute, they twice had to move and found the difficulties of their living conditions insurmountable."[22]

18 Marietta Blau: Discoverer of the Cosmic Ray "Stars"

LEOPOLD E. HALPERN

Of all the forgotten pioneer women atomic scientists, Marietta Blau is among the most deserving of recognition. Her development of a photographic method for the study of cosmic rays played a major role in this field and in the wider study of particle physics, while the discovery of the Blau-Wambacher "stars" produced on photographic plates exposed to cosmic radiation[1] initiated the study of nuclear fragmentation. Her work, spanning forty-five years, resulted in over sixty-five publications related to aspects of radiation and to the improvement in photographic emulsions. To those in the field of nuclear physics, Marietta Blau's research was quite well known, but in the wider scientific realm her contributions have received little recognition.[2]

I first met Blau in 1953[3] and was immediately struck by the effects of the radiation to which she had been exposed throughout her career – the cataracts in her eyes and the damage to her hands. Her pleasant personality showed in her accounts of the past, in which she was sympathetic yet objective about people and events and often humorous. Conversations with her longtime friend and former colleague Herta Leng attested to Blau's legendary helpfulness.

Blau was born in Vienna on 29 April 1894. Her father was the Jewish lawyer Markus Blau.[4] At the time anti-Semitism was on the rise again in Vienna, and although the emperor Franz Joseph espoused tolerance, the mayor of the city and other local politicians made anti-Semitic remarks. Zola's warnings in France about the corrupting political and social effects of anti-Semitism were lost on the Viennese.

After graduating from the Private Girl's High School for Continued Education of Women in 1915, Blau studied physics at the University of Vienna and then obtained her doctorate with a thesis on the absorption of gamma rays in 1919.[5] She worked briefly at the Central x-ray Institute and then from October 1920 to June 1921 at the Radium Institute (Institut für Radiumforschung) in Vienna. After a short period at an industrial plant in Berlin making x-ray tubes, she obtained an assistantship at the University of Frankfort am Main. During this time (1921–23) she worked on x-ray absorption and, with Kamillo Altenburger, made a theoretical study of such processes.[6]

For family reasons she returned to Austria in 1923, where she undertook research at the Radium Institute and at the Second Physical Institute of the University of Vienna until 1938. She worked first on the decay of Radium A (polonium-218), then with Elizabeth Róna on the ionization caused by proton rays.[7] Next, she joined a group led by Hans Pettersen, who suggested that she devise a method of following radiation tracks using a photographic emulsion. After a lengthy period of research, Blau finally succeeded and her method was applied to reactions involving α-particles and recoil protons.

Marietta Blau did not have a paid position at the Vienna Institute. Throughout her time in Vienna she was supported by her family and therefore had no pension rights after the war. When her work in Vienna proved successful, Blau asked whether she could finally become an assistant professor, or at least obtain some form of paid employment. The answer was: "You are a woman and a Jew and together this is too much."[9] This response was probably made less out of enmity than to make her aware of the political realities of the time. However, in 1932 she received a one-year grant from the Austrian Association of University Women. She used the funds to work with Professor Pohl at the University of Göttingen on crystal physics and then to perform research at the Curie Institute in Paris on the study of the artificial disintegration of atoms using a photographic method.

During the 1930s, the study of cosmic rays had become crucial in the development of atomic physics. Victor Hess, an Austrian physicist, had shown in 1912 that the air was more strongly ionized at high altitude than at sea level.[10] From this finding he deduced that part of the radiation observed at the earth's surface came from outer space. The existence of rays from space became accepted during the 1920s but the visual imaging of these rays by means of photographic film came later. However, the primitive photographic emulsions of the time were too insensitive for cosmic ray studies. At the time, physicists needed the data from measurements on cosmic rays to test their theoretical studies in quantum electrodynamics and particle physics.[11] It was Blau's work, upon her return to Vienna, on improvements in sensitivity of the photographic emulsion that resulted in enhanced images of the tracks.[12] This research was partially done alone but also with some collaborators, particularly Herta Wambacher. Wambacher, who became a long-term associate of Blau, had studied law for a while before her interest shifted to natural science. Blau had invited her with typical generosity to join her research program.

The photographic method developed by Blau had the advantage of making the observation of nuclear processes into a simple procedure, and it allowed for the detection of rare processes by increased exposure time. As a result, it became one of the most important tools of nuclear research. Blau decided to make use of these advantages to search for nuclear and α-particle tracks caused by cosmic radiation. Until 1935, such tracks were observed only occasionally in cloud chambers. Blau and Wambacher exposed their emulsions on the Hafelekar mountain near Innsbruck at an altitude of 2,300 metres. When they developed the plates they found unexpectedly long tracks. These indicated extremely high energy cosmic ray particles, greater than anything known before. For their work in identifying the different tracks of α-rays, protons, and neutrons in photographic emulsions, Blau and Wambacher received the Ignaz L. Lieben prize of the Austrian Academy of Sciences in 1937.

Blau and Wambacher then worked on a more exact study of the tracks and on their elimination by radioactive contamination. In these later studies, they observed a set of tracks like a star. This had never been seen before.[13] Additional exposures at Hafelekar enabled them to find more such tracks, some of which had as many as twelve branches. They concluded that the stars were a result of the cosmic rays colliding with atoms of silver or bromine in the emulsion, the stars themselves being the tracks of the fragments from the disintegrations of the atoms.[14] It is this work, the discovery of the Blau-Wambacher stars, that gave the pair worldwide acclaim.

I first heard about the work of Blau and Wambacher from a lecture series held by Berta Karlik during the late 1940s. Karlik had presented the contribution of the two as equal and stated that, after Blau's departure, Georg Stetter and Wambacher had successfully continued the research with balloons to higher altitudes. Out of curiosity, I asked the institute's laboratory technician, A. Schwella, about the relative contributions of the two women. Schwella, who was well respected for his honesty and courage during the Nazi period, knew everything that had happened at the institute over the preceding decades. He became excited by the question and said it was well known that Blau had been the real researcher and that Wambacher had been a mere assistant. Conversations with others and with Blau herself confirmed this. Yet in some articles, the discovery of the stars are ascribed to Wambacher and Blau.[15] How did this change in attribution take place?

To answer this question, we must consider an individual who came to wield a great deal of authority at the institute – Georg Stetter. Stetter had been hired many years earlier by Karl Przibram, an impartial scholar. Both Stetter and Wambacher became very early members of the then-illegal Nazi party,[16] and they developed a sexual relationship. Stetter did all that he could to discourage Blau from staying at the institute and to encourage her to hand over her proposed research to Wambacher, a "true Aryan." When the Nazis occupied Austria, Stetter was appointed as head of the institute, a position that enabled him to amend the perspective of the Blau-Wambacher saga.

Blau and Wambacher planned to expand their work of exposure of photographic emulsions to measurements at higher altitudes. Stefan Meyer helped them to obtain a grant from the Academy of Sciences to cover the expenses of balloon flights for this purpose. These promising developments were stopped when Hitler invaded Austria in 1938. Fortunately, Blau had left Austria right before the *anchluss* to visit Niels Bohr's institute in Copenhagen. It was providential that she had left Austria, for the treatment of Jewish professionals by the Austrians was far more sadistic than anything seen in Germany up to that time.[17] She did not return to Vienna and instead accepted an invitation by Ellen Gleditsch to visit Oslo to work at the University Chemical Institute, Blindern. Gleditsch also helped Blau to bring her mother out of Austria.

While in Oslo, Blau reported on the examination of some photographic plates that she had brought with her from Austria. These plates had been exposed to cosmic rays for five months at an altitude of 3,400 metres at the Institute for Cosmic-ray Studies on the Jungfrau mountain. The emulsions showed some new phenomena: very short tracks. She ascribed the tracks to cosmic rays that had collided with atoms to produce secondary radiation that, in turn, collided with atoms in the emulsion. In the paper she noted, "I wish to express my sincerest gratitude to Prof. Ellen Gleditsch for her kind hospitality that has made it possible for me to continue my investigation."[18] Also in Oslo, she published a paper on the α-radiation of an unknown element (later identified as an isotope of samarium).[19]

Unfortunately, there was no possibility of a permanent stay in Scandinavia as the small quota of refugees allowed by Denmark and

Norway was exhausted. However, Albert Einstein had been trying to find a position for Blau and he wrote to the American Association of University Women asking for their help: "She has done outstanding original scientific work and is estimated very highly by her colleagues."[20] Einstein also recommended Blau for a professorship of physics at the Technical University of Mexico City.[21] Possibly as a result of Einstein's enthusiastic reference, a position was offered to her in Mexico, which she accepted in 1939.

In view of the threat of war, Blau looked for the quickest way across the Atlantic: a German airship flight via Hamburg. During the short stop in Hamburg, several officials came on board and asked her to descend with her baggage from the airship cabin. This "invitation" must have been extremely frightening for her. The officials knew exactly what they were looking for, confiscating from the baggage all of her important scientific material – particle tracks on emulsions and a draft of future research plans. Then they let her continue the flight. It is of note that, according to Blau, the later studies of Stetter and Wambacher showed a conspicuous resemblance to Blau's research proposals that were confiscated at Hamburg.

At the Laboratory for Radioactivity in Mexico, Blau studied the levels of radioactivity in minerals, oil, and water samples.[22] In her usual helpful way, she tried while in Mexico to place a refugee scientist in a position at the Bacteriological Institute. When the two women visited the institute's laboratory, it was noticed that a graduate student had left a culture of typhus bacteria exposed. The research personnel were already immunized and the visitors were immediately inoculated as the disease can be contracted by inhalation, but both women became infected. Blau, who lived alone, remained helpless for days until worried colleagues sent someone to her home to see why she hadn't appeared at work. Fortunately, she made a full recovery.

Blau moved to the United States in May 1944. Her first work at the International Rare Metals Refinery in New York involved a study of multiplier phototubes in radioactivity measurements.[23] Then she worked on industrial applications of radioactivity for the Gibbs Manufacturing and Research Corporation.[24] Her next position was with the Canadian Radium and Uranium Corporation in their New York laboratories, where she obtained several patents and a publication on the measurement of radioactivity.[25] In 1948 she was invited by the Atomic Energy Commission to work as a research physicist at Columbia University. Her work at Columbia, aided by a number of students, related to the study of high-energy radiation from particle accelerators using her photographic method. In addition, she devised a new method of developing emulsions[26] and continued her studies on cosmic radiation.[27] A semiautomatic machine that she developed together with co-workers for the determination of tracks proved to be very useful in her next position.[28]

The Atomic Energy Commission invited her in 1950 to work at Brookhaven where a very high energy accelerator, the Cosmotron, was available. This machine produced a range of exotic particles such as mesons and hyperfragments that previously could only be observed using cosmic rays. Now, however, they could be produced at will. This was a fruitful period for Blau, matching her successes in Vienna.[29]

In 1955 she resigned her position at Brookhaven and accepted an appointment as associate professor at the University of Miami, where she taught several branches of physics. She was also awarded a US Air Force grant, which she used for the acquisition of equipment. Together with co-workers and students, she constructed an apparatus for the measurement of ionization parameters in photographic emulsions; this device proved to be particularly useful in particle physics.[30] Her well-known expertise in this field led to an invitation to write several parts of the prestigious work *Methods of Experimental Physics – Nuclear Physics*, including the whole chapter on photographic emulsions.[31]

While in Miami Blau wrote to Berta Karlik at the Vienna Institute:

> Florida reminds me a little of Mexico, naturally the mountains are missing and everything is artificial and arranged for the tourist industry. In the season that has just ended people dressed awfully and loud. Miami Beach, only about 30 km from here, I saw only once but even if I had the time I would not go down there for a second time. Due to the many highly elegant hotels one cannot see the ocean. Near to the house where I am living, there is a bridge to a lonely island, and I was there one evening. There are palm trees on the shore, the moon was rising and the long shadows of the palm trees on the emerald-green water were of wondrous beauty. In the garden where I live there are coconut palms and mango and avocado trees and beautiful birds are flying about. In spite of all the beauty, I am homesick and am thinking of the nice days I spent in Vienna.[32]

During this time, a cataract diminished her eyesight to the point where she felt unable to continue her regular work without an operation. Her short period of work in the United States left her, after retirement, with an income from Social Security of only two hundred dollars a month. This was not sufficient to cover living and medical expenses in America. Medical expenses in Austria were still relatively low at the time, so she decided to have the operation there. The Austrian eye surgeon who examined her found her general state of health so weak that she had to wait until it improved before the operation could be risked. This waiting time extended over years. M. Jakobi, an old acquaintance who had become a town councillor in Vienna, helped her to rent a small apartment at a low rate from the City of Vienna. While she was waiting, her health was just good enough to take on some part-time unpaid tasks at the Radium Institute that included the supervision of several research dissertations. She hoped that once she regained her eyesight, she could return to her research in the United States.

About this time, Marietta Blau finally received some international recognition. Erwin Schrödinger, the Nobel laureate, had twice nominated Blau for a Nobel Prize on the basis of her cosmic ray discoveries,[33] but each time the nomination was rejected. However, in 1962 she was awarded the prestigious Schrödinger Prize. Erwin Schrödinger himself had first suggested her for this highest of Austrian distinctions, and Karl Przibram had nominated her for it on this particular occasion. The prize, which involved a cash grant of about fifteen hundred dollars, was awarded by the academy to Blau and, posthumously, to Wambacher – an unusual arrangement as deceased persons were not usually considered for the award. The names of Marietta Blau and Hertha Wambacher were also engraved on the honour board of the University of Vienna.

That same year the cataract operation was performed successfully and Blau felt able to work, though she had developed a heart condition. She invited me to her home to discuss her plans. She could no longer survive in Austria on her US Social Security income, but she was allowed to earn an additional hundred dollars per month tax free and she hoped to find this money through professional work, such as editing and refereeing for scientific journals or even scanning cosmic ray photographs.[34]

Blau was explicit that she did not want to ask for a salaried position at the institute in Vienna. Upset with her treatment over the years by members of the institute, there was one particular cause of her bitterness: Stetter had been reappointed after the war. When the Allies had reached Vienna at the end of the war, Stetter fled, leaving behind a "black list" in which he had classified all the members of the institute according to their (Nazi) political reliability. In spite of this, the reappointed head of the Theoretical Division of the institute, Hans Thirring, had used his influence to invite Stetter back as head of the Experimental Division. This appointment was presented as a generous act of reconciliation.[35] Yet there had been no similar gesture towards Blau, who had been the victim of Nazi persecution.

Blau was unable to obtain financial recompense from the Austrian government. Austria claimed that it had been an occupied country rather than a member of the Axis powers and thus refused to offer restitution to the few returning refugees like Blau (at the time of the *anchluss*, the Nazi government of Austria had confiscated seventy-five percent or more of the funds of those fleeing the country). However, Blau needed a position more for psychological reasons than for the income. She asked me to help her find a suitable activity. The International Laboratory at Cern, Switzerland, seemed to be promising. It had abundant funding and Blau was amply qualified. They refused to consider her, however, though no reason was given. The International Atomic Energy Commission in Vienna was also contacted as a potential employer but made no response. It is tragic that these massive, rich organizations were unwilling to make a token gesture of good will by hiring one of the pioneers in the field as a consultant at the rate of one hundred dollars per month.

When I saw Blau again, her heart condition had worsened to such an extent that she could hardly work. As it happened, the most helpful individual proved to be Otto Frisch.[36] He, too, came from Vienna and he seemed to know Blau quite well, commenting on her helpfulness and modest reticence. He remarked that among Blau's many fields of interest, her work on photomultiplier technology during her years in the United States had come to be of great importance in nuclear research. Frisch convinced the two companies that produced the photographic emulsions perfected by Blau to make a one-time grant of one hundred pounds sterling in lieu of royalties. Blau accepted one of the offers but declined the other as it required her to give permission for them to use her name in advertising.

The following year Blau appeared so weak that one feared for her life. With her was a young woman who, Blau said, was enthusiastic to hear about her discoveries and visited her often. When I telephoned Blau's home in 1969, her brother answered and announced that she was in hospital in an intensive-care unit. This was the beginning of the end. Blau died in Vienna on 27 January 1970. A telephone conversation and a letter from her brother after her death brought more distressing information: the young woman who cared for Blau during her last year was known to associate with very sick people with the intention of seizing their possessions when they died.

During her life, the retiring Blau never received the recognition that her discoveries warranted. Yet in Einstein's opinion she was a gifted nuclear physicist. Schrödinger nominated her twice for a Nobel Prize for her discovery of the "stars" in the photographic plates. More recently, I have tried unsuccessfully to do the same. Why has she been overlooked? This is an interesting question, and one that has no easy answer. Her shyness and the fact that she was not part of the male network of researchers probably contributed to the lack of recognition. We may never know whether Stetter's anti-Semitic influence blighted her prospects in Europe even to the extent of Cern's failing to offer her a position. In any event, it is hoped that this biographical essay will, at last, give some recognition to Blau's life and work.

19 Elizaveta Karamihailova: Bulgarian Pioneer of Radioactivity

SNEZHA TSONEVA-MATHEWSON,
MARELENE F. RAYNER-CANHAM,
and
GEOFFREY W. RAYNER-CANHAM

As discussed earlier, the University of Vienna was a magnet for many of the pioneer women in radioactivity: Meitner began her studies there in 1901, Horovitz in 1907, and Blau in 1915. The next woman to follow this path was Elizaveta Karamihailova, or Elizabeth Kara-Michailova, as she called herself in English. She was born on 3 September 1897 in Vienna.[1] Her father, Dr Ivan Karamihailov, was a famous Bulgarian surgeon, while her English mother, Mary Slade, came from Minster Lovell, Oxfordshire. Karamihailova spent her childhood in Vienna but the family moved to Sophia, Bulgaria, in 1909, settling in a spacious house on Dimiter Polyanov Street in keeping with her father's status. Believing that his skills could be put to better use in Bulgaria, Dr Karamihailov organized the construction of the Red Cross Hospital in Sophia and became its unpaid director. Both Mary, a musician, and her sister-in-law, Elena Karamihailova, a famous Bulgarian artist, were influential in the intellectual life of Sophia, and Elizaveta grew up in a scholarly environment.

Karamihailova finished secondary school in Sophia and then entered the University of Vienna in 1917, from which she was awarded a DPhil (Vienna) in physics and mathematics in February 1922.[2] Her thesis work, "electric figures on different materials," was undertaken at the Institut für Radiumforschung (Institute for Radium Studies) under the direction of Professor Karl Przibram. It became the subject of her first research paper.[3] From 1922 until 1935, she continued her research at the institute in Przibram's group, but she became much more involved in radioactivity studies; in particular, the release of light during radioactive change (radioluminescence). Some of this work was published in collaboration with Hans Petterson[4] and Berta Karlik.[5] Karamihailova participated on joint projects with Marietta Blau[6] on the radiation from polonium and with Ernst Föyn and Elizabeth Róna on the neutron bombardment of thorium.[7]

In 1935 Karamihailova obtained a position at Cambridge University as a Yarrow Scientific Research Fellow. The Yarrow fellowships were endowed by Sir Alfred Yarrow to enable gifted women scientists to perform academic research at Girton College, Cambridge, for up to five years.[8] By the 1920s, when the awards were first presented, the ability of women to perform high-quality scientific research had been recognized and Yarrow's was one of the first endowments aimed specifically at women scientists. Karamihailova obtained some financial support as well from the Rockefeller Foundation and from the International Federation for University Women, thus her earlier work must have been rated highly.

Karamihailova spent four years at Cambridge working at the Cavendish Laboratory. Initially she studied the energy of gamma rays released from actinium. In the publication containing her results, it is interesting to note that she expresses "my gratitude for the permission given to me by the late Lord Rutherford to work in the Cavendish Laboratory."[9] Later, she studied the ionization of gases under high pressure, a topic of considerable importance for the development of radiation detectors.[10]

Dame Mary Cartwright, a mathematician and the previous Yarrow research fellow (1930–34), remembered her well: "E.K-M was violently anti-communist, had a horror of gunfire, and used to come out in spots about every three weeks after her experiments."[11] The last comment is consistent with Elizabeth Róna's view that Karamihailova suffered from a hysterical fear of radiation. According to Róna, "E.K-M" claimed that she had trouble swallowing and had a swelling of her tongue as soon as she entered the room in which one gram of radium was kept.[12]

During her stay at Cambridge Karamihailova applied for a post at Halle University in Germany, but she was denied the position on political grounds. In 1939 she returned to Bulgaria to help with the development of Bulgarian science. She was awarded the position of associate professor in atomic physics at Sophia University, but, tragically, her arrival in Bulgaria coincided with the beginning of the war. As a result, she had to organize the first course in atomic physics and the associated laboratory work from scratch. She donated her own equipment to the university and many instruments had to be hand made. As well, she turned her office at the university into a laboratory for the students. It is true to say that Karamihailova founded the study of radioactivity in Bulgaria.

Karamihailova tried to continue the research on multiple ionization that she had started in England, but without the sophisticated equipment she had little success. More promising was her research into cosmic rays using photographic plates, following on the work of her former colleague, Marietta Blau. When she began that line of research she had only one microscope, a darkroom, and a notebook.

Karamihailova was quite unassuming: no task was beneath her. One of her students, Hristo Hristov, later recalled an event that made her popular, yet she would never boast about it. At one point a radioactive preparation had disappeared from the Alexandrov Hospital. Professor Sahatliev and his assistants turned to Karamihailova for help. She and Hristov set out in search of it "like Sherlock Holmes and Dr Watson."[13] They wandered all over the place with a Geiger-Müller counter mounted on a wooden stretcher. After two or three days of fruitless searching Hristov began to lose hope, but Karamihailova would not give up. Finally, the needle jumped, and then again! Underneath a pile of waste they found the lost sample and put it back in the lead-lined container.

Hristov also reminisced about the social gatherings at Professor Karamihailova's home in the 1940s. Sometimes she would invite the young graduate and undergraduate students to her home, explaining to the students that it was the tradition at Cambridge and Vienna to do this. Despite rationing, she would offer them tea with biscuits and jam. The students enjoyed these relaxed gatherings where they could talk about their work and muse about science and life.

The Karamihailov house had peeling walls and big trees in the garden. Because of the housing shortage, it sheltered many tenants. From the once-numerous Karamihailov family there remained only Elizaveta, her father (a pensioner by now), and her Aunt Elena, the artist, who lived together on part of the ground floor. The spacious sitting-room contained threadbare furniture, haphazardly arranged, and many objects reminiscent of the sophisticated cultural atmosphere in which Elizaveta had been brought up: her mother's piano, her aunt's paintings, and her father's polyglot library.

In 1945 a separate department of atomic physics was established, with Karamihailova as its first chairperson. Following in the traditions of Vienna and Cambridge, she taught not only general courses on atomic physics but also specialized courses on spectroanalysis, luminescence, radioactivity, and nuclear physics. When the communist regime took power, the anti-communist Karamihailova was harassed by the authorities and the publication of her results was forbidden. Moreover, her correspondence with the Western world was stopped and she was not permitted to travel abroad.

Ten years later she was transferred to the Physics Institute of the Bulgarian Academy of Sciences and given the post of senior researcher and head of the radioactivity and spectroscopy laboratory; whether the move was voluntary or not is unclear. About this time, Karamihailova began to study the radioactivity in the environment, measuring radioactivity levels in soil types, rocks, and mineral spring waters throughout the country.

She was interested in the medical condition endemic nephritis and the possibility that its occurrence might be related to locally high levels of natural radioactivity. Together with a medical team, she undertook the task of measuring the level of natural radioactive elements in the soil and water. This involved collecting an enormous number of samples from different parts of Bulgaria, a task that she pursued year-round and without regard to extremes of climate. In the end the credit went to the medical team, not to her. However, in the 1960s she was allowed to publish two of the reports on her study of environmental radioactivity,[14] and another was published in 1975 after her death;[15] the others remained for official use only.[16] Possibly as a result of the first two of these publications, she was promoted to full professor in 1962.

During her lifetime, Professor Karamihailova was the author or co-author of more than forty scientific works, though none of her research proved to be of great significance in itself. Her lonely toil ended with her death on 22 May 1968, and, like so many others, her final illness was probably radiation-related. She left all her property to the Bulgarian Academy of Sciences. Like many of the pioneer women of radioactivity, Karamihailova dedicated her life to the cause of research, yet this devotion was never recognized. Hristov commented: "Professor Karamihailova's life was not an easy one, but she passed through the hardships with dignity and stoicism. She did not receive the warmth she deserved. Yet she fought her loneliness. She always offered her colleagues her attention and consideration, and it was a pity that we often forgot to pay her back."[17]

20 Elizabeth Róna: The Polonium Woman

MARELENE F. RAYNER-CANHAM
and
GEOFFREY W. RAYNER-CANHAM

The expert on the separation of polonium-210 from radioactive ores was Elizabeth Róna, and her skills were put to use in the Second World War. Starting as an organic chemist, she worked in the field of radioactivity for most of her life. Though she was not involved in any great discovery, she worked with some of the biggest names in the field. She held positions in seven different countries and performed research until she was eighty-two years old.

For an adventurous life, few could equal Róna. Born in Budapest on 20 March 1890,[1] her mother, Ida Mahler, had little influence on her interests; it was her father, Samuel Róna, a medical doctor, who fostered her love of science. Róna recalled how he had taken her to a Budapest hospital and shown her the new and exciting x-ray machine. Of course, Róna did not realize at the time how important a role x-rays and radioactivity were to play in her life. She had a wide-ranging love of science and during the summers she would stay with her grandmother in the country and explore the wonders of nature.[2]

After graduating from high school, she attended the University of Budapest. She wanted to become a doctor, but her father thought that medical studies were too difficult for a woman. He died during her sophomore year but she continued to respect his wishes, choosing instead to study physics, chemistry, and geochemistry up to the DrPhil level. Her thesis work on the rate of organic bromination reactions was undertaken with Dr Stefan Bugarszky and the results subsequently published.[3]

At the age of twenty-one Róna graduated from university and travelled to Berlin, where she worked at the chemical division of the Animal Physiology Institute and at the Kaiser Wilhelm Institute, Berlin-Dahlem. Her research involved the use of yeast as a chemical reagent.[4] Then in October 1913, she transferred to the Technical University of Karlesruhe. Her initial plans were to work with the prominent physical chemist George Bredig, but her fellow students advised her to choose Kasimir Fajans, who was working in the new and exciting field of radioactivity. Two other factors influenced Róna in her choice of supervisor. First, Bredig was a traditional authoritarian professor, while Fajans fired his students with his enthusiasm and made them feel comfortable in his presence. Second, when she was invited to one of Bredig's parties, she was expected to join the wives, where she felt totally out of place during discussions of children and recipes. At Fajan's parties, however, she was welcomed like any other member of his research group. She must have had a positive relationship with Fajans as she corresponded with him throughout his life.

Róna stayed in Karlesruhe from October 1913 to the spring of 1914. She spent the summer in England working at University College, London,[5] possibly as a member of Ramsay's research group. It is unclear whether Róna left London because the position was short term or because she wanted to return to her homeland at the outbreak of war. In any event, she travelled across Europe to Budapest, where she stayed until 1921. She joined the research group of George von Hevesy, who worked in the field of physical chemistry,[6] and it was in his laboratory that Róna undertook her first major research project in the field of radioactivity.[7] She was involved in the earliest use of radioisotopes as a means of following chemical reactions, work for which Hevesy was awarded a Nobel Prize in 1943.[8] Róna recalled the pleasant atmosphere of the group: "There was no pressure, and, although I did not have much experience in radioactivity, Hevesy let me use my own imagination; there was a free flow of ideas. He had brought to Budapest from his stay in Manchester [with Rutherford] the habit so dear to the English of five o'clock tea. Across from the laboratory was the tea room Gerbaud, one of the best pastry shops in the city. We had a cup of tea and some delightful pastries there in the afternoon, discussing our experiments or theories, or engaging in just plain talk, which pleasantly relieved the pressure of daily work."[9] During her time with Hevesy Róna also started to correspond with Eva Ramstedt,[10] who was working in the same field. It is unclear how contact was first established, but Ramstedt became one of her life-long friends.

Hevesy left Budapest in 1918, and Róna was offered a challenging job by Francis Tangl, a well-known biochemist and physiologist at the University of Budapest. Her task was to set up the necessary chemistry courses and laboratory work for medical students, whom Tangl considered deficient in chemical knowledge. Róna was flattered by this offer. Given her youth and sex, she took the job and also managed to finish up some of the biochemical research she'd been engaged in at Berlin-Dahlem.[11]

In 1919 the communists took over in Hungary, instituting a reign of violence and fear. The revolutionaries seized the apartment that Róna and her mother lived in and they took refuge with an aunt. The revolution was followed by an equally violent counterrevolution (the White Terror) during which anyone with communist sympathies, particularly at the university, became a target.[12] With the resulting depletion in staff, Róna had to bear an increasing share of the workload. She was able to resign in 1921 to take up an offer of a grant to return to the Kaiser Wilhelm Institute in Berlin-Dahlem to work with Otto Hahn and Lise Meitner in the radioactivity department.

Once more, Róna was blessed with enjoyable working conditions. She considered Hahn to be "gay and self-confident with a pleasant sense of humour" and Meitner to be "an introvert, shy and reserved."[13] Her initial research was on the separation of ionium from uranium ores.[14] The work was important since at the time ionium was considered to be a new element, though we now know it to be thorium-230. However, as inflation began to soar in Germany, only grants related to economically essential work were permitted, so Róna was transferred to the Textile Institute of the Kaiser Wilhelm Institute. Salaries were paid daily and the money had to be spent immediately. As Róna commented, "if one went by subway in the morning to a seminar without a return ticket, one could not buy one in the evening because the value of the mark had diminished so much during the day."[15]

The political turmoil in Hungary having abated, Róna returned there in 1923 and obtained a research position with a textile manufacturer. When it became apparent that her work was not being appreciated, she resigned. Fortunately, two days later Stefan Meyer offered her a position at the Vienna Radium Institute, which she accepted. She was given the only office available, that of the retired director, Franz Exner. One condition of occupancy was that nothing be changed or moved, including Exner's prominent display of tobacco pipes, which Róna thought must have amused her visitors.

To Róna, "the atmosphere at the institute was most pleasant. We were all members of one family. Each took an interest in the research of others, offering help in the experiments and ready to exchange ideas. Friendships developed that have lasted to the present day. The personality of Meyer and that of the associate director, Karl Przibram, had much to do with creating that pleasant atmosphere."[16]

Róna's most important work in Vienna followed on the discovery that bombarding the atoms of an element with alpha particles could lead to transmutation of that element. There was a consequent need for highly concentrated sources of alpha rays; Róna found that polonium-210 was an ideal source of this radiation and she devised a way of separating and concentrating the isotope from spent radon needles that had been used by hospitals.[17]

Both the Vienna Radium Institute and the Cavendish Laboratory were interested in Róna's talents. Through James Chadwick, who was visiting Vienna, Rutherford offered her a position at Cambridge but she declined, deciding to stay in Austria. She felt, however, that she should spend 1926 at the Curie Institute to learn how to prepare radiation sources properly. Róna was thrilled to meet the famous Madame Curie, though she found her very reserved. Róna also tried to meet André Debierne, but Debierne had become a near recluse as a result of his experiences in the First World War.

While working with Curie, Róna was involved in a dangerous mishap:

> One day she [Curie] asked me to come to the laboratory the following Saturday morning but not to tell anybody. The project was to open a flask containing a solution of a strong radium salt, which had been closed for many years. It is well known that under the intense radiation the solvent water is decomposed and hydrogen peroxide accumulates, so that, if proper precautions are not taken, there is a violent explosion. That is just what happened on that memorable afternoon. After Mme Curie scratched the neck of the glass flask with a file and approached a narrow flame, a violent explosion scattered glass all over. It is a miracle that we were not hurt or highly contaminated. Mme Curie was not a highly skilled experimenter at that time: that was probably due to her severely burned fingers, from her long work with radioactive materials. At that time she prepared secondary standards, weighing the radium on a microbalance, without any protection against the intense radiation. I still wonder that she lived to be 67 before succumbing to pernicious anemia.[18]

Róna was assigned to work with Irène Curie, who was the expert on polonium. While most people viewed Irène as arrogant because of her extreme seriousness, reserve, and disregard for her appearance, Róna found her to be a "warm, candid, even romantic person."[19] Irène

Curie had developed a method of extracting polonium-210 from lead-210 and bismuth-210 by means of electrolysis. Róna tried this method, but she could only produce a single layer of polonium atoms on the electrode surface. Róna devised a new method whereby the polonium atoms were sublimed from a heated electrode to a cooled piece of palladium foil.[20] When she returned to the Vienna Radium Institute, Róna continued to work with polonium, starting a fruitful collaboration with Berta Karlik[21] on the radiation from polonium and on the decay products of actinium x (radium-233).[22]

Like most of the research laboratories studying radioactivity, the Radium Institute of Vienna treated the hazards of the work very lightly. Róna commented upon an incident that occurred when she and Gustav Ortner were asked to open some sealed tubes containing radium salts. Unless they had been correctly heat treated, she knew that they might explode. Her caution was probably related to her earlier experience with Curie. "I went to Stefan Meyer, director of the Radium Institute, and asked for gas masks to protect us from such an eventuality. He laughed and tried to assure me that no danger was involved. However, I was not convinced and bought two gas masks with my own money. When we tried to open the first tube, it exploded, and the same thing happened with the second, scattering radioactive material all around. The gas masks saved us from severe damage. The basement room was closed permanently because it was impossible to get rid of the contamination."[23]

In 1934, the Joliot-Curies were invited to the Vienna Institute to describe their discovery of artificial radioactivity. Berta Karlik and Róna were delegated to entertain the visitors and the four of them spent hours hiking through the Vienna forests. Irène slowly relaxed during these excursions and became quite talkative, though Róna concluded that she was "a compassionate but somewhat naive person."[24]

Before the Joliot-Curies left Vienna, they invited Róna to work again for a time at the Curie Institute and she accepted their offer that same year. While in Paris, she made the short journey across the English Channel to meet Ernest Rutherford at Cambridge, staying with Karamihailova in Girton College.[25] She commented on "his booming voice, his liveliness, his sparkling blue eyes," and she noted that "a talk with him was a refreshing experience."[26] Although radioactivity was usually the major problem in nuclear science laboratories, she thought that electrocution by the dangling high-voltage wires was the greatest danger at Cambridge.

Karamihailova was anxious that Róna follow the proper college etiquette. Róna recounted an incident at dinner: "As a guest, I was seated at high table. Mutton was served. Because it was very dry, I started to reach for the mustard. My friend [Karamihailova] nervously whispered into my ear: 'Only a moron eats mustard with mutton. Mint sauce is alright, naturally.'"[27]

After her return to the Vienna Radium Institute, Róna started a study on the effect of bombarding radioactive isotopes with neutrons. She was able to produce a number of completely new isotopes, which were of special interest as they had the longest half-lives of any artificially created isotopes at that time. This was dangerous work as the neutron sources that Róna used were mixtures of radium or polonium with beryllium powder. Not only was there the danger from radiation but inhalation of the toxic beryllium powder could cause severe lung damage. During this time, Róna was awarded the Academy of Sciences' Haitinger Prize for outstanding work in physics.[28]

With the Austrian *anschluss* in March 1938, Róna, being of Jewish origin, had to leave Vienna. Curiously, her autobiography makes no mention of her whereabouts during the 1938–39 period, nor was the historian of Austrian scientists, Wolfgang Reiter, able to find any indication as to where she might have voyaged.[29] We do know that Ellen Gleditsch invited her to spend the winter of 1939–40 at the University of Oslo, where she taught in the Department of Chemistry and joined Gleditsch in some radiochemical research.[30] Otto Hahn visited Norway during her stay and she later wrote to Fajans of her admiration for his open anti-Nazi views.[31] Unable to return to either Austria or Germany, Róna travelled to Hungary where she spent the 1940–41 year working at the Radium-Cancer Hospital of Budapest.[32] In 1941 she decided to flee to the United States, arriving in New York on a visitor's visa. Unfortunately, she left no account of the journey.

Through a chance encounter at a meeting of the American Physical Society, she was offered a teaching position at Trinity College, a Catholic college for women in Washington, DC. About the same time, she was invited to work on the uranium content of seawater for the Geophysical Laboratory of the Carnegie Institute.[33] She was able to combine the two, teaching in the mornings and then, during the afternoons, analyzing samples provided by the Woods Hole Oceanographic Institute, Massachusetts.

In 1942, despite her visitor's status, Róna was asked by the Office of Scientific Research and Development (OSRD) to devise a route for the extraction of large quantities of polonium from radon sources. Although she was still regarded as a security risk (her family still lived in Nazi-controlled Hungary), she was given clearance due to her unique knowledge of polonium. Róna performed this work at the University of Rochester, New York, and she agreed to pass on the method to the Canadian Radium and Uranium Company in New York for mass production without any financial compensation. It is interesting that her old friend Blau was working with the Canadian Radium and Uranium Company at the same time.

OSRD suggested that a student be hired to help Róna with the laboratory work but that the student should not be a chemist or a physicist, given the highly secret nature of the work Róna found a conscientious French major to help her. Almost certainly, the student never became aware that Róna was helping to develop a nuclear weapon.

Róna took up a position at the Argonne National Laboratory in 1947, working on uranium reactions.[34] Three years later she obtained a position as senior scientist at the Special Training Division of the Oak Ridge Associated Universities (ORAU). There was a need to train people in the use and handling of radioactive isotopes, and her unique expertise made her a valuable teacher. In addition, when the courses were opened to foreign students in 1954, her fluency in a number of European languages increased her usefulness even further. At the same time, she continued research work that she had started at the Argonne Laboratories on a method to determine the thorium isotope ratios in minerals.[35]

Róna loved nature and her house in Oak Ridge possessed a beautiful rose garden. She made many friends there, and she enjoyed showing off her grasp of American slang, however imperfect. One of her former co-workers, Edith Wilson, recalled: "One day Dr Róna was going to ask a lady from Norway who was a participant in one of the courses to talk to some group who had asked her [Róna] to speak. She [Róna] turned around to me, sort of shrugged, and said, 'This is what the Americans call "passing the bucket."'"[36]

Oceanography had always been one of Róna's interests. On her summer vacations between 1928 and 1940 she had travelled to an oceanographic research station in southern Sweden, analyzing water samples and marine sediments for radium and uranium content. Friends would come and visit such as Berta Karlik, Ellen Gleditsch, and the well-known oceanographer Hans Pettersson.[37] She decided to renew this interest and start a research program in geochronology and geophysics of marine samples. This work was carried out at ORAU in collaboration with Texas A & M University.[38]

In 1965, at the age of seventy-five, she retired from ORAU and then immediately took a position as professor of chemistry with the Institute of Marine Sciences of the University of Miami.[39] It seems more than a coincidence that her good friend Blau had been a professor of physics at the University of Miami until a few years earlier. Probably she had first visited the university while Blau was there. In a letter to Fajans from her new home on Key Biscayne, she described her arrival: "I came here immediately after hurricane Betsy, the island where I live was still flooded. I found out today that even without a hurricane the island can be flooded, it rains so hard here, real tropical rain, that it is hard to get around or drive. It is a pretty place to live, [a] few very nice apartment buildings, a few private homes and motels. Most of the island is uninhabited. A 10 minute drive through a park and a bridge, and I am in the laboratory. I keep my house in Oak Ridge and will spend summers there, because the heat in the summer here is not for me."[40]

The research program that she started involved the absolute dating of sediment cores and her preferred method was to use the ratios of the radioactive isotopes thorium-230 and protactinium-231 for this purpose.[41] Although it is often assumed that old age is a handicap in frontier research, Róna continued to publish prolifically.

In 1972, at the age of eighty-two, she announced her second retirement; she found that her apartment in Key Biscayne was becoming too expensive and the annual migration between Florida and Tennessee too arduous. As she wrote to Fajans, the Key Biscayne area was becoming "spoiled more and more from the highrises, etc,"[42] so she made a permanent move back to Oak Ridge.

Even at the age of eighty-four Róna was physically active. In the same letter to Fajans, she said that she had gone in the spring of 1974 to the Galapagos Islands and to Quito, Ecuador. This was followed by a summer visit to a niece in New Hampshire. In the same year, at the urging of her former colleagues at ORAU, she started writing her memoirs. This kept her busy until 1978. She continued to have an active social life, though a broken hip necessitated the use of a walker.[43] Róna was taken ill on Christmas Day 1980, and she remained in hospital until her death in July 1981. In her obituary, the former director of ORAU described Róna as a "fascinating spritely woman" whose "facility for languages was an aid in President Eisenhower's Atoms for Peace program."[44] Her friend Edith Wilson commented: "In spite of Dr Róna's brilliance and prestige, she was one of the nicest and most humble people I have ever known. She said that Eve Curie in the biography of her mother wrote, 'She never learned to be famous.' I think this also applied to Dr Róna ... she never learned to be famous either."[45] As with her friend and former colleague Marietta Blau, it was probably her humble nature that contributed to the lack of recognition of her work. Her contributions to the war effort, for example, for which she waived any payment, were never acknowledged. Nor was she ever recognized for her pioneering work in radioactivity over a sixty-year span.

21 Ida Noddack: Proposer of Nuclear Fission

FATHI HABASHI

Ida Noddack (1896–1978) was a geochemist, not a nuclear scientist, but she startled the top nuclear physicists and radiochemists of her time by proposing the concept of atomic fission. Ridiculed by the most eminent chemists and physicists at the time, she was later proved to be correct. In his autobiography, Otto Hahn wrote: "Her [Noddack's] suggestion was so out of line with the then accepted ideas about the atomic nucleus that it was never seriously discussed."[1] She is best known as the codiscoverer of rhenium, the last naturally occurring element to be identified. She was twenty-eight when she made this great discovery and this was the first achievement of her career.

I first met Ida Noddack and her husband, Walter Noddack, in 1955 in Karlesruhe, Germany, at a private dinner at the home of F.A. Henglein (1893–1986), professor of technology at the Technische Hochschule. A subsequent visit to Köln in 1969, by which time I had become fluent in German, resulted in the start of a friendship with Ida Noddack that continued until her death.

Ida Eva Tacke was born on 25 February 1896 in Lackhausen, a small village near Wesel on the Rhein, north of Köln.[2] Her father, Adelbert Tacke, was a lacquer and varnish manufacturer in Lackhausen and her mother was Hedwig Danner. She went to school in Wesel (1902–12) and then in Aachen (1912–15). After completing her high school diploma, she studied chemistry at the Technische Hochschule (now Technische Universität) in Charlottenburg, obtaining the Diplom-Ingenieur degree in 1919, together with first prize in chemistry and metallurgy, and the Doktor Ingenieur in 1921 under Professor D. Holde. Her doctoral thesis was on the anhydrides of high molecular weight fatty acids,[3] a topic that she subsequently dropped. Upon graduation, she obtained industrial chemistry positions in Berlin, first at the Allgemeine Elektrizität Gesellschaft from 1921 to 1923, and then at Siemens-Halske from 1924 to 1925.

In 1925 she joined the Physikalische Technische Reichsanstalt (Imperial Physico-Technical Research Office), a government laboratory in Berlin. Tacke commenced work with Dr Walter Noddack, the head of the chemical laboratory. Tacke and Noddack developed an interest in some of the missing elements from the periodic table. Until the end of the nineteenth century, elements were discovered essentially by accident. It was in 1896 that the Russian chemist Dimitri Ivanovich Mendeléev proposed that ordering the elements according to their atomic weight led to a pattern of repeating properties among elements placed in the same column (group). In constructing this periodic table, Mendeléev found that the pattern only held if he skipped a number of spaces in the table. He proposed that the spaces corresponded to hitherto missing elements. Using his concept of periodicity, he was able to predict the probable properties of many of these missing elements by interpolation. That is, by taking the properties of the elements above and below the missing ones, he assumed that the properties of the unknown element must fall somewhere about the mean between those of the two known neighbouring members of the group.

As a result, many of the missing elements were discovered over the following decade. According to Mendeléev, this still left two elements to be discovered in the manganese group, which he provisionally named eka-manganese (EM) and dvi-manganese (DM). As the elements followed one another, interpolation of properties was impossible. Only the first member of the group, manganese itself, was known. Moreover, it was questioned whether the spaces existed at all.

It was a British scientist, Henry Moseley, one of Rutherford's former students, who provided proof of the vacancies in 1912. He showed that the wavelengths of the X-ray emission spectra from an element depended upon an integer value. This integer correlated with the ordering of the periodic table and the number was referred to as the atomic number.[4] Elements 43 and 75, the spaces under manganese, were among the integers for which no corresponding element was known; hence there was no doubt that these two elements existed. It was now a question of deciding where to look for them.

Tacke and Noddack focused their attention on the mystery of the two missing elements, making a tedious and systematic study of properties of the elements near the two gaps. They found that, although usually there was a gradual change in properties in the main group elements, among the transition metals the subsequent members of the group often bore little resemblance in their properties to the first member of the group. This explained why previous investigators had failed to discover the missing elements – they were searching for them in manganese ores on the assumption that the missing elements would resemble manganese in chemical properties. Tacke and Noddack predicted that these elements would have properties different from manganese and would instead be similar to their horizontally occurring neighbours. Hence they concentrated on ores containing minerals of the metals molybdenum, tungsten, ruthenium, and osmium, the horizontal neighbours of eka- and dvi-manganese. Their examination of trends in terrestrial abundance also led them to believe that the missing elements would be found in extremely low concentrations, even less abundant than ruthenium and osmium.

They prepared more than four hundred enriched samples from different ores for examination. It was Moseley's discovery, the periodicity of the wavelengths of the x-ray emission lines of the elements, that Tacke and Noddack utilized as their main technique in looking for the unknown elements. In June 1925, with the help of Otto Berg, an x-ray specialist at Siemens-Halske in Berlin, they identified in a Norwegian columbite ore, a new element that they called rhenium in honour of the River Rhein. This was the missing element dvi-manganese, element 75.[5] A year after the discovery of rhenium, the Noddacks prepared the first gram of the metal from 660 kilograms of molybdenite ore and, later, wrote numerous papers on its chemistry. The papers were produced by three author sets: Noddack and Tacke, Tacke and Berg, and Tacke alone, indicating that she had played a leading role in the work.

The discovery of this metal, which had eluded so many other researchers, should have brought worldwide fame to Tacke and Noddack, and indeed Ida Tacke did give a major address to the Society of German Chemists on their discoveries. The president of the society thanked Fraulein Tacke for the address, noting that she was the first woman ever to have spoken before the gathering, and hoping that other *Chemikerinnen* (lady chemists) might soon follow her example.[6]

However, Tacke and Noddack's reception outside Germany was influenced by their claim to have found the other missing member of the group, eka-manganese, element 43.[7] They said they had identified some weak x-ray lines belonging to this element concurrently with the discovery of rhenium. They named it masurium after

Masuria, a district in eastern Prussia (now in Poland) famous for its lakes where Walter Noddack had come from. The inability of other chemists to obtain any trace of masurium when they repeated the extraction method of Tacke and Noddack has been described by Kenna.[8] Subsequently, it was found that only radioactive isotopes of element 43 could exist; Tacke and Noddack had been mistaken in their conclusions. An attempt was made in 1988 by van Assche to rehabilitate the reputation of Tacke, Noddack, and Berg.[9] He re-examined their old experimental results and argued that they may indeed have discovered element 43. However, Kuroda showed that the concentrations of the element were far below the levels that could have been detected by the methods that were used in the 1920s.[10]

The name masurium itself sparked a later controversy. The British chemist J. Newton Friend argued that while rhenium was appropriately named "in honour of their national river," "masurium" was not a benign choice but in fact represented a nationalistic political agenda. It was in Masuria during World War I that the Russians had suffered a devastating defeat by the German forces. The battle (6–15 September 1914) became known in history as the Battle of the Masurian Lakes. Newton Friend criticized the choice of name as "a stupid psychological blunder which no civilized scientist would make."[11] Aaron Ihde, in his classic 1964 work on the history of chemistry, considered both names to be overtly nationalistic, commenting that the Rhine and the Masurian marshes represented the barriers to the Allied advances in World War I.[12] Whether the criticism was valid or not, it contributed to a negative image of Tacke and Noddack by scientists in other countries.

In 1926 Ida Tacke married Noddack, becoming a "wife chemist." This "wife chemist" held nothing more than research positions, usually unpaid, over the course of her forty-two-year married career. In order to continue her research and overcome the lack of workspace and equipment, Ida Noddack often laboured in makeshift laboratories, using equipment borrowed from Walter Noddack and his associates. Her activity evidently depended largely upon Walter Noddack's employment situations, and professionally she remained very much in the shadow of her husband. It was this lack of formal rank that kept her later proposals on atomic fission from being taken seriously.

The employment of married women was a problem even during the era of the Weimar Republic. The republic had guaranteed equality for both sexes but there was an ongoing debate about female *Doppelverdiener* (second-income earners), one of the main arguments being that the high unemployment problem would be alleviated if married women left their jobs and returned home, strengthening the family unit in the process.[13] When the Reich government came to power, it formalized this position with the Law on the Legal Position of Female Public Servants, which required the dismissal of females upon marriage. Moreover, women were barred from senior positions and from many professions such as law. The hostility towards women intellectuals reflected two concerns of Nazi administrators at the time: that women were usurping men's rights to high-paying jobs and thereby increasing unemployment among men; and that the birth rate was rapidly dropping, threatening the future of the German race.[14] With these political pressures, Noddack's circumstances as a married, childless woman would have precluded any possibility of her obtaining a recognized position.

Even though the Noddacks were not directly affected by the rise to power of the National Socialists, life for scientists had become very difficult partly because of the promotion of "German physics" and "Aryan physics" by two Nobelists, Philipp Lenard and Johannes Stark.[15] Lenard attacked the theory of relativity newly formulated by Albert Einstein, while Stark was attacking the new quantum mechanics developed by Werner Heisenberg. The attacks – among other things a clash between old and new – appeared to be scientific in nature but in reality were racial and political, in keeping with the policies of the ruling party. Beyerchen, in his review of scientists under Hitler, commented that most German scientists (including, one would assume, the Noddacks) decided upon at least minimal cooperation with the regime as the price of preserving German science.[16] In fact, the question of political affiliation influenced the entire scientific community of pre-war Germany, although it affected individual scientists in different ways. Membership in the Nazi party provided certain advantages – in obtaining professional promotions and gaining institutional support for one's projects, for example – but it was scorned by a disapproving network of scientists with anti-Nazi inclinations.

The Noddacks' work on eka-manganese (masurium) had been a joint project, though with several of the publications bearing only Ida's name and most of the remainder citing her first, she must have been the main researcher. After about 1933, Ida Noddack developed her own independent lines of research, the first being an intensive study of the periodic table. In this work she discussed the possibility of discovering the transuranium elements, the elements beyond uranium in the periodic table.[17]

Noddack is included in this compilation because of her next endeavour, which originated with some research in nuclear physics started by Enrico Fermi in Rome. In 1934 Fermi had discovered that neutrons could be captured by atoms and that the frequency of capture increased when they were slowed down by passing them through a hydrogen-rich material such as paraffin or water. He was thus able to produce atoms of higher atomic number than those bombarded. For example, on bombarding cobalt with neutrons he was able to produce nickel.

However, when Fermi and his co-workers bombarded uranium with neutrons, they obtained more than one radioactive product. Thinking along the lines of his previous experiments, Fermi suggested that one of these products was formed by neutron capture, that is, it was a transuranium element (element 93).[18] He assigned the new element a position under rhenium in the periodic table and called it eka-rhenium. Fermi's paper attracted Ida Noddack's attention because it dealt with another element in the manganese group and she studied it carefully.

Soon afterward, she published a paper showing that Fermi's experimental evidence was incomplete.[19] She was critical of his conclusions, arguing that all elements in the periodic system would have to be eliminated before one could claim to have found a transuranium element. She went further and suggested that "when heavy nuclei are bombarded by neutrons, it would be reasonable to conceive that they break down into numerous large fragments which are isotopes of known elements but are not neighbours of the bombarded elements." In this statement Ida Noddack conceived, before anybody else, the idea of nuclear fission. Her argument was as follows: when atoms are bombarded by protons or alpha particles, the nuclear reactions that take place involve the emission of an electron, a proton, or a helium nucleus and the mass of the bombarded atom suffers little change. When, however, neutrons are used, new types of nuclear reaction should take place that are completely different from those previously known.

Fermi's experiments were repeated by Otto Hahn and his co-workers in Berlin. They agreed with his conclusions and published a series of papers on extensive radiochemical separations of the so-called transuranium elements.[20] The results, however, became so contradictory that after five years of intensive research and extensive publication, the concept of transuranium elements had to be abandoned. In their paper dated 6 January 1939, Hahn and Strassmann wrote: "As chemists we must actually say that the new particles do not behave like radium but in fact like barium; as nuclear physicists we cannot make this conclusion which is in conflict with all experience in nuclear physics."[21]

At that time Hahn was fifty-five and already director of the Kaiser Wilhelm Institute for Chemistry (now the Max Planck Institute). A well-established scientist, he had travelled abroad on numerous scientific missions; he spoke excellent English; he had discovered protactinium with his associate Lise Meitner in 1918; and he had written a textbook on radiochemistry. Yet he could not accept the new idea that the uranium atom was split into two fragments. Hahn noted later: "It took four years for scientists to explain what actually takes place when uranium is bombarded with neutrons. Because of this delay in finding the true interpretation, the ideas that Lise Meitner, Fritz Strassmann, and I advanced about the artificial isotopes all turned out to be wrong. We thought them to be transuranium elements – elements with an atomic number higher than 92. Everyone else who had worked on the problem had arrived at the same conclusion; there seem to be no other." [22] Ida Noddack later recalled: "When in 1935 or 1936 my husband suggested to Hahn by word of mouth that he should make some reference, in his lectures and publications, to my criticism of Fermi's experiments, Hahn answered that he did not want to make me look ridiculous as my assumption of the bursting of the uranium nucleus into larger fragments was really absurd."[23]

Ida Noddack wrote a short article in the same journal as Hahn and Strassman in which she reminded Hahn of her suggestion five years earlier that the uranium atom might have undergone splitting; she ended by telling him regretfully that he never cited her paper on this matter although she had once explained her view to him personally. The editor of the journal asked Hahn to comment, but he refused. As a result, the editor added a note to the article (my translation): "Editor's remark. Otto Hahn and Fritz Strassman informed us that they have neither the time nor the interest to answer the preceding note. They think that they would rather renounce commenting, as the possibility of breaking down a heavy atom into smaller fragments – an idea already expressed by many others, cannot be concluded without experimental evidence. They have their own opinion on the correctness of the views of Frau Ida Noddack and the way she expressed them to the peers."[24] The deliberate downplaying of Noddack's contribution by Hahn has been addressed in detail elsewhere.[25]

Ida Noddack's lack of institutional support and the publication of her work only in German journals were factors that prevented her suggestion of uranium fission from receiving any serious consideration. The prevailing theory in nuclear physics, moreover, precluded the possibility of splitting the uranium nucleus with a single slow-moving neutron. Noddack's status as a chemist further separated her from the mainstream of discourse regarding what seemed to be a problem of physics.

The masurium controversy of the 1920s had had a negative influence on the Noddacks' reputation and was present in the memories of many scientists when Ida Noddack made her 1934 suggestion of fission.[26] The adverse publicity outweighed recognition for the Noddacks' discovery of rhenium, though studies of the new substance revealed the skill required to isolate this very rare element. Ida Noddack's lack of English and the fact that she never travelled to the English-speaking world meant that few outside Gremany recognized her talents.

It is true to say that, although the experiments that led to the discovery of the fission phenomenon were conducted by very capable men, the interpretation came mainly from two women scientists. Ida Noddack was nominated for the Nobel Prize in chemistry in 1933, 1935, and 1937,[27] but it was awarded to those who made the experiments, not to those who interpreted the results. The physics prize for 1938 was awarded to Enrico Fermi "for identification of new radioactive elements and discovery of nuclear reactions effected by slow neutrons," work he accomplished in 1934. The chemistry prize for 1944 went to Otto Hahn "for work on atomic fission" that he published in 1939. The interpretation of Fermi's work was postulated as a hypothesis by Ida Noddack in 1934, while that of Otto Hahn and his young collaborator Fritz Strassmann was argued on a theoretical basis by Lise Meitner and her nephew, Otto Frisch.

In 1935, shortly after Ida Noddack had proposed the fission hypothesis, the Noddacks moved to Freiburg in Baden, where Walter Noddack was appointed professor of physical chemistry at the University of Freiburg and Ida Noddack obtained an actual formal position as a research associate. They stayed there for seven years until 1942, when Walter Noddack was appointed to Strasbourg University where the couple remained until 1944, when the city was returned to France. To have received such an appointment implied that Noddack harboured Nazi sympathies. After the war, however, the "Public Prosecution of the Denazification Court" absolved him of any guilt. Ida Noddack was not investigated, one assumes, because she held no academic rank at the university.

The "masurium" incident continued to plague the Noddacks. As Fritz Paneth pointed out in 1947, they had never withdrawn their claim to have found element 43 despite the complete lack of any elaboration and confirmation of the proposal. Paneth concluded his harsh attack with the comment: "During the war, W. Noddack was appointed professor of inorganic chemistry in Strasbourg by the occupying power: when in 1945 the French chemists returned, they found the symbol Ma painted on the wall of the main lecture theatre in a large representation of the Periodic Table."[28]

Between 1944 and 1956 Ida Noddack was unemployed. During this period the Noddacks spent some time in Turkey. In 1956 they moved to the newly founded Staatliche Forschungs Institut für Geochemie (State Research Institute for Geochemistry) in Bamberg, Germany. Ida Noddack worked on a number of projects related to the rare earth elements. Walter Noddack died in 1960 and Ida Noddack retired six years later.

Because of her important contribution to the periodic table, Ida Noddack was invited by the USSR Academy of Sciences to the Mendeléev periodic table centennial held in Leningrad in October 1969. She was then the only chemist alive to have discovered a naturally occurring element. Although illness prevented her from attending, her paper entitled "The Periodic System and the Search for Ekamanganese" was translated into Russian and read at the conference; this happened to be her last scientific contribution to chemistry. In this paper, Ida Noddack told the circumstances of her discovery and bitterly recounted the controversy that took place during her scientific career. She spent her last years in Wohnstift Augustinum, an apartment building for elderly people in Bad Neuenahr on the Rhein near Bonn. She died on 24 September 1978. The Noddacks had no children.

Ida Noddack received a number of honours during her life: the Leibig Medal of the German Chemical Society in 1931, the Scheele Medal of the Swedish Chemical Society in 1934, and the High Service Cross of the German Federal Republic in 1966. She was awarded honorary memberships in the Spanish Society of Physics and Chemistry (1934) and the International Society of Nutrition Research (1963) and received an honorary doctorate in science from the University of Hamburg (1966). In a more permanent gesture of recognition, her name was given to a street in her home town of Lackhausen. However, she was never recognized by any English-speaking scientific organization. Perhaps this was because of the "masurium affair" or because she was a German who had not openly rejected the Nazi regime. Whatever the reason, there is no doubt that she was indeed the first to propose the concept of nuclear fission.

22 … And Some Other Women of the Austro-German Group

MARELENE F. RAYNER-CANHAM
and
GEOFFREY W. RAYNER-CANHAM

Some of the women who worked in scientific research seemed to vanish; perhaps they married or returned to their families, or took on a profession more "suitable" for a single woman such as school teaching. Among the Austro-German group, one of those who completely disappeared from the record was Hélène Towara, who spent the summer of 1914 working with Kasimir Fajans at the University of Karlesruhe, Germany. Towara was born on 21 March 1889 in Maikop, Russia, to Greek Orthodox parents; her father was a merchant.[1] She obtained an advanced high school diploma from the Moscow Commerce Institute. As we saw in the case of Catherine Chamié, Russian women were not permitted to enter university at the time. Towara thus chose to continue her education in Germany. Her work with Fajans involved the study of decay products of radium.[2] In a letter to Elizabeth Róna Fajans commented, "With the Russian lady Towara we thought to find a radioactive isotope of Bi but her results were not correct."[3] No trace of her could be found after that summer.

Among the other research groups, George von Hevesy recruited at least two women after the departure of Elizabeth Róna. One in particular, Hilde Levi, performed a number of important studies of artificial radioactivity between 1934 and 1936.[4] In an account of the major discoveries of Hevesy, one of his biographers, D. Lockner, commented: "Hevesy discovered experimentally that neutron irradiation of scandium and calcium gives K-42 and described several radioactive isotopes of rare-earth metals produced by neutron irradiation. This work was done in collaboration with Hilde Levi, as was the discovery of neutron activation analysis in 1936, a technique first used to determine the content of europium in preparations of gadolinium. It was more than 20 years before scientists really started to use this important method."[5]

During the interwar period, Vienna continued to be a haven for women in atomic science. As happened at the Curie Institute, many of the women published one paper and then vanished from the research scene. A few, such as Margarete Hoffer, Elizabeth Neuninger, and Herta Scheichenberger (all of whom worked with Roná), published more extensively. The key figure at Vienna was Berta Karlik, yet her name is totally unknown in the English-speaking world.[6] Though born four years into the twentieth century and therefore technically beyond the period of our study, it is important to mention her existence and contributions.

BERTA KARLIK

Karlik was born on 24 January 1904 in Vienna, where she enrolled at the university in 1923.[7] She completed her degree in physics and mathematics in 1928. As a student she had undertaken research with Stefan Meyer and her first paper, on scintillations from zinc sulfide, appeared in 1927.[8] She started work at the Institut für Radiumforschung in 1929. For the 1930–31 academic year, she travelled to England with a Crosby Hall stipend from the British Federation of University Women. She spent part of the year with Sir William Bragg at the Royal Institution in London. During her stay, she worked on crystallographic methods of studying hydrocarbons[9] with Helen Gilchrist and she co-authored a book of crystallography data tables with another of Bragg's protégées, Isabel Ellie Knaggs.[10] The remainder of her time in England was spent studying radium therapy at Hammersmith Hospital and working with Ernest Rutherford at the Cavendish. Then she travelled to Paris, becoming a researcher at the Institut Curie for the last part of 1931, before returning to Vienna. Her other lengthy absence from Austria was in 1938 when she lived in Sweden for one year studying the radioactivity of seawater with Föyn and Róna.[11] Karlik published prolifically in German-language scientific journals,[12] her most important research being the discovery of the element astatine in nature.[13] This work, which was performed with one of the next generation of women at Vienna, Traude Bernert, was overlooked by many as it was performed during the Second World War.

But Karlik was an important character in another way: she corresponded with many of the other women researchers such as Meitner,

Gleditsch, Róna, and Karamihailova; thus helping to maintain the "invisible college" among women atomic scientists that Gleditsch had first developed. In addition, when Blau and Róna fled the institut it was Karlik who, despite her anti-Nazi sentiments, managed to remain at her post throughout the war, enabling the research in radioactivity to survive into the postwar period.[14] In 1947 she was promoted to head of the institute upon the retirement of Stefan Meyer. Karlik was also a well-respected teacher and she was known by her many admiring students as "Aunt Berta." Within Austria, she received numerous awards for her contributions to science. She died on 4 February 1990 at the age of eighty-six – yet another of the researchers to survive to an exceptional age.[15]

23 Epilogue: The End of an Era and a New Generation

MARELENE F. RAYNER-CANHAM
and
GEOFFREY W. RAYNER-CANHAM

This compilation encompasses the work of women researchers in radioactivity who were born before 1900. Thus, we have looked at the lives and work of women who entered academic life at a time when the universities were just opening up to them, particularly in the sciences. Though most of their research was accomplished in the first two or three decades of the twentieth century, some of our subjects, such as Blau, Joliot-Curie, Meitner, and Róna, continued their work into the 1940s and beyond. Yet as a group they represented a unique generation – perhaps we might call them the "innocent" generation – the first group to enter the hitherto male-only institutions of higher learning. To them, there were no limits to their potential progress; they would give their all to science. In those early years the discrimination they faced was generally an expression of individual male traditionalists rather than part of an organized phenomenon.

Thus, it might be said that the 1900–20 era was in some respects a "golden age" for women researchers in science. In this closing chapter we show how the situation changed for the next cohort, and how the milieu of atomic science became less welcoming to women researchers.

As we have seen, few of the women who studied radioactivity became famous, but they did make significant contributions. In the classic work *Discovery of the Elements,*[1] for example, Harriet Brooks, Irène Curie, Marie Curie, Stefanie Horovitz, Lise Meitner, and Jesse Slater, together with Yvette Cauchois (see below), are referenced as having made contributions to the discovery of the radioactive elements. Yet the chapter on the later discoveries of synthetic elements (from 1940 onwards) only mentions Lise Meitner and Marguerite Perey. What happened after the early "golden" years? There appear to be two trends: a diminishing rate of participation of women in science, and various factors relating specifically to the study of radioactivity.

THE PARTICIPATION OF WOMEN IN SCIENCE

During the late nineteenth and early twentieth centuries, the sciences had become popular pursuits for women,[2] and within the sciences women's colleges in the United States often had larger enrolments in physics courses than in chemistry courses.[3] But this trend was not to continue. We have shown that a decline in participation by women in chemistry occurred in the 1920s[4] and this decline was apparent across the sciences. Vetter noted in the context of PhDs in the United States: "Once before, women made a significant dent in the sturdy armor of the male science community. During the decade of the 1920's, women earned twelve of every one hundred PhDs awarded in science and engineering, but this was a higher proportion than they ever would again until 1975."[5] And Kistiakowsky has reported specific data for physics doctorates showing that the proportion of women declined from about four percent in the 1920s to a minimum of less than two percent in the 1950s, with a recovery occurring in the 1970s.[6]

Vetter attributes the decrease to "the depression of the 1930's, World War II in the 1940's and the GI bill for returning veterans."[7] However, Barnard, who noted the phenomenon of decreasing participation among all academic women, considered the drop to reflect an increasing preference among women for children and family life over academic qualifications.[8] More recently, it has become apparent that the end of the First World War marked a turning point in male society's views on the intellectual abilities of women. In our opinion, this was a key factor in the drop in women's enrolments. Before the Great War, women were tolerated in male professions, perceived more as a curiosity than a threat. During the war, women filled a significant portion of skilled occupations. Then, with the end of the war, as Roberts comments: "They [women] were often regarded with open hostility by men who had realised for the first time that women were fully capable of carrying out jobs previously perceived as men's, thus presenting a real challenge."[9] This was certainly true for industrial positions; as we have seen, Leslie and Hitchins were among the thousands of women who lost their wartime posts. Rossiter noted the same phenomenon in the United States, quoting an announcement in 1919 by the Illinois Steel Company that "the women chemists of the Illinois Steel Company not only made good as chemists but showed their fine spirit by resigning in order to make places for the men returning from war work."[10]

The anti-feminist attitude became prevalent worldwide during the interwar period. In Italian universities one popular student song contained the line: "We don't want women at school, we want them naked laid out on a couch."[11] Similar sentiments obtained in Canada. The historian Strong-Boag has reviewed the rise in prejudice against Canadian academic women in the 1930s, noting that some of the newly instituted university scholarships were reserved exclusively for men.[12] The hostility against women students was so widespread at the University of Alberta that the head of the Woman-Hater's Club was elected president of the student union in 1935. This misogynistic trend was related by Strong-Boag to the change in the image of women from the independent "flapper" of the twenties to the dependent "moppet" of the thirties. She, too, remarked on the critical decline in the percentage of women enrolled in male-dominated fields, including the sciences, between 1920 and 1940: "Explanations for this decline are not yet clear, but it owed something to the absence in these years of an articulate feminist movement that had nurtured the pioneers who had breached male monopolies in the nineteenth century."[13] Moreover, she commented, for the women entering university an arts degree opened the doors to women-accessible careers such as a teacher, socialworker, librarian, clerical worker, saleswoman, and wife, whereas there were few acceptable employment possibilities with a science degree.

Faludi, examining the situation in the United States, echoed the view that women's retreat was not voluntary: "Post-feminist sentiments first surfaced, not in the 1980s media, but in the 1920s press. Under this barrage, membership in feminist organizations soon plummeted, and the remaining women's groups hastened to denounce the Equal Rights Amendment or simply converted themselves to social clubs."[14] Faludi added that the 1920s had also eroded a decade of growth for female professionals, and that, in the United States, there were fewer female doctors by 1930 than in 1910.

In Germany, the "threat" of young women choosing intellectual rather than family-rearing activities was taken very seriously by the Nazi government. A law was passed in 1934 requiring female high school graduates to first perform a six-month "labour service" before they could apply to a university. This was intended primarily as an obstacle that would curb, or at least delay, women's enrolments at the institutions of higher learning. As Pauwels remarked: "Moreover, the lives of the 'labor maids' (*Arbeitsmaiden*) in the camps was far from pleasant ... The prospect of six months in a Spartan labor camp far from home, of hard work in the fields, inexorable discipline, and minimal comfort was undoubtably a powerful deterrent to many young women who would normally have embarked on an academic career."[15]

THE SITUATION IN ATOMIC SCIENCE

Even though women's participation in academia declined worldwide, women did not disappear completely from the ranks of the researchers in atomic science. For example, Rutherford had a number of women working with him in later years, including G.I. Harper and E. Salaman,[16] who vanished into obscurity after 1930. Katharine Blodgett, who also worked with Rutherford, did not vanish but obtained a PhD in 1926, the first woman to obtain this qualification at Cambridge.[17] She moved into the field of surface chemistry and her name is immortalized in the phenomenon of Langmuir-Blodgett films. Women continued to join the Paris and Vienna groups as well. The most famous of the later Paris workers was Marguerite Perey,[18] who became the first woman to be elected *correspondante* of the Académie des Sciences, an honour denied to Marie Curie and to Irène Joliot-Curie. The highlight of Perey's career was the discovery of the radioactive element francium before she, too, succumbed to cancer. Another prominent member of the Paris group was Yvette Cauchois.[19]

But the proportion of women in atomic science declined and there were none to join the ranks of the "movers and shakers" such as Curie, Joliot-Curie, and Meitner. Together with the more general societal reasons discussed earlier, there were special factors involved in atomic science. First, the growth of early atomic science had paralleled that of biochemistry. Creese remarked: "In the case of the biochemists, their field at the time was a developing discipline. The entry paths and entry qualifications of its practitioners were not well defined. This lack of prestige, due to the slowness of academic chemists to recognise the full power and potential of research in the field, offers one explanation for its relative openness to women."[20] However, as both biochemistry and atomic science became more professional and mainstream, women became more marginalized. Many women in the early years had begun research work with a bachelor's degree or less. In later years, doctoral degrees became a necessity to enter the research community – a much harder commitment for women at a time when marriage and children were still expected to be their main goal in life.

Then, too, in the early years the study of radioactivity occupied the middle ground between chemistry and physics, bridging the two. With the concepts of transmutation and isotopes well understood,[21] atomic science had become a branch of pure physics – the male-dominated field of theoretical physics where Bohr, Heisenberg, Schrödinger, and so on held sway. No longer the "orphan" field that it initially was,[22] it had become a sphere in which Nobel Prizes were to be won and thus it attracted the brightest of the new generation. As Slater commented: "Here was a decade [the 1920s] in which the most momentous physics discoveries in a century were being made, and here were fifty or more ambitious young men, entering a field with a smaller number of older and very distinguished workers, all trying to be in on the exciting discoveries that all were convinced were going to be made."[23]

Along with the change in emphasis came the change and expansion of locations. The United States became one of the new foci – a centre that had no tradition of encouraging women researchers. In fact, the contrary was true; women were simply not welcome in this environment. In the United States, Jane M. Dewey was unable to find a physics post at a regular university in 1929 in spite of her excellent credentials. Most of the physics faculty at Princeton were not even prepared to allow her to take up a research fellowship.[24] Though Goeppert-Mayer had an outstanding background in quantum mechanics and was to become a Nobel laureate in physics, upon her arrival in the United States in the early 1930s, she was treated not as a brilliant academic but as "an academic wife, whose husband has a fairly secure position, so that the university considers her captive, who can be hired with no guarantees, for as little as possible, and dropped whenever it becomes expedient."[25] This exclusion of women from the centre of research continued into the Second World War, and the women involved in the Manhatten Project were completely marginalized in its "men's club" atmosphere.[26]

Of equal importance, the style of the research process had changed. The supportive atmosphere of the early twentieth century with friendly supervisors such as Rutherford and Meyer had become competitive and aggressive. When Joan Freeman arrived at the Cavendish in 1946, she was told: "Progress would be painful and intellectually demanding ... Worthwhile results would not come easily. Far too many people were flocking into the field ... It would be a case of survival of the fittest."[27] Traweek studied the situation in the 1980s. Her conclusions were that "the traits required for gaining entry into this exclusive community [high-energy physics] – aggressive individualism, haughty self-confidence, and a sharp competitive edge – are traits typically defined as masculine by our society."[28] Thus atomic science, once a haven for women, turned into the exact opposite, an environment where, as Traweek stated, "extreme masculinity" became the only basis for success, excluding in the process the half of the population that was once so welcome.

Appendix

DATES OF SELECTED EVENTS IN THE HISTORY OF ATOMIC SCIENCE (1895–1940)

Date	*Person*	*Event*
1895	H. Bequerel	Natural radioactivity observed
1897	J.J. Thomson	Discovery of electron as particle
1898	P. Curie and M. Curie	Polonium and radium discovered
1902	E. Rutherford and F. Soddy	Theory of atomic disintegration
1906	P. Curie	Died
1910	F. Soddy	Theory of isotopes
1912	E. Rutherford	Nuclear model of atom
1917	O. Hahn and L. Meitner	
	F. Soddy and J. Cranston	Protactinium discovered
1919	E. Rutherford	Artificial transmutation reported
1925	I. Tacke and W. Noddack	Rhenium discovered
1932	J. Chadwick	Neutron discovered
1934	I. Curie and F. Joliot	Artificial radioactivity discovered
1934	M. Curie	Died
1937	E. Rutherford	Died
1938	O. Hahn and F. Strassmann	
	I. Curie and F. Joliot	Nuclear fission observed
1938	L. Meitner and O. Frisch	Nuclear fission explained
1939	M. Perey	Francium discovered

Contributors

E. Tina Crossfield, Okotoks, Alberta

Grete P. Grzegorek, John Abbott College, Ste Anne-de-Bellevue, Quebec

Fathi Habashi, Department of Extractive Metallurgy, Laval University, Quebec City, Quebec

Leopold Ernst Halpern, Department of Physics, Florida State University, Tallahassee, Florida, US

Miruna Popescu, Bucuresti, Romania

Helena M. Pycior, Department of History, University of Wisconsin, Milwaukee, Wisconsin, US

Geoffrey W. Rayner-Canham, Department of Chemistry, Sir Wilfred Grenfell College, Memorial University, Corner Brook, Newfoundland

Marelene F. Rayner-Canham, Department of Physics, Sir Wilfred Grenfell College, Memorial University, Corner Brook, Newfoundland

Snezha Tsoneva-Mathewson, Plovdiv, Bulgaria

Sallie Watkins, Pueblo West, Colorado, US

Anne-Marie Weidler Kubanek, John Abbott College, Ste-Anne-de-Bellevue, Quebec

Stephanie Weinsberg-Tekel, North Bay, Ontario

Notes

ABBREVIATIONS

AIC	Archives of the Institut Curie, Paris
BCA	Barnard College Archives, New York
BHL	Bentley Historical Library, University of Michigan
BMC	Bryn Mawr College, Pennsylvania
BN	Bibliothèque Nationale, Paris
CCA	Churchill College Archives, Cambridge University
CU	Catholic University of America, Washington, DC
CUA	Cambridge University Archives
FA	Fajans Archives, University of Michigan
GCA	Goucher College Archives, Maryland
MPG	Bibliothek und Archiv der Max-Planck-Gesellschaft, Berlin-Dahlem
OUA	Oxford University Archives
OUL	Oslo University Library
UL	Leeds University Library

CHAPTER ONE

1 Cited in Alex Keller, *The Infancy of Atomic Physics: Hercules in His Cradle* (Oxford: Clarendon Press, 1983), 9.
2 Russell Moseley, "Tadpoles and Frogs: Some Aspects of the Professionalization of British Physics 1870–1939," *Social Studies on Science* 7 (1977): 423–46.

3 R. Lucas, *Bibliographie der radioactiven Stoffe* (Hamburg and Leipzig: Verlag von Leopold Voss, 1908).
4 Lawrence Badash, "Radium, Radioactivity, and the Popularity of Scientific Discovery," *Proceedings of the American Philosophical Society* 122 (1978): 145–54.
5 Gerald L. Geison, "Scientific Change, Emerging Specialties, and Research Schools," *History of Science* 19 (1981): 20–40; and Gerald L. Geison, "Research Schools and New Directions in the Historiography of Science," *Osiris* 8 (1993): 227–38.
6 Diane Crane, *Invisible Colleges* (Chicago: University of Chicago Press, 1972).
7 These three main schools were first described by Robert W. Lawson, "The Part Played by Different Countries in the Development of the Science of Radioactivity," *Scientia* 30 (1921): 257–70.
8 Keller, *The Infancy of Atomic Physics,* 3.
9 Ibid., 5.
10 Thomas S. Kuhn, 2d ed., *The Structure of Scientific Revolutions* (Chicago: University of Chicago Press, 1970).
11 J.L. Davis, "The Research School of Marie Curie in the Paris Faculty, 1907–1914," *Annals of Science* 52 (1995): 321–55.
12 Dong-Won Kim, "J.J. Thomson and the Emergence of the Cavendish School, 1885–1890," *British Journal of the History of Science* 28 (1995): 191–226.
13 Lewis Pyenson, "The Incomplete Transmission of a European Image: Physics at Greater Buenos Aires and Montreal, 1890–1920," *American Philosophical Society, Proceedings* 122 (1978): 92–114.
14 Lawrence Badash, *Radioactivity in America: Growth and Decay of a Science* (Baltimore: Johns Hopkins University Press, 1979), 272.
15 Elizabeth Crawford, *Nationalism and Internationalism in Science, 1880–1939* (Cambridge, UK: Cambridge University Press, 1992), 99.
16 Lawrence Badash, "The Suicidal Success of Radioactivity," *British Journal for the History of Science* 12 (1979): 245–56.
17 Lawrence Badash, "British and American Views of the German Menace in World War I," *Notes and Records of the Royal Society of London* 34 (1979): 91–121; P. Foreman, "Scientific Internationalism and the Weimar Physicists," *Isis* 64 (1973): 151–80; D.J. Kevles, " Into Hostile Political Camps: The Reorganization of International Science in World War I," *Isis* 62 (1971): 47–60.
18 Otto R. Frisch, "How It All Began," *Physics Today* (November 1967): 43–8.
19 G.E.M. Jauncey, "The Early Years of Radioactivity," *American Journal of Physics* 14 (1946): 226–41; Lawrence Badash, "Radioactivity before the Curies," *American Journal of Physics* 33 (1965): 128–35; Lawrence

Badash, "Bequerel's 'Unexposed' Photographic Plates," *Isis* 57 (1966): 267–9.

20 G.E.M. Jauncey, "The Birth and Early Infancy of x-rays," *American Journal of Physics* 13 (1945): 362–79.

21 Lawrence Badash, "The Discovery of Thorium's Radioactivity," *Journal of Chemical Education* 43 (1966): 219–20.

22 Thaddeus J. Trenn, "Rutherford on the Alpha-Beta-Gamma Classifactions of Radioactive Rays," *Isis* 67 (1976): 61–75.

23 S.B. Sinclair, "J.J. Thomson and Radioactivity," *Ambix* 35 (1988): 91–104, 113–26.

24 A. Pais, "Radioactivity's Two Early Puzzles," *Reviews of Modern Physics* 49 (1977): 925–38.

25 Alfred Romer, "The Transformation Theory of Radioactivity," *Isis* 49 (1958): 3–12.

26 Thaddeus J. Trenn, "Rutherford and Soddy: From a Search for Radioactive Constituents to the Disintegration Theory of Radioactivity," *Rete* 1 (1971): 50–70.

27 Marjorie Malley, "The Discovery of Atomic Transmutation: Scientific Styles and Philosophies in France and Britain," *Isis* 70 (1979): 213–23.

28 George B. Kauffman, "The Atomic Weight of Lead of Radioactive Origin: A Confirmation of the Concept of Isotopy and the Group Displacement Laws," *Journal of Chemical Education* 59 (1982): 3–8, 119–23.

29 Eri Yagi, "On Nagaoka's Saturnian Atomic Model," *Japanese Studies in the History of Science* 3 (1964): 29–47.

30 Thaddeus J. Trenn, "The Geiger-Marsden Results and Rutherford's Atom, July 1912 to July 1913, The Shifting Significance of Scientific Evidence," *Isis* 65 (1974): 74–82; John L. Heilbron, "The Scattering of a and b particles and Rutherford's Atom," *Archive for History of Exact Sciences* 4 (1968): 247–307.

31 Guiseppe Bruzzaniti and Rita Quaglia, "p-e Model and Anomolies of 'Nuclear Electrons,'" *Rivista di Storia della Scienza* 2(1) (1994): 75–126.

32 Bernd Kröger, "On the History of the Neutron," *Physis* 22 (1980): 175–90.

33 Günter Herrmann, "Five Decades Ago: From the 'Transuranics' to Nuclear Fission," *Angewandte Chemie, International Edition* 29 (1990): 481–508.

34 Roger H. Stuewer, "The Origin of the Liquid Drop Model and the Interpretation of Nuclear Fission," *Perspectives on Science* 2 (1994): 76–129.

35 Kurte Starke, "The Detours Leading to the Discovery of Nuclear Fission," *Journal of Chemical Education* 56 (1979): 771–5.

36 Henri Becquerel and Pierre Curie, "Action physiologique des rayons du radium," *Comptes Rendus* 132 (1901): 1289–91.

37 May Sybil Leslie to Arthur Smithalls, 8 June 1911, Leeds University Archives, Smithalls Collection.
38 Thomas Powers, *Heisenberg's War* (New York: Alfred A. Knopf, 1993).
39 Gordon F. Hull, "The New Spirit in American Physics," *American Journal of Physics* 11 (1943): 23–30.
40 Otto Hahn, *Otto Hahn: My Life*, trans. Ernst Kaiser and Eithne Wilkins (New York: Herder and Herder, 1970), 110.
41 Elizabeth Róna, "Laboratory Contamination in the Early Period of Radiation Research," *Health Physics* 37 (1979): 723–7.
42 C. Richard Cothern and James E. Smith, eds., *Environmental Radon* (New York: Plenum Press, 1989), 54–7.
43 Samuel Hellman, "Curies, Cure and Culture," *Perspectives in Biology and Medicine* 36 (1992): 39–45; Eve Curie, *Madame Curie: A Biography*, trans. Vincent Sheean (New York: Da Capo Press, 1986 [1937]), 199, 302.
44 Sonya Cotelle's accident and severe radiation-caused symptoms are described in a letter from Irène Curie to Marie Curie, 5 August 1927, in *Curie Correspondence (1905–1934)* (Paris: Les Éditeurs Français Réunis). See also Moïse Haïssinsky, *Le Laboratoire Curie et son apport aux sciences nucleaires* (Paris: Laboratoire Curie, 1971), 33.
45 Rosalynd Pflaum, *Grand Obsession: Madame Curie and Her World* (New York: Doubleday, 1991), 234–5; Robert Reid, *Marie Curie* (New York: E.P. Dutton, 1974), 271–9.
46 Robert Alvarez, "Radiation Workers: The Dark Side of Romancing the Atom," *Science for the People*, 18 (March/April 1986): 6–11.
47 Bryce DeWitt, "Cécile Andrée Paule DeWitt-Morette (1922-)," in Louise S. Grinstein, Rose K. Rose, and Miriam H. Rafailovich, eds., *Women in Chemistry and Physics: A Biobibliographic Sourcebook* (Westport, CT: Greenwood Press, 1993), 150.
48 Róna, "Laboratory Contamination," 723–7.

CHAPTER TWO

1 Pierre and Marie Curie (together) obtain their traditional recognition and sometimes one finds a reference to Lise Meitner and Irène Joliot-Curie, but it is very rare to find mention of any of the other women pioneers. See, for example, Henry A. Boorse, Lloyd Motz, and Jefferson Hane Weaver, *The Atomic Scientists: A Biographical History* (New York: Wiley Science, 1989).
2 Ruth Hubbard, "Facts and Feminism," *Science for the People* 18 (March/April 1986): 16–20, 26.
3 Derek J. de Solla Price, *Little Science, Big Science ... and Beyond* (New York: Columbia University Press, 1986), 238.

4 Deborah J. Warner, "Women Astronomers," *Natural History* 88 (1979): 12–26; Andrea K. Dobson and Katherine Bracher, "A Historical Introduction to Women in Astronomy," *Mercury* 21 (1992): 4–15; Andrew Fraknoi and Ruth Freitag, "Women in Astronomy: A Bibliography," *Mercury* 21 (1992): 46–7; John Lankford and Rickey L. Slavings, "Gender and Science: Women in American Astronomy, 1859–1940," *Physics Today* 43 (March 1990): 58–65.

5 Maureen M. Julian, "Women in Crystallography," in G. Kass-Simon and Patricia Farnes, eds., *Women of Science: Righting the Record* (Bloomington: Indiana University Press, 1990), 335–83.

6 Maureen N. Julian, "Profiles in Chemistry: Rosalind Franklin, From Coal to DNA to Plant Viruses," *Journal of Chemical Education* 60 (1983): 660–2; Anne Sayre, *Rosalind Franklin and DNA* (New York: W.W. Norton, 1975).

7 Maureen N. Julian, "Profiles in Chemistry: Dorothy Crowfoot Hodgkin, Nobel Laureate," *Journal of Chemical Education* 59 (1982): 124–5.

8 Maureen N. Julian, "Profiles in Chemistry: Isabella Kahle and a Mathematical Breakthrough in Crystallography," *Journal of Chemical Education* 63 (1986): 66–7.

9 Maureen N. Julian, "Kathleen Lonsdale and the Planarity of the Benzene Ring," *Journal of Chemical Education* 58 (1981): 365–6; Maureen N. Julian, "Profiles in Chemistry: Kathleen Lonsdale 1903–1971," *Journal of Chemical Education* 59 (1982): 965–6.

10 Robert K. Merton, "The Matthew Effect in Science II," *Isis* 79 (1988): 606–23.

11 Margaret W. Rossiter, "The 'Matthew' Matilda Effect in Science," *Sage* 23 (1993): 325–41.

12 H. Brooks to E. Rutherford, Spring 1903 (n.d.), CUA.

13 Marcia M. Bonta, *Women in the Field: America's Pioneering Women Naturalists* (College Station: Texas A &M Press, 1991), xii-xiii.

14 Ruth Hubbard, "Feminism in Academia: Its Problematic and Problems," in Anne M. Briscoe and Sheila M. Pfafflin, eds., *Expanding the Role of Women in the Sciences* (New York: New York Academy of Sciences, 1979), 251. See also Joan N. Burstyn, "Education and Sex: The Medical Case against Higher Education for Women in England, 1870–1900)," *Proceedings of the American Philosophical Society* 117 (1973): 79–89; Louise M. Newman, ed., *Men's Ideas/Women's Realities: Popular Science, 1870–1915* (New York: Pergamon Press, 1985), 54.

15 Roy MacLeod and Russell Moseley, "Fathers and Daughters: Reflections on Women, Science, and Victorian Cambridge," *History of Education* 8 (1979): 321–33.

16 Lydia E. Becker, "On the Study of Science by Women," *Contemporary Review* 10 (1869): 386–404.
17 Roberta Wein, "Women's Colleges and Domesticity, 1875–1918," *History of Education Quarterly* 14 (1974): 31–48.
18 Jill Conway, "Stereotypes of Femininity in a Theory of Sexual Evolution," in Martha Vicinus, ed., *Suffer and Be Still: Women in the Victorian Age* (Bloomington: Indiana University Press, 1972), 141.
19 W.H. Brock, ed., *H.E. Armstrong and the Teaching of Science 1880–1930* (Cambridge, UK: Cambridge University Press, 1973).
20 Linda M. Fedigan, "The Changing Role of Women in Models of Human Evolution," in Gill Kirkup and Laurie S. Keller, eds., *Inventing Women: Science, Technology, and Gender* (Cambridge, UK: Polity Press, 1992), 103–22.
21 Sue Zschoche, "Dr Clarke Revisited: Science, True Womanhood, and Female Collegiate Education," *History of Education Quarterly* 29 (1989): 545–69.
22 Barbara M. Solomon, *In the Company of Educated Women* (New Haven, CT: Yale University Press, 1985), 63.
23 Sharon B. McGrayne, *Nobel Prize Women in Science* (New York: Birch Lane Press, 1992), 6.
24 Ute Frevert, *Women in German History* (Oxford: Berg, 1989), 124.
25 Claire Goldberg Moses, *French Feminism in the Nineteenth Century* (Albany: State University of New York Press, 1990), 175.
26 Linda Harriet Edmondson, *Feminism in Russia, 1900–17* (Stanford: Stanford University Press, 1984).
27 Judith Fingard, "College, Career, and Community: Dalhousie Coeds, 1881–1921," in Paul Axelrod and John G. Reid, eds., *Youth, University and Canadian Society* (Montreal/Kingston: McGill-Queen's University Press, 1989), 26.
28 M. Cary Thomas, "Present Tendencies in Women's College and University Education," *Educational Review* 35 (1908): 64–85; reproduced in J. Stacey et al., eds., *And Jill Came Tumbling After – Sexism in American Education* (New York: Laurel, 1974), 276.
29 Martha Vicinus, *Independent Women: Work and Community for Single Women 1850–1920* (Chicago: University of Chicago Press, 1985), 138.
30 C.M. Derick, "In the 80's," *Old McGill 1927*, 200.
31 Elizabeth A. Irwin, "Women at McGill," *McGill News* 1 (December 1919): 41.
32 Frevert, *Women in German History*, 121–2.
33 Ibid., 123–4.
34 Margaret W. Rossiter, "'Womens Work' in Science, 1880–1910," *Isis* 71 (1980): 381–98; also in Margaret W. Rossiter, *Women Scientists in America* (Baltimore: Johns Hopkins University Press, 1982), 52.

35 Joyce Antler, "'After College, What?': New Graduates and the Family Claim," *American Quarterly* 32 (Fall 1980): 409–34.
36 Rossiter, *Women Scientists in America*, 53–7.
37 Williamina Fleming, "A Field for 'Women's Work' in Astronomy," *Astronomy and Astrophysics* 12 (1893): 688–9.
38 John Lankford and Rickey L. Slavings, "Gender and Science: Women in American Astronomy, 1859–1940," *Physics Today* 43 (March 1990): 58–65.
39 F.H. Portugal and J.S. Cohen, *A Century of DNA* (Cambridge, MA: MIT Press, 1977), 267.
40 Cited in Lawrence Badash, "Nuclear Physics in Rutherford's Laboratory before the Discovery of the Neutron," *American Journal of Physics* 51 (1983): 884–9.
41 Vivian Gornick, *Women in Science* (New York: Simon & Schuster, 1983), 15.
42 Ann Gibbons, "Key Issue: Mentoring," *Science* 255 (1992): 1368; Clarice M. Yentsch and Carl J. Sindermann, *The Woman Scientist: Meeting the Challenges for a Successful Career* (New York: Plenum, 1992).
43 Vivienne Mayes, "Lee Lorch at Fisk: A Tribute," *American Mathematical Monthly* 88 (1981): 708–11.
44 Julian, "Women in Crystallography," 335–83.
45 Marelene F. Rayner-Canham and Geoffrey W. Rayner-Canham, "Women's Fields of Chemistry: 1900–1920," *Journal of Chemical Education* 73 (1996): 136–8.
46 Mary R.S. Creese, "British Women of the Nineteenth and Early Twentieth Centuries Who Contributed to Research in the Chemical Sciences," *British Journal for the History of Science* 24 (1991): 275–305.
47 Anna Sayre to Maureen Julian, cited in Julian, "Women in Crystallography," 339–40.
48 J. Needham, "Sir F.G. Hopkins' Personal Influence and Character," in J. Needham and E. Baldwin, eds., *Hopkins and Biochemistry* (Cambridge, UK: Cambridge University Press, 1949), 114–15.
49 George B. Kauffman, "The Misogynist Dinner of the American Chemical Society," *Journal of College Science Teaching* 12 (1983): 381–3.
50 Joan Mason, "A Forty Years' War," *Chemistry in Britain* 27 (1991): 233–8.
51 Otto Hahn, *Otto Hahn: A Scientific Autobiography*, trans. and ed. Willy Ley (New York: Charles Scribner's Sons, 1966), 65.
52 Margaret W. Rossiter, "Mendel the Mentor," *Journal of Chemical Education* 71 (1994): 215–19.
53 David Wilson, *Rutherford: Simple Genius* (Cambridge, MA: MIT Press, 1983), 46.
54 P.L. Kapitza, "Recollections of Lord Rutherford," *Royal Society (Great Britain), Proceedings*, series A, 294 (1966): 123–37.

55 Ernest Rutherford, 2d ed., *Radioactivity* (Cambridge, UK: Cambridge University Press, 1905); Ernest Rutherford, *Radioactive Substances and Their Radiations* (Cambridge, UK: Cambridge University Press, 1913).
56 Ellen Gleditsch to Ernest Rutherford, 1 November 1915, CUA.
57 Lise Meitner, "Looking Back," *Bulletin of the Atomic Scientists* 20 (November 1964): 5.
58 Rutherford, *Radioactivity*; and Rutherford, *Radioactive Substances*.
59 Meitner, "Looking Back," 5.
60 Katherine Haramundanis, ed., *Cecelia Payne-Gaposchkin* (Cambridge: Cambridge University Press, 1984), 118.
61 A. Phillips, *A Newnham Anthology* (Cambridge: Cambridge University Press, 1979), 120.
62 See chapter 3, below.
63 Eva Ramstedt to Marie Curie, 14 December 1911, BN.
64 Ernest Rutherford to Bertram Boltwood, 20 November 1911, CUA.
65 Margaret Ashton to Ernest Rutherford, 11 March 1908, CUA.
66 Ernest Rutherford to Kasimir Fajans, 3 December 1923, CUA.
67 Elizabeth Róna, *How It Came About* (Oak Ridge: Oak Ridge Associated Universities, 1978), 4.
68 Text of Speech (n.d.), Soddy Collection, OUA.
69 Otto Hahn to Ernest Rutherford, 19 December 1913, CUA. Hahn made some similar comments to Rutherford in a letter of 12 October 1912.
70 R.L. Sime, "Belated Recognition: Lise Meitner's Role in the Discovery of Fission," *Journal of Radioanalytical and Nuclear Chemistry* 142 (1990): 13–26; Ruth L. Sime, *Lise Meitner. A Life in Physics* (Berkeley: University of California Press, 1996).
71 Hahn, *A Scientific Autobiography*, 51.
72 Lise Meitner to Patricia Alison, 16 November 1954, Meitner Collection, CCA.
73 N.a., *A History of the Cavendish Laboratory 1871–1910* (London: Longmans, 1910), 331.
74 Lord Rayleigh, *The Life of Sir J.J. Thomson* (Cambridge, UK: Cambridge University Press, 1942), 28.
75 Bertram Boltwood to Ernest Rutherford, 12 September 1913, CUA.
76 Mary Rutherford to Bertram Boltwood, 6 October 1913, CUA.
77 We thank Monique Bordry, Archivist, Institut Curie, for permission to view the early files of the institute.
78 Maxine F. Singer, "Heroines and Role Models," *Science* 253 (1991): 249; T.M. Semkow and K.W. Semkow, "In Defense of Marie Curie," *American Journal of Physics* 59 (1991): 871–2.
79 Rosalynd Pflaum, *Grand Obsession: Madame Curie and Her World* (New York: Doubleday, 1991), 82.

80 Eve Curie, *Madame Curie: A Biography*, trans. Vincent Sheean (New York: Da Capo Press, 1986 [1937]); (New York: Doubleday, 1937); Robert Reid, *Marie Curie* (New York: E.P. Dutton, 1974); Françoise Giroud, *Marie Curie – A Life* (New York: Holmes and Meier, 1986); Pflaum, *Grand Obsession;* Susan Quinn, *Marie Curie. A Life* (New York: Simon & Schuster, 1995).
81 Marie Curie, *Pierre Curie*, trans. Charlotte and Vernon Kellogg (New York: Macmillan, 1963 [1923]), 197.
82 Pflaum, *Grand Obsession*, 134.
83 Ernest Rutherford to Bertram Boltwood, 14 December 1918, CUA.
84 Personal communication, Wolfgang L. Reiter, 8 December 1991. See also "Appendix: Verzeichnis der Mitarbeiter (1910–1950)," *Festschrift des Institutes für Radiumforschung: Anlässlich seines 40Jährigen Bestandes (1910–1950)*, in *Sitzungberichte Akademie der Wissenschaften Wein Math-naturwissenschaften*, abt. 2A, 159 (1950).
85 M. Blau, "Bericht über die Entdeckung der durch kosmische Strahlung erzeugten 'Sterne' in photographischen Emulsionen," *Festschrift des Institutes für Radiumforschung Anlässlich seines 40Jährigen bestandes (1910–1950)*, in *Sitzungberichte Akademie der Wissenschaften Wein Math-naturwissenschaften*, abt. 2A, 159 (1950): 53–7.
86 F. A. Paneth, "Prof. Stefan Meyer," *Nature* 165 (1950): 548–9.
87 R.W. Lawton, "Prof. Stefan Meyer," *Nature* 165 (1950): 549.
88 Patsy A. McLaughlin and Sandra Gilchrist, "Women's Contributions to Carcinology," in Frank Truesdale, ed., *History of Carcinology* (Rotterdam: A.A. Balkema, 1993), 204.
89 Barbara Lotze, ed., *Making Contributions: An Historical Overview of Women's Role in Physics* (College Park, MD: American Association of Physics Teachers, 1984), 22.
90 Eva Ramstedt to Marie Curie, 11 August 1925, BN.
91 Jessie Barnard, *Academic Women* (University Park: Pennsylvania State University Press, 1964), 207. The "marriage question" is also discussed by Penina Glazer and Miriam Slater, *Unequal Colleagues: The Entrance of Women into the Professions, 1890–1940* (New Brunswick: Rutgers University Press, 1986), 57–64.
92 de Solla Price, *Little Science, Big Science*, 119–34; Diane Crane, *Invisible Colleges* (Chicago: University of Chicago Press, 1972).
93 The Meitner Collection, CCA, contains considerable correspondence between these individuals.
94 Eva Ramstedt to Lise Meitner, 17 June 1922, CCA.
95 Berta Karlik to Lise Meitner, 31 May 1953, CCA.
96 Eva Ramstedt to Lise Meitner, 17 June 1920, CCA.
97 Ellen Gleditsch to Lise Meitner, 16 June 1926, CCA.

98 Stefan Meyer to Ernest Rutherford, 22 May 1921, CUA; Róna, *How It Came About*, 37.
99 Eva Ramstedt to Marie Curie, 11 August 1925, BN.
100 Elizabeth Róna to Kasimir Fajans, 26 September 1965, FA.
101 Berta Karlik and Elizabeth Kara-Michailova, "Zur Kenntnis der Szintillationsmethode," *Zeitschrift für Physik* 48 (1928): 765–83.
102 E. Föyn, E. Kara-Michailova, and E. Róna, "Zur Frage der künstlichen Umwandlung des Thoriums durch Neutronen," *Naturwissenschaften* 23 (1935): 391–2.
103 Marietta Blau and Elizabeth Kara-Michailova, "Über die durchdringende Strahlung des Poloniums," *Sitzungberichte Akademie der Wissenschaften Wein Math-naturwissenschaften*, abt. 2A, 140 (1931): 615–22.
104 Marietta Blau and Elizabeth Róna, "Ionization durch H-Strahlen," *Sitzungberichte Akademie der Wissenschaften Wein Math-naturwissenschaften*, abt. 2A, 135 (1926): 573–85.
105 Ernst Föyn, Berta Karlik, Hans Pettersson, and Elizabeth Róna, "Radioactivity of Sea Water," *Nature* 143 (1939): 275–6.
106 Ellen Gleditsch and Elizabeth Róna, "Beitrag zur Untersuchung des Ionen-austausches zwischen Salzen und gesättigten Lösungen," *Chemisches Zentralblatt* 112 (1941): 1117; also *Archiv for Mathematik og Naturvidenskab* 44 (1941): 53–62.
107 Marietta Blau, "Photographic Tracks from Cosmic Rays," *Nature* 142 (1938): 613.
108 Fanny Cook Gates to Ernest Rutherford, 9 April 1904, CUA.
109 Lawrence Badash, ed., *Rutherford and Boltwood, Letters on Radioactivity* (New Haven, CT: Yale University Press, 1969), 294.
110 Jadwiga Szmidt to Ernest Rutherford, 29 December 1915, CUA.
111 Ellen Gleditsch to Ernest Rutherford, 1 November 1915, CUA.
112 May Sybil Leslie to Frederick Soddy, 12 November 1922, OUA.
113 M. Curie to E. Gleditsch, 24 December 1924, University of Oslo Archives.
114 L. Blanquies, "Comparaison entre les rayons a produits par differentes substances radioactives," *Comptes rendus* 148 (1909): 1753–6.
115 J.M.W. Slater, "On the Excited Activity of Thorium," *Philosophical Magazine*, series 6, vol. 9 (1905): 628–44.
116 May Sybil Leslie, "Le thorium et ses produits de désagrégation," *Le Radium* 8 (1911): 356–63.
117 Frederick Soddy and Ruth Pirret, "The Ratio between Uranium and Radium in Minerals," *Philosophical Magazine*, series 6, vol. 20 (1910): 345–9; Ruth Pirret and Frederick Soddy, "The Ratio between Uranium and Radium in Minerals II," *Philosophical Magazine*, series 6, vol. 21 (1911): 652–8.
118 Ellen Gleditsch, "Sur le radium et l'uranium contenus dans les mineraux radioactifs," *Le Radium* 6 (1909): 165–6.

119 Marietta Blau and Elizabeth Róna, "Anwendung der Chamié'schen photographischen Methode zur Prüfung des chemischen Verhaltens von Polonium," *Sitzungsberichte Akademie der Wissenschaften Wein Math-naturwissenschaften*, abt. 2A, 139 (1930): 275–9.

CHAPTER THREE

1 For consistency, all future references will be to Marie Curie, although her original name was Maria Sklodowska. There are two recommended biographies of Curie: Eve Curie (trans. Vincent Sheean), *Madame Curie* (Garden City, N.Y.: Doubleday, 1940); and Robert Reid, *Marie Curie: A Biography* (New York: E.P. Dutton, 1974). An autobiography appears in Marie Curie (trans. Charlotte and Vernon Kellogg), *Pierre Curie*, (New York: Macmillan, reprint 1963). The present essay relies on these works for the details and general themes of Curie's life. But a particular biography is cited only where details or themes peculiar to it are presented, or where it is directly quoted.
2 Marie Curie, *Pierre Curie*, 78.
3 Ibid., 79.
4 Quoted in Eve Curie, *Madame Curie*, 36.
5 Marie Curie, *Pierre Curie*, 78.
6 Ibid., 83. There are some striking similarities between the Sklodowska sisters' close relationship and that of Sofia Kovalevskaia and her sister, Aniuta, not the least of which is the strong influence of progressive ideologies on the two pairs of sisters. On Kovalevskaia's early relationship with Aniuta and their interest in Russian nihilism, see Ann Hibner Koblitz, *A Convergence of Lives: Sofia Kovalevskaia, Scientist, Writer, Revolutionary* (Cambridge, Mass.: Birkhäuser, 1983), esp. 31–79.
7 Marie Sklodowska to Kazia Przyborovska, 25 October 1888, in Eve Curie, *Madame Curie*, 78.
8 Marie Sklodowska to Henrietta Michalovska, December 1886, in Eve Curie, *Madame Curie*, 72.
9 Ibid.
10 Bronia Sklodowska to Marie Curie, March 1890, in Eve Curie, *Madame Curie*, 83–84.
11 Marie Sklodowska to Bronia Sklodowska, 12 March 1890, in Eve Curie, *Madame Curie*, 84.
12 Marie Curie, *Pierre Curie*, 83.
13 Marie Sklodowska to Joseph Sklodowski, 15 September 1893, in Eve Curie, *Madame Curie*, 115.
14 For details on Pierre Curie's life and work through 1894, see Marie Curie, *Pierre Curie*, 11–33; and Jean Wyart, "Pierre Curie," in Charles C. Gillispie (editor) *Dictionary of Scientific Biography* (New York: Scribner, 1970–80) 3: 503–508.

15 This is Pierre Curie's description of the collaboration, quoted from one of his letters to Marie Sklodowska shortly before their marriage. See Marie Curie, *Pierre Curie,* 23.
16 Eve Curie, *Madame Curie,* 120.
17 The terms "anti-natural path" and "anti-natural thought" were Pierre Curie's. As he explained, they referred to a life so dominated by science that, although he was "obliged to eat, drink, sleep, laze, love," he would not "succumb" to these distractions. See Eve Curie, *Madame Curie,* 120, 126.
18 The only indication that Pierre shared some domestic responsibilities is Cunningham's claim that visitors to the young couple's apartment "found Pierre sweeping the floor and his wife cooking the meals." See Marion Cunningham, *Madame Curie (Sklodowska) and the Story of Radium* (London: n.p., n.d.), 32.
19 Marie Curie, *Pierre Curie,* 86.
20 Eve Curie, *Madame Curie,* 207.
21 Marie Curie to Joseph Sklodowski, 23 November 1895, in Eve Curie, *Madame Curie,* 145–146.
22 Ibid., 144–145.
23 Marie Curie taught her laboratory students "that a research bench should always be neat and tidy, that 'a good researcher' did not wait until the end of an experiment to clear up, but did so as he went along." See Maurice Goldsmith, *Frédéric Joliot-Curie: A Biography* (London: Lawrence and Wishart, 1976), 35.
24 Marie Curie, *Pierre Curie,* 88.
25 Ibid. Marie Curie, however, seems to have spent more than her usual amount of time at home immediately before and after the birth of her second daughter, Eve. See Eve Curie, *Madame Curie,* 225–227.
26 Lawrence Badash, "Radioactivity before the Curies," *American Journal of Physics* 33 (1965): 128.
27 G.C. Schmidt independently made the same discovery in 1898. See Lawrence Badash, "The Discovery of Thorium's Radioactivity," *Journal of Chemical Education* 43 (1966): 219–220.
28 Marie Curie's 1903 thesis contains a lucid account of her early work on radioactivity. See Marie Curie, *Radioactive Substances* (Westport, Conn.: Greenwood Press, 1971). This edition is a reprint of the translated version of the thesis that appeared in *Chemical News* of 1903.
29 For Marie Curie's own account of the extraction of radium salt, see Marie Curie, *Radioactive Substances,* 19–31.
30 James Christie, "The Discovery of Radium," *Journal of the Franklin Institute* 167 (May 1909): 361.
31 Marie Curie, "Note," *Comptes rendus* 126 (1898): 1101.
32 See, for example, Marie Curie, *Radioactive Substances,* 93–94. Also, when Gösta Mittag-Leffler (who had been so supportive of Sofia Kovalevskaia)

informed Pierre of his consideration for the Nobel Prize of 1903, he responded with a private latter emphasizing Marie's part in the couple's research on radioactivity and thus the appropriateness of her sharing the prize. See Elisabeth Crawford, *The Beginnings of the Nobel Institution: The Science Prizes, 1901–1915* (Cambridge: Cambridge University Press, 1984), 141.

33 Marie Curie, *Pierre Curie*, 94.

34 Ibid., 95.

35 Ibid.

36 André Debierne (1874–1949), a student of Pierre Curie, was an early collaborator and close friend of both Curies and, after 1906, of Marie Curie. He was one of the select group admitted to Curie's home, where he apparently spent many Sundays trying to entertain the Curie girls; he also saw Curie off on her trips abroad. Upon Curie's death, Debierne succeeded her as director of the Laboratoire Curie. For details of the Curie-Debierne friendship, see Eve Curie, *Madame Curie*, 269, 280, and Reid, *Marie Curie*, 278–279.

37 For a basic discussion of this work, see Reid, *Marie Curie*, 137–140.

38 David Wilson, *Rutherford: Simple Genius* (Cambridge, Mass.: MIT Press, 1983), 253.

39 Marie Curie to Ernest Rutherford, September 1910, Add. Ms. 7653 Ernest Rutherford Papers, Syndics of Cambridge University Library (CUA).

40 Reid, *Marie Curie*, 144.

41 Wilson, *Rutherford*, 253.

42 Besides Gleditsch (and Curie's famous daughter Irène), Curie's women students included May Leslie and Marguerite Perey. The Norwegian Gleditsch, who began a fruitful career devoted to radioactivity in Curie's laboratory in 1907, eventually returned to her native country as a professor of chemistry at the University of Oslo. Leslie worked with Curie during 1910–11, after which time Curie arranged for her admission to Rutherford's laboratory from which she published a long paper on the coefficients of diffusion of thorium and actinium emanations. Marguerite Perey (1909–1975), who served as an assistant at the Laboratoire Curie beginning in 1929, eventually became a professor of nuclear chemistry at the University of Strasbourg and, like her mentor, discovered a new element, francium. See Chapter 8, Ellen Gleditsch: Professor and Humanist.

43 Bertram Boltwood to Ernest Rutherford, 12 September [1913], Bertram Boltwood Papers, Yale University Library; Mary Rutherford to Bertram Boltwood, 6 October [1913], CUA. The cited letters are also found in Lawrence Badash (editor), *Rutherford and Boltwood: Letters on Radioactivity* (New Haven, Conn.: Yale University Press, 1969), 285–287.

44 Of non-French scientists, Rutherford – a major pioneer of radioactivity along with the Curies – seems to have been Marie Curie's closest and

most supportive peer. For discussions of the Rutherford-Curie relationship, see Reid, *Marie Curie,* 141–145, and Wilson, *Rutherford,* 254–267. Both take very positive views of Rutherford's treatment of Curie.

45 A.S. Eve, *Rutherford: Being the Life and Letters of the Rt. Hon. Lord Rutherford, O.M.* (New York: Macmillan, 1939), 213.

46 Lise Meitner, "Looking Back," *Bulletin of the Atomic Scientists* 20 (November 1964): 5.

47 Otto Hahn (trans. by Ernst Kaiser and Eithne Wilkins), *Otto Hahn: My Life. The Autobiography of a Scientist* (New York: Herder and Herder, 1970), 89.

48 E. Rutherford to B. Boltwood, 11 January 1909, in Badash, *Rutherford and Boltwood,* 206.

49 Ernest Rutherford, "Radium Standards and Nomenclature," *Nature* 84 (1910): 430.

50 E. Rutherford to B. Boltwood, 20 November 1911, in Badash, *Rutherford and Boltwood,* 258.

51 E. Rutherford to B. Boltwood, 22 April 1912, in Badash, *Rutherford and Boltwood,* 270.

52 Eve Curie, *Madame Curie,* 266.

53 Marie Curie to Ernest Rutherford, 18 April 1919, Ernest Rutherford Papers, Syndics of Cambridge University Library.

54 Adrienne R. Weill-Brunschvicg, "Paul Langevin," in Charles C. Gillispie (editor) *Dictionary of Scientific Biography* (New York: Scribner, 1970–80) 8: 9–14.

55 For a reconstruction of the Curie-Langevin relationship and details of the scandal, see Reid, *Marie Curie,* 162–182.

56 Henry Smith Williams, "The Case of Madame Curie," *World To-Day* 21 (1912): 1632–1635. Williams was a distinguished American physician who specialized in mental diseases and pioneered nonspecific protein therapy. See "Dr. H.S. Williams Dies on the Coast," *New York Times* (5 July 1943): 15.

57 "Editors in Duel over Mme. Curie," *New York Times* (24 November 1911): 3.

58 For some details of the address, see Reid, *Marie Curie,* 184.

59 Drawing these two women together were their shared experiences of rejection by major scientific societies. Whereas the Royal Society refused Ayrton admission on the grounds that she was married, in early 1911 the French Academy of Sciences had turned down Curie's bid for membership primarily because she was a woman.

60 E. Rutherford to B. Boltwood, 20 November 1911, in Badash, *Rutherford and Boltwood,* 258–259.

61 Reid, *Marie Curie,* 167; Eve, *Rutherford,* 224.

62 Marie Curie to Ernest Rutherford, 25 February 1919, CUA.

63 See, for example, Marie Meloney, "The 'New Woman,'" *Delineator* (November 1922): 1; and Marie Meloney to Charles Eliot, 9 December

1920, Marie M. Meloney Papers, Rare Book and Manuscript Library, Columbia University.

64 For an insightful discussion of the effects of Curie's visit and the "Madame Curie strategy" on American women scientists, see Margaret W. Rossiter, *Women Scientists in America: Struggles and Strategies to 1940* (Baltimore: Johns Hopkins University Press, 1982), esp. 122–128, 156–159.

65 Charles Eliot to Marie Meloney, 18 December 1920, Meloney Papers.

66 M. Meloney to Charles Eliot, 24 December 1920, Meloney Papers.

67 Charles Eliot, "The Woman That Will Survive," *Delineator* (August 1914): 5.

68 "Mme. Curie in Boston," *New York Times* (19 June 1921), sec. 2, 1.

69 B. Boltwood to E. Rutherford, 14 July 1921, in Badash, *Rutherford and Boltwood,* 346.

70 Ibid.

71 Eve Curie, *Madame Curie,* 337.

72 Marie Curie to Bronia Dluska [née Sklodowska], 12 April 1932, in Eve Curie, *Madame Curie,* 358.

CHAPTER FOUR

1 Thorleiv Kronen and Alexis C. Pappas, *Ellen Gleditsch. Et liv i forskning og medmenneskelighet* (Oslo: Aventura Forlag, 1987), 133. Much of the material in this chapter has been taken from the book, and we are indebted to the authors for their kind permission to extract the information. We are also grateful to Marianne Ainley and Tina Crossfield for helpful advice and suggestions.

2 Kronen and Pappas, *Ellen Gleditsch,* 14.

3 In 1902 the Institute of Pharmacology was still a separate institution and only became part of the Faculty of Science and Mathematics at the University of Oslo in 1931.

4 The name of Norway's capital, Kristiania, was changed back to Oslo in 1925. For the purpose of clarity, Oslo has been used throughout this chapter.

5 Interview with E. Gleditsch in *Urd* (Oslo), 14 January 1911, in Kronen and Pappas, *Ellen Gleditsch,* 18.

6 E. Gleditsch, "Sur quelques derivés d'amylbenzene tertiaire," *Bulletin de la société chimique de France* 35 (1907): 1094–7.

7 E. Bodtker to M. Curie, 23 June 1907, in Kronen and Pappas, *Ellen Gleditsch,* 208.

8 Kronen and Pappas, *Ellen Gleditsch,* 34.

9 E. Bodtker to E. Gleditsch, 9 July 1907, in Kronen and Pappas, *Ellen Gleditsch,* 209.

10 M. Curie to E. Gleditsch, 26 July 1907, OUL.

11 E. Gleditsch, "Marie Sklodowska Curie," *Nordisk Tidsskrift*, 1959: 417, in Kronen and Pappas, *Ellen Gleditsch*, 49.
12 Interview with E. Gleditsch in *Urd*, (Oslo), 14 January 1911, in ibid., 35.
13 Kronen and Pappas, *Ellen Gleditsch*, 37.
14 M. Curie and E. Gleditsch, "Action de l'émanation du radium sur les solutions des sels de cuivre," *Comptes rendus* 147 (1908): 345–9.
15 Ellen Gleditsch, "Sur le radium et l'uranium contenus dans les minéraux radioactifs," *Le Radium* 6 (1909): 165–6; Ellen Gleditsch, "Sur de rapport entre l'uranium et le radium dans les minéraux radioactifs," *Le Radium* 8 (1911): 256–73.
16 Interview with E. Gleditsch in *Dagbladet*, 13 June 1964, in Kronen and Pappas, *Ellen Gleditsch*, 36.
17 E. Gleditsch's speech at fourth IFUW congress, Amsterdam, 1926, in ibid., 112.
18 E. Gleditsch, "Marie Sklodowska Curie," *Tidsskrift for Kemi*, (1912).
19 B. Boltwood to E. Gleditsch, 11 September 1913, OUL.
20 B. Boltwood to E. Rutherford, 12 September 1913, in B. Boltwood to E. Rutherford, 12 September 1913, in Lawrence Badash, ed., *Rutherford and Boltwood: Letters on Radioactivity* (New Haven, CT: Yale University Press, 1969), 285.
21 Lawrence Badash, *Radioactivity in America: Growth and Decay of a Science* (Baltimore: Johns Hopkins University Press, 1979), 153–8.
22 It is possible that Gleditsch was nominated for the degree by Gladys Anslow of Smith College, as Anslow had an interest in nuclear physics. See George Fleck, "Gladys Amelia Anslow (1892–1969)," in Louise S. Grinstein et al., eds., *Women in Chemistry and Physics: A Bio-bibliographic Sourcebook* (Westport, CT: Greenwood Press, 1993), 9–17.
23 B. Boltwood to E. Gleditsch, 2 July 1914, OUL.
24 E. Gleditsch to B. Boltwood, 1 November 1914, in Kronen and Pappas, *Ellen Gleditsch*, 225.
25 Interview with Ellen Gleditsch in *Adresseavisen* (Trondheim), 10 May 1930, in Kronen and Pappas, *Ellen Gleditsch*, 63.
26 Suffrage for women in Norway was introduced in 1913.
27 Interview with E. Gleditsch in *New York Press*, 26 October 1913, in Kronen and Pappas, *Ellen Gleditsch*, 64.
28 E. Gleditsch to B. Boltwood, 17 July 1914, in Kronen and Pappas, *Ellen Gleditsch*, 223. In the same letter Gleditsch mentions her meeting with Jadwiga Szmidt; see chapter 13.
29 E. Gleditsch to B. Boltwood, 14 March 1915, in Kronen and Pappas, *Ellen Gleditsch*, 228.
30 Ellen Gleditsch, "The Life of Radium," *American Journal of Science* 41 (1916): 112–24.
31 B. Boltwood to E. Gleditsch, 23 November 1915, OUL.

32 B. Boltwood to E. Gleditsch, 28 June 1915, OUL.
33 B. Boltwood to E. Rutherford, 18 January 1916, in Badash, *Rutherford and Boltwood*, 316.
34 T.W. Richards and C. Wadsworth, "Further Studies of the Atomic Weight of Lead of Radioactive Origin," *Journal of the American Chemical Society* 38 (1916): 2613–22.
35 E. Gleditsch to E. Rutherford, 1 November 1915, CUA.
36 E. Gleditsch to M. Curie, 10 November 1915, in Kronen and Pappas, *Ellen Gleditsch*, 236.
37 *Kvinnelige studenter 1882–1932* (1932): 245, in ibid., *Ellen Gleditsch*, 68.
38 M. Curie to E. Gleditsch, 22 June 1916, OUL.
39 E. Rutherford to E. Gleditsch, 15 November 1916, OUL.
40 Thorstein Hiortdahl, *Kortfattet Laerebok i anorganisk Kjemi*, 6th ed. (Oslo: Cammermeyer, 1917).
41 E. Gleditsch to B. Boltwood, 1 March 1916, in Kronen and Pappas, *Ellen Gleditsch*, 240.
42 Th. Hiortdahl to E. Gleditsch, 6 January 1917, in ibid., 244.
43 Ellen Gleditsch and Eva Ramstedt, *Radium og de radioaktive processer* (Kristiania: Aschehoug Forlag, 1917). The preface and acknowledgement ends with the words, "In closing, we want to express our deeply felt gratitude to our common teacher, Mme Curie, in whose laboratory we both received our grounding in the studies of radiochemistry."
44 M. Curie to E. Gleditsch, 27 January 1918, OUL.
45 E. Rutherford to E. Gleditsch, 14 June, 1919, OUL.
46 F. Soddy to E. Gleditsch, 6 May 1920, OUL. Rutherford's letter to Gleditsch concerning the visit of the trio is also in the archives: E. Rutherford to E. Gleditsch, 14 June 1914, OUL.
47 Postcard from E. Gleditsch, S. Leslie, and E. Ramstedt to M. Curie, 12 June 1920, in Kronen and Pappas, *Ellen Gleditsch*, 254.
48 F. Soddy to E. Gleditsch, 17 November 1922, OUL.
49 M. Curie to E. Gleditsch, 7 July 1920, OUL.
50 Ellen Gledisch, "L'âge des minéraux d'après la théorie de la radioactivité," *Bulletin de la Société chimique de France* 31 (1922): 351–72.
51 E. Gleditsch to M. Curie, 20 November 1922, in Kronen and Pappas, *Ellen Gleditsch*, 260.
52 Kronen and Pappas, *Ellen Gleditsch*, 86.
53 M. Curie to E. Gleditsch, 7 August 1921, OUL; M. Curie to E. Gleditsch, 23 August 1921, OUL.
54 E. Gleditsch to M. Curie, 18 August 1921, in Kronen and Pappas, *Ellen Gleditsch*, 255.
55 E. Gleditsch to M. Curie, 28 February 1922, in ibid., 258.
56 E. Gleditsch to M. Curie, 20 November 1922, in ibid., 260.; M. Curie to E. Gleditsch, 29 November 1922, in ibid., 261.
57 E. Gleditsch to M. Curie, 15 December 1924, BN.

58 M. Curie to E. Gleditsch, 24 December 1924, OUL.
59 M. Curie to E. Gleditsch, 5 May 1926, OUL.
60 E. Gleditsch to M. Curie, 26 May 1926, BN.
61 S. Dedichen to M. Curie, 1 October 1926, in Kronen and Pappas, *Ellen Gleditsch*, 278.
62 M. Curie to S. Dedichen, 7 October 1926, in ibid., 278.
63 Ellen Gleditsch and Liv Gleditsch, "Contribution a l'étude des isotopes. Sur le poids atomique du chlore dans les sels de potasse d'Alsace," *Journal de chimie physique* 24 (1927): 238–44; Ellen Gleditsch and Liv Gleditsch, "Le conductivité électrique des solutions aqueuses de radon," *Journal de chimie physique* 25 (1928): 290–3.
64 E. Gleditsch, *Radioaktivitet og grundstofforvandling* (Oslo: O. Norlis Forlag, 1924). It was also published in English: Ellen Gleditsch, *Contribution to the Study of Isotopes* (Oslo: J. Dybwad, 1925).
65 E. Gleditsch speech at IFUW meeting in Madrid, in Kronen and Pappas, *Ellen Gleditsch*, 113.
66 Interview with E. Gleditsch in *Dagbladet* (Oslo), 7 October 1929, in ibid., 119.
67 "College Women Greets a Leader," *New York Times*, 17 March 1929.
68 M. Curie to K. Bonnevie, 11 May 1929, in Kronen and Pappas, *Ellen Gleditsch*, 283.
69 K. Bonnevie to M. Curie, 25 May 1929, in ibid., 284.
70 "Blir lille Ellen professor?" *Fikenbladet* (Oslo), 17 May 1929, in ibid., 130.
71 E. Gleditsch to M. Curie, 26 June 1929, in ibid., 285.
72 "Den nye professor i kjemi. Froken Ellen Gleditsch utnevnt i gaar," *Tidens Tegn* (Oslo), 22 June 1929, in ibid., 133.
73 Kronen and Pappas, *Ellen Gleditsch*, 309.
74 Ibid., 157.
75 E. Gleditsch speech to IFUW congress, Geneva, 8 August 1929, in ibid., 115.
76 Bergljot Qviller Werenskiold is the co-author with Gleditsch of the publication *Investigation of Uranothorites from the Arendal District, Norway* (1932); see Kronen and Pappas, *Ellen Gleditsch*, 309.
77 Bergljot Q. Werenskiold, personal interview, 1984, in Kronen and Pappas, *Ellen Gleditsch*, 141.
78 Kronen and Pappas, *Ellen Gleditsch*, 68.
79 Ellen Gleditsch and Elizabeth Róna, "Contribution to the Study of the Exchange of Ions between Salts and their Saturated Solutions," *Archiv for Mathematik og Naturvidenskab* 44 (1941): 53–62.
80 Henning Sinding-Larsen, "UNESCO. Ellen Gleditsch foreslar revisjon av laereboker," *Aftenposten*, 23 November 1946.
81 E. Gleditsch to Irène Joliot-Curie, 26 February 1947, in Kronen and Pappas, *Ellen Gleditsch*, 295.
82 Frédéric Joliot-Curie to E. Gleditsch, 26 August 1947 and 24 September 1947, in ibid., 296.

83 Ellen Gleditsch and Aamund Salvesen, "Sur les échanges de ions dans un sel et sa solution saturée," *Bulletin de la société chimique de France* (1952): 523–7.
84 Kronen and Pappas, *Ellen Gleditsch,* 311.
85 Ibid., 190 and 307–11.
86 A. Rosenquist, "Ellen Gleditsch in Memoriam," July 1968, International Federation of University Women, press release.
87 E. Gleditsch to M. Curie, 20 November 1922, in Kronen and Pappas, *Ellen Gleditsch,* 260.
88 Abraham Pais, *Inward Bound* (New York: Oxford University Press, 1986), 56.
89 M.F. Rayner-Canham and G.W. Rayner-Canham, "Pioneer Women in Nuclear Science," *American Journal of Physics* 58 (1990): 1036–43.
90 The authors of this biography both grew up in Scandinavia. One of the co-authors studied chemistry during the sixties for an undergraduate and graduate degree at Uppsala University, Sweden, where there were no women professors in any branch of chemistry. As a young woman entering the field of experimental science, it would have been tremendously encouraging and inspiring to know that a woman had been a professor of chemistry for sixteen years at the nearby University of Oslo. However, not once was Ellen Gleditsch's name heard.
    The other co-author went through the school system in Norway up to "examen artium" (matriculation exam), at a time when Gleditsch was still very much at the height of her activity. Yet she was mentioned neither in the science, humanities, nor pre-med courses. At the same time there was no lack of information about Marie Curie and radium, a subject covered liberally in textbooks and assignments. It was easy to get the impression that Curie was the only famous woman scientist at the time. How interesting it would have been to have learnt about the "Norwegian Marie Curie."
91 Marianne Gosztonyi Ainley has used the term "invisible" to describe the life and careers of women scientists whose contributions to science for various reasons have been ignored by the scientific community. M.G. Ainley, "Introduction," in M.G. Ainley, ed., *Despite the Odds. Essays on Canadian Women and Science* (Montreal: Véhicule Press, 1990), 20.
92 Kronen and Pappas, *Ellen Gleditsch,* 82.
93 Odd Holaas, ed., *Norge under Haakon VII, 1905–57* (Oslo: J.W. Cappelens Forlag, 1957).

CHAPTER FIVE

1 Biographical details were provided by P.S. Morrish, Sub-librarian, MSS. and Special Collections, Leeds University Library; F. Boyle,

Senior Information Scientist, Royal Society of Chemistry; and by Anne Barrett, College Archivist, Imperial College, London.

2 H.M. Dawson and M.S. Leslie, "Dynamics of the Reaction between Iodine and Acetone," *Transactions of the Chemical Society* 95 (1909): 1860–70.

3 John Sichel, personal communication, 31 August 1993; F.S. Dainton, "Reputable Memories," *Chemistry in Britain* 29 (1993): 573.

4 A. Smithalls to M. Curie, 22 April 1909, AIC. We thank Monique Bordry for access to these records.

5 M.S. Leslie to M. Curie, 14 June 1909, AIC.

6 Gaston Dupuy, "Notice sur la vie et les travaux de André Debierne (1874–1949)," *Bulletin de la Société chimique de France* (1950): 1024–6.

7 M.S. Leslie to A. Smithalls, 30 November 1909, Smithalls Collection, UL.

8 Ibid.

9 M.S. Leslie to A. Smithalls, 10 June 1910, UL.

10 M.S. Leslie to A. Smithalls, 30 November 1909, UL.

11 M.S. Leslie to A. Smithalls, 8 June 1911, UL.

12 M.S. Leslie, "Sur le poids moléculaire de l'emanation du thorium," *Comptes rendus* 153 (1911): 328–30; M.S. Leslie, "Le thorium et ses produits de désagrégation," *Le Radium* 8 (1911): 356–63; M.S. Leslie, "Sur la périod du radiothorium et le nombre des particules a données par le thorium et ses produits," *Le Radium* 9 (1912): 276–7.

13 M.S. Leslie to A. Smithalls, 7 July 1910 and 8 June 1911, UL.

14 M.S. Leslie to A. Smithalls, 8 June 1911, UL.

15 M. Curie to E. Rutherford, 24 May 1911, CUA.

16 M.S. Leslie, "A Comparison of the Coefficients of Diffusion of Thorium and Actinium Emanations, with a Note on their Periods of Transformation," *London, Edinburgh and Dublin Philosophical Magazine* 24 (1912): 637–47.

17 E. Rutherford to B. Boltwood, 18 March 1912, CUA.

18 H.M. Dawson and M.S. Leslie, "Ionization in Non-aqueous Solvents," *Transactions of the Chemical Society* 99 (1913): 1601–9.

19 J.A. Hall, A. Jaques, and M.S. Leslie, "Nitric Acid Absorption Towers," *Journal of the Society of Chemical Industry, Review* 41 (1922): 285–93.

20 Edward S.F. Rogans, personal communication, 31 August 1993.

21 Edward S.F. Rogans, "Reputable Memories," *Chemistry in Britain* 29 (1993): 573.

22 These unpublished reports were entitled "Experiments in the Design of Absorption Towers for Nitrous Fumes" (October 1916) and "Experiments in the Design of Absorption Towers for Nitrous Fumes: Supplementary Report" (8 November 1916), Archives Section, Edward Boyle

Library, University of Leeds. We thank Peter Towse, University of Leeds, for this information.

23 We thank J. Mary D. Forster, University Archive, University of Leeds, for a copy of this report.

24 See, for example, *Home Office Report: Substitution of Women in Non-munitions Factories during the War* (London, UK: His Majesty's Stationery Office, 1919).

25 Charles S. Whewell, personal communication, 27 March 1993.

26 "Obituary: Arthur Hamilton Burr," *Institute of Chemistry of Great Britain and Ireland, Journal and Proceedings* (1934): 68.

27 M.S. Burr, "A Preliminary Note on Solvate Formation," *Proceedings of the Leeds Philosophical and Literary Society* 1 (1926): 74–80.

28 M.S. Burr (née Leslie), *The Alkaline Earth Metals*, vol. 3, part 1 of J. Newton Friend, ed., *A Textbook of Inorganic Chemistry* (London: Charles Griffin, 1925).

29 J.C. Gregory and M.S. Burr (née Leslie), *Beryllium and Its Congeners*, vol. 3, part 2 of J. Newton Friend, ed., *A Textbook of Inorganic Chemistry* (London: Charles Griffin, 1926).

30 Postcard from E. Gleditsch, S. Leslie, and E. Ramstedt to M. Curie, 12 June 1920, cited in Thorleiv Kronen and Alexis C. Pappas, *Ellen Gleditsch. Et liv i forskning og medmenneskelighet* (Oslo: Aventura Forlag, 1987), 254.

31 F. Soddy to E. Gleditsch, 12 June 1920, OUL.

32 E. Rutherford to E. Gleditsch, 14 June 1920, OUL.

33 E. Gleditsch, E. Ramstedt and M.S. Burr (Leslie) to M. Curie, 12 July 1932, Curie Archives, BN.

34 May Sybil Leslie to Frederick Soddy (several items), Soddy Archives, Oxford University.

35 A.H. Burr (deceased) and M.S. Burr, "The Nature of the Dyeing Process in Cellulose Acetate Rayon and the Hydrolytic Constant of Chrysodine G," *Journal of the Society of Dyers Colourists* 50 (1934): 42–7.

36 M.S. Burr and H.M. Dawson, "Analysis of the Complex Group of Reactions Involved in the Final Stages of the Idealised Hydrolysis of Aqueous Solutions of Sodium Bromo-acetate," *Proceedings of the Leeds Philosophical and Literary Society* 3 (1937): 293–9.

37 Margaret W. Rossiter, *Women Scientists in America: Struggles and Strategies to 1940* (Baltimore: Johns Hopkins University Press, 1982), 71.

38 We thank Frances Boyle, Royal Society of Chemistry Information Services, for this information.

39 *Yorkshire Post*, 6 July 1937, 5.

40 H.M. Dawson, "May Sybil Burr. 1887–1937," *Journal of the Chemical Society* (1938): 151–2.

CHAPTER SIX

1 The early biographical information is taken from a curriculum vitae in the AIC. We thank Monique Bordry, Chargée des Archives, Institut Curie, for access to the files.
2 Ann Hibner Koblitz, *A Convergence of Lives: Sofia Kovalevskaia, Scientist, Writer, Revolutionary* (Cambridge, MA: Birkhäuser, 1983).
3 C. Steinberg, "Yulya Vsevolodovna Lermontova (1846–1919)," *Journal of Chemical Education* 60 (1983): 757–8.
4 Her thesis work was published as: A. Schidlof and Mlle Catherine Chamié, "Influence de la rapidité des variations du champ magnétisant sur l'hystérésis alternative," *Archives des sciences physiques et naturelles* 36 (1913): 13–40.
5 C. Chamié to Marie Curie, 3 January 1921, AIC.
6 Reference for C. Chamié, from the Faculté des Sciences, Université de Genève, 12 July 1921, AIC.
7 Reference for C. Chamié, from the Groupe Académique Russe, Paris, 18 July 1921, AIC.
8 Reference for C. Chamié, from the Société d'Oevres d'Enseignement pour les Réfugiés de Russe, 14 July 1921, AIC.
9 M. Yovanovitch and Mlle Chamié, "Préparation du sel étalon radifère," *Comptes rendus* 175 (1922): 266–8.
10 Mlles Ellen Gleditsch and C. Chamié, "Contribution à l'étude des propriétés chimiques du mésothorium 2 et de l'actinium," *Comptes rendus* 182 (1926): 380–1. It was Curie who proposed that Gleditsch work with Chamié during 1925. See M. Curie to E. Gleditsch, 24 December 1924, University of Oslo.
11 Mlles Irène Curie and C. Chamié, "Sur la constante radioactive du radon," *Comptes rendus* 178 (1924): 1808–10.
12 See chapter 12, below.
13 Mlle C. Chamié, "Sur l'ionisation produite par l'hydration du sulfate de quinine," *Journal de physique et le Radium* 7 (1926): 204–14.
14 Mlle C. Chamié, "Étude du phénomène des groupements d'atoms des radioéléments," *Journal de physique et le Radium* 10 (1929): 44–8; Mlle C. Chamié, "Formation du corps solide par les radioéléments," *Journal de physique et le Radium*, series 8, vol. 7 (1946): 345–9.
15 Irène Joliot-Curie, *Les radioéléments naturels* (Paris: Hermann, 1946).
16 M. Blau and E. Róna, "Anwendung der Chamié'schen photographischen Methode zur Prüfung des chemischen Verhaltens von Polonium," *Sitzungberichte Akademie der Wissenschaften Wein Math-naturwissenschaften*, abt. 2a, 139 (1930): 275–9.
17 Elizabeth Róna, *How It Came About* (Oak Ridge: Oak Ridge Associated Universities, 1978), 28.

18 Copy of a reference for C. Chamié by M. Curie, 28 November 1930, AIC. The purpose of this reference is not indicated.
19 C. Chamié to Marie Curie, 23 July 1926, AIC.
20 C. Chamié to M. Camine, 12 April 1932, AIC.
21 C. Chamié to Agnès Noecker, 26 January 1932, BN.
22 M. Curie to Monsieur le Doyen, 14 June 1929, AIC.
23 Le Doyen de la Faculté des sciences to M. Curie, 20 June 1929, AIC.
24 M. Curie, memorandum, June 1931, AIC. The covering letter is missing though the destination would presumably be the dean or whoever controlled the research funding.
25 Mlles C. Chamié and H. Filçakova, "Sur l'absorption du polonium par le mercure dans différentes solutions," *Journal de chimie et de physico-chimie* 46 (1949): 174–5.
26 Mlle C. Chamié and Mmes H. Faraggi and B. Marques, "Sur les activités en profondeur de l'argent irradié par les deutons," *Comptes rendus* 229 (1949): 358–60.
27 Mlle C. Chamié, "Sur le phénomène des activités singulières provoquées par les deutons dans la masse de l'argent," *Journal de physique et le Radium* 11 (1950): 77–9.
28 Mlle C. Chamié, M. Caillet, and M. G. Fournier, *Tables relatives à la décroissance et à l'accumulation du radon ou émination du radium* (Paris: Gauthier-Villars, 1930).
29 C. Chamié, *Principes nouveax de psychologie* (Paris: Hermann, 1937).
30 C. Chamié, *Psychologie de savoir* (Paris: Hermann, 1950).
31 We thank Ginette Gablot, archivist, Archives Curie et Joliot-Curie, for this search.
32 Moïse Haïssinsky, *Le Laboratoire Curie et son apport aux sciences nucléaires* (Paris: Laboratoire Curie, 1971), 33.
33 Elizabeth Róna, "Laboratory Contamination in the Early Period of Radiation Research," *Health Physics* 37 (1979): 723–7.

CHAPTER SEVEN

1 Most of the biographical information on Maracineanu was obtained from C.G. Berdreag, *The Bibliography of Romanian Physics* (Bucharest: Technical Publishing House, 1957), and Edmond Nicolau and J.M. Stefan, *100 Romanian Researchers and Inventors* (Bucharest: Jon Creanja Publishing House, 1987).
2 S. Maracineanu, "Recherches sur la constante du polonium," *Comptes rendus* 176 (1923): 1879–81.
3 S. Maracineanu, "Sur une méthode de mesure pour un fort rayonnement," *Comptes rendus* 177 (1923): 682–5.

4 Mme Monique Bordry, Institut Curie, is thanked for this information. The thesis was later published as Stéphanie Maracineanu, *Recherches sur la constante du polonium et sur la pénétration des substances radioactives dans les métaux* (Paris: Presses universitaires de France, 1924).

5 M. Curie to the Faculty of Science, Bucharest, 1 August 1924, Library of the Romanian Academy.

6 S. Maracineanu, "Recherches sur la pénétration des substances radioactives dans les métaux," *Comptes rendus* 177 (1923): 1215–17; S. Maracineanu, "Recherches sur l'effet du soleil au point de vue radioactif," *Bulletin de la Section Scientifique de l'Academie Roumaine* 9 (1924): 51–61.

7 S. Maracineanu, "Actions spéciales du Soleil sur la radioactivité du plomb et de l'uranium," *Comptes rendus* 181 (1925): 774–6.

8 A. Nodon, "Sur l'action photogènique des ultraradiations," *Comptes rendus* 174 (1922): 1061–2.

9 A. Nodon, "Relations entre la radioactivité du radium et l'activité des radiations solaires," *Comptes rendus* 176 (1923): 1705–7.

10 S. Maracineanu, "Recherches sur la radioactivité du plomb, qui a été-soumis pendant longtemps au rayonnement solaire," *Comptes rendus* 184 (1927): 1322–5; S. Maracineanu, "Recherches sur la radioactivité de la matière exposée pendant longtemps au rayonnement solaire," *Comptes rendus* 184 (1927): 1547–9; S. Maracineanu, "Effet spécial du polonium, du rayonnement solaire et de la haute tension sur le plomb," *Comptes rendus* 185 (1927): 122–4.

11 H. Deslandres, "Remarques sur la communication précédente," *Comptes rendus* 184 (1927): 1549–50; H. Deslandres, "Remarques sur la communication précédente," *Comptes rendus* 185 (1927): 124–5.

12 F. Behounek, "Zum Ursprung der durchdringenden Strahlung der Atmosphäre," *Physikalische Zeitschrift* 27 (1926): 8–10; F. Behounek, "Einige Bemerkungen zum Ursprung der durchdringenden Strahlung der Atmosphäre," *Physikalische Zeitschrift* 27 (1926): 536–9; F. Behounek, "Zur Erwiderung von Herrn Kolhörster in Heft 17 dieser Zeitschrift," *Physikalische Zeitschrift* 27 (1926): 712–3.

13 W. Kolhörster, "Erwiderung zu der Arbeit von Herrn Behounek in Heft 16 dieser Zeitschrift," *Physikalische Zeitschrift* 27 (1926): 555–6.

14 See chapter 20 below.

15 E. Rona and E.A.W. Schmidt, "The Penetration of Polonium into Metals," *Sitzungberichte Akademie der Wissenschaften Wein Math-naturwissenschaften,* abt. 2A, 136 (1927): 65–73.

16 See chapter 11, below.

17 S. Maracineanu, "Sur la phénomènes, semblables à ceux des corps radioactifs, présentés par les métaux," *Comptes rendus* 186 (1928): 746–8.

18 S. Maracineanu, "L'effet du rayonnement solaire sur les phenomenes de radioactivité et de transmutation," *Bulletin de la Section Scientifique de l'Academie Roumaine* 12 (1929): 5–9.
19 Ch. Fabry and E. Dubreuil, "Sur une prétendue transformation du plomb par l'effet du rayonnement solaire," *Comptes rendus* 190 (1930): 91.
20 S. Maracineanu, "Remarques sur une note de MM. Fabry and Dubreuil intitulée: sur une prétendue transformation du plomb par l'effet du rayonnement solaire," *Comptes rendus* 190 (1930): 373–4.
21 G.I. Pokrowski, "Über eine möglische Wirkung kurzwelliger Strahlung auf Atomkerne," *Zeitschrift für Physik* 63 (1930): 561–73.
22 See, for example, J. Reboul, "Sur l'émission probable d'un rayonnement peu pénétrant par certains métaux," *Comptes rendus* 196 (1933): 1596–8; J. Reboul, "Sur l'action exercée par les métaux ordinaires sur la plaque photographique et sur l'electromètre," *Comptes rendus* 202 (1936): 1920–2.
23 A. Smits and Mlle C.H. MacGillavry, "Remarques sur la note de Mlle. Maracineanu," *Comptes rendus* 190 (1930): 635–7.
24 A. Boutaric and Mlle Madeleine Roy, "Sur la radioactivité de divers métaux provenant de toitures anciennes," *Comptes rendus* 190 (1930): 483–5.
25 A. Lepape and M. Geslin, "Sur la radioactivité acquise par les matériaux exposés à l'action des agents atmosphériques," *Comptes rendus* 190 (1930): 676–8.
26 A. Boutaric and Mlle Madeleine Roy, "Sur la radioactivité des matériaux provenant de toitures anciennes," *Comptes rendus* 190 (1930): 1410–2.
27 S. Maracineanu, "Remarques sur les notes de MM. Smits, Boutaric et Lepape," *Bulletin de la Section Scientifique de l'Academie Roumanie* 13 (1930): 55–8.
28 H. Behounek, "Ein Beitrag zu den Versuchen über die Beeinflussung des radioaktiven Zerfalles," *Physikalische Zeitschrift* 31 (1930): 215–24; S. Maracineanu, "Bemerkungen zu Behounek: Ein Beitrag zu den Versuchen über die Beeinflussung des radioactiven Zerfalls," *Physikalische Zeitschrift* 31 (1930): 1032–6; F. Behounek, "Erwiderung auf die Bemerkungen von Frl. St. Maracineanu," *Physikalische Zeitschrift* 31 (1930): 1036–8; Stéphanie Maracineanu, "Bemerkungen zur Erwiderung des Herrn Behounek," *Physikalische Zeitschrift* 31 (1930): 1038–9.
29 S. Maracineanu to L. Meitner, 12 March 1936, Meitner Collection, CCA.
30 *Neues Wiener Zeitschrift*, 5 June 1934.
31 Unidentified and untitled publication, S. Maracineanu, Library of the Romanian Academy.

32 Records of the Romanian Academy meeting, 29 November 1936, Library of the Romanian Academy.
33 S. Maracineanu, "Les substances radioactives sous l'influence du rayonnement solaire provoquent la pluie," *Cultura Nationala* 5 (1930): 3–5.
34 Letter of transit for Mlle Maracineanu, Governor General of Algeria, 20 August 1934, Library of the Romanian Academy.
35 Report on weather observations in the Touggourt Territory of Algeria, 28 August to 1 September, Library of the Romanian Academy.
36 Souchier (representative of the Governor General) to Stéphanie Maracineanu, 2 September 1934, Library of the Romanian Academy.
37 S. Maracineanu, "La radioactivité du globe, les radiations et les tremblements de terre," *Comptes rendus des séances de l'Académie des Sciences de Roumanie* 6 (1942): 72–5.

CHAPTER EIGHT

1 Most of the biographical details were obtained from Dorabialska's autobiography: Alicja Dorabialska, *Jeszcze jedno zycie* (Warsaw, Poland: Pax, 1972). This book, "Still Another Life" in English, is dedicated to the memory of her mentor, Wojciech Swientoslawski. There are also the following obituaries, both of which include a comprehensive list of her publications: Jerzy Kroh and Wladyslaw Reimschüssel, "Wspomnienie o prof. Dr Alicji Dorabialskiej," *Wiadomosci Chemiczne* 31 (1977): 315–24; and Czeslaw Wronkowski, "Alicja Dorabialska (1897–1975)," *Wiadomosci Chemiczne* 34 (1980): 479–90.
2 A. Dorabialska, "Badaia termochemiczne nad oksymami," (Thermochemical Researches on Oximes) *Roczniki Chemii* 1 (1921): 424–47; 448–67.
3 For example, D.K. Yovanovitch and A. Dorabialska, "Sur une méthod nouvelle pour mesurer l'absorption du rayonnement beta et gamma de corps radioactifs," *Comptes rendus* 182 (1926): 1459–61.
4 Robert Reid, *Marie Curie* (New York: E.P. Dutton, 1974), 281.
5 W. Swientoslawski and A. Dorabialska, "Microcalorimètre adiabatique pour recherches radiologique," *Comptes rendus* 185 (1927): 763–5.
6 A. Dorabialska, "Prof. Dr Wojciech Swietoslawski – Uczony i Czlowiek," *Roczniki Chemii* 18 (1938): 289–302.
7 Gilette Ziegler, ed., *Curie Correspondence: Choix de lettres (1905–1934)* (Paris: Les Éditeurs Français Réunis, 1974), 307, 312.
8 See chapter 4, above.
9 This information was obtained from *The General Encyclopaedia* (Warsaw, Poland: PWN 3, 1963), 109. It is unclear whether Dorabialska travelled into Warsaw to give the classes or whether the underground polytechnic had moved outside of Warsaw to Lwow.

10 Some of this work is summarized in H. Bem, A. Dorabialska, W. Reimschüssel, H. Sugier, and W. Swiatkowski, "Kierunki prac radiochemicznych katedry chemii fizycznej politechniki Lodzkiej," (Radiochemical Research at the Department of Physical Chemistry, Technical University of Lodz) *Nukleonika*, Suppl. 10 (1965, published 1966): 339–46. Her publications continued to appear until 1975.

11 We thank Maria Golonko, Technical University of Wroclaw, for this information.

CHAPTER NINE

1 This is not surprising, considering that estimates of Marie Curie's weekly exposure to damaging radiation was roughly one rem per week. The current safety standard is approximately one hundredth this amount. Robert Reid, *Marie Curie* (New York: William Collins and Sons, 1974), 130.

2 Rosalynd Pflaum, *Grand Obsession: Marie Curie and Her World* (New York: Doubleday, 1989), 96–7.

3 Pflaum, *Grand Obsession*, 101.

4 Marie Curie to Henriette Perrin, 1905, cited in Eve Curie, *Madame Curie. A Biography*, trans. Vincent Sheean (New York: Da Capo Press, 1986 [1937]), 224.

5 Curie, *Madame Curie*, 123.

6 Ibid., 234.

7 Pflaum, *Grand Obsession*, 99.

8 Ibid., 322.

9 Robin McKowan, *She Lived for Science: Irène Joliot-Curie* (New York: Julian Messner, 1992), 29.

10 Curie, *Madame Curie*, 272.

11 McKowan, *She Lived for Science*, 30.

12 Françoise Giroud, *Marie Curie. A Life*, trans. Lydia Davis (New York: Holmes and Meier, 1986), 152.

13 Maurice Goldsmith, *Frédéric Joliot-Curie. A Biography* (London: Lawrence and Wishart, 1976), 30.

14 Gilette Ziegler, *Curie Correspondance: Choix de lettres (1905–1934)* (Paris: Les Éditeurs Français Réunis, 1974), 32.

15 Pflaum, *Grand Obsession*, 174.

16 Giroud, *Marie Curie*, 186.

17 Ziegler, *Curie Correspondance*, 55.

18 Pflaum, *Grand Obsession*, 320.

19 See also: Joan Mason, "Hertha Ayrton: A Scientist of Spirit," in Gill Kirkup and Laurie S. Keller, eds., *Inventing Women: Science, Technology and Gender* (Cambridge, UK: Polity Press, 1992), 168–77.

20 Ziegler, *Curie Correspondance*, 84.
21 Pflaum, *Grand Obsession*, 191.
22 Ziegler, *Curie Correspondance*, 61.
23 Pflaum, *Grand Obsession*, 189.
24 The villa at l'Arcouest was so well known to a certain group of French scientists and intellectuals that it was dubbed "Port Science" by the locals. Irène and Fred continued to live there with their children, upholding a tradition lasting three generations.
25 Ziegler, *Curie Correspondance*, 131.
26 Ibid., 134.
27 Ibid., 141.
28 Ibid., 153.
29 Giroud, *Marie Curie*, 208.
30 Ibid., 207.
31 Pflaum, *Grand Obsession*, 201.
32 Ibid., 206.
33 Francis Perrin, "Irène Joliot-Curie," in C.C. Gillespie, ed., *Dictionary of Scientific Biography* (New York: Scribner's and Sons, 1973), 7: 157–9.
34 Giroud, *Marie Curie*, 207.
35 Pflaum, *Grand Obsession*, 211.
36 Giroud, *Marie Curie*, 220.
37 I. Curie, "Sur le poids atomique du chlore dans quelques minéraux," *Comptes rendus* 172 (1921): 1025–8.
38 Pflaum, *Grand Obsession*, 221.
39 Ibid., 227.
40 Ibid., 238.
41 Ibid., 255.
42 I. Curie, "Recherches sur les rayons a du polonium. Oscillation de parcours, vitesse d'émission, pouvoir ionisant," *Annales de physique* 2 (1925): 403.
43 Pflaum, *Grand Obsession*, 319.
44 Interview: Sonja Hanneborg by Anne Marie Weidler-Kubanek, 27 October 1992, Oslo. Hanneborg was a student in the Curie Institute from 1924 to 1925.
45 Pflaum, *Grand Obsession*, 259.
46 Goldsmith, *Frédéric Joliot-Curie*, 28. Marie was shocked to discover that Joliot had neither his baccalaureate nor his license, which is equivalent to a first degree and a master's. Until these credentials were completed, Joliot could not supplement his meagre salary with a teaching appointment.
47 Ibid., 25.
48 Ibid., 31.
49 Ibid., 32.

50 Pflaum, *Grand Obsession*, 261.
51 Goldsmith, *Frédéric Joliot-Curie*, 31.
52 Ibid., 33.
53 Ibid., 34.
54 Ibid., 41.
55 Angèle Pompëi, *Europe* 38 (1961): 234.
56 Pflaum, *Grand Obsession*, 235.
57 Goldsmith, *Frédéric Joliot-Curie*, 37.
58 I. Curie and F. Joliot, "Sur la nature du rayonnement pénétrant excités dans les noyaux légers par les particules alpha," *Comptes rendus* 194 (1932): 1229–32.
59 James Chadwick, "Possible Existence of a Neutron," *Nature* 129 (1932): 312; "The Existence of a Neutron," *Proceedings of the Royal Society, Part A*, 136 (1932): 692–708.
60 Goldsmith, *Frédéric Joliot-Curie*, 50.
61 I. Curie and F. Joliot, "Électrons de matérialisation et de transmutation," *Journal de physique et le Radium* 4 (1933): 494–500.
62 I. Curie and F. Joliot, "Mass of the Neutron," *Nature* 133 (1934): 721.
63 I. Curie and F. Joliot, "La complexité du proton et la masse du neutron," *Comptes rendus* 197 (1933): 237–8.
64 Pflaum, *Grand Obsession*, 302.
65 I. Curie and F. Joliot, "Artificial Production of a New Kind of Radioelement," *Nature* 133 (1934): 201–2.
66 I. Curie and F. Joliot, "Chemical Separation of the New Elements That Emit Positive Electrons," *Comptes rendus* 198 (1934): 559–61.
67 Artificial radioactivity was a term Fred never liked because he considered their radioactivity to be just as natural as the one studied by Marie and Pierre Curie; Goldsmith, *Frédéric Joliot-Curie*, 53.
68 Noëlle Loriot, *Irène Joliot-Curie* (Paris: Presses de la renaissance, 1991), 9.
69 Pflaum, *Grand Obsession*, 329.
70 Goldsmith, *Frédéric Joliot-Curie*, 65.
71 I. Joliot-Curie and P. Savitch, "Sur le radioélément de période 3,5 h. formé dans l'uranium par les neutrons," *Comptes rendus* 206 (1938): 1643–4.
72 Pflaum, *Grand Obsession*, 335.
73 Ibid., 335.
74 McKowan, *She Lived for Science*, 123.
75 Pflaum, *Grand Obsession*, 336.
76 See chapter 16, below.
77 O. Hahn and F. Strassmann, "Über den Nachweis und das Verhalten der bei der Bestrahlung des Urans mittels Neutronen entstehenden Erdalkalimetalle," *Naturwissenschaften* 27 (1939): 11–15; L. Meitner and O. Frisch, "Disintegration of Uranium by Neutrons: a New Type of Nuclear Reaction," *Nature* 143 (1939): 239.

78 F. Joliot, "Preuve expérimentale de la rupture explosive des noyaux d'uranium et de thorium sous l'action des neutrons," *Comptes rendus* 208 (1939): 341–3.
79 Pflaum, *Grand Obsession*, 323.
80 Wilfred Knapp, *France, Partial Eclipse. From the Stavinsky Riots to the Nazi Conquest* (New York: Macdonald/American Heritage Press, 1972), 43.
81 Ibid., 84.
82 Women would not obtain the vote in France until after the war, in 1945.
83 Pflaum, *Grand Obsession*, 326.
84 Ibid., 326.
85 Ibid., 327.
86 McKowan, *She Lived for Science*, 115.
87 Goldsmith, *Frédéric Joliot-Curie*, 77.
88 Ibid., 96.
89 Pflaum, *Grand Obsession*, 376.
90 Solomon's wife, Hélène, was arrested and deported to Auschwitz along with Jacques's mother on 23 January 1943. She survived to be liberated by the Soviet army. Unfortunately, her mother-in-law died in the concentration camp.
91 Pflaum, *Grand Obsession*, 389.
92 Since there were no official diplomatic channels between occupied France and Switzerland, Gentner's wife acted as an intermediary for Irène and Fred's correspondence.
93 Pflaum, *Grand Obsession*, 382.
94 I. Joliot-Curie and S.T. Tsien, "Parcours des rayons a de l'ionium," *Journal de physique et le radium* 6 (1945): 162–3.
95 Irène Curie to Ellen Gleditch, 1 February 1947, cited in Torleiv Kronen, and Alexis C. Pappas, *Ellen Gleditch Et liv i forskning og medmenneskelighet* (Oslo: Aventura, Forlag, 1987).
96 Pflaum, *Grand Obsession*, 408.
97 Zoë. "Z" Stood for energy zero, or the state of France's economy after the war, "o" stood for oxide of uranium, the source of Zoe's energy, and "ë" stood for eau lourde, or heavy water, the reactor's moderator.
98 McKowan, *She Lived for Science*, 158.
99 Ibid., 159.
100 George B. Kauffman and J.P. Adloff, "Marguerite Catherine Perey (1909–1975)," in Louise S. Grinstein, Rose K. Rose, and Miriam H. Rafailovich, eds., *Women in Chemistry and Physics: A Biobibliographic Sourcebook* (Westport, CT: Greenwood Press, 1993), 470–5.
101 Irène Curie to Ellen Gleditch, 30 April 1951, cited in Kronen and Pappas, *Ellen Gleditch*.

102 Irène Curie to Ellen Gleditch, 8 February 1951, cited in ibid.
103 Pflaum, *Grand Obsession*, 447.
104 Ibid., 453.
105 Ibid., 457.
106 Ibid., 421.
107 Ibid., 421.
108 McKowan, *She Lived for Science*, 165.
109 Irène Curie to Ellen Gleditch, 5 July 1953, cited in Kronen and Pappas, *Ellen Gleditch*.
110 Pflaum, *Grand Obsession*, 463.
111 Pompëi, *Europe*, 16.
112 Irène Curie, "Sur une nouvelle méthode pour la comparaison précise du rayonnement des ampoules de radium," *Journal de physique et le radium* 15 (1954): 790–5.
113 Pflaum, *Grand Obsession*, 462.
114 Kronen and Pappas, *Ellen Gleditch*; 99.
115 Pflaum, *Grand Obsession*, 295.

CHAPTER TEN

1 We thank Mme Monique Bordry, Institut Curie, for this information.
2 L. Blanquies, "Comparaison entre les rayons α produits par différentes substances radioactives," *Comptes rendus* 148 (1909): 1753–6; L. Blanquies, "Comparaison entre les rayons α produits par différentes substances radioactives," *Le Radium* 6 (1909): 230–2.
3 Mlle L. Blanquies, "Sur les constituants de la radioactivité induite de l"actinium," *Comptes rendus* 151 (1910): 57–60; L. Blanquies, "Sur les constituants de la radioactivité induite de l"actinium," *Le Radium* 7 (1910): 159–62.
4 May Sybil Leslie to Arthur Smithalls, 30 November 1909, Smithalls Collection, UL.
5 Speeches and Guest List, *Cinquantenaire du Premier Cours de Marie Curie à la Sorbonne*, 12 January 1957, Paris.
6 See, for example, J.S. Lattès and Antoine Lacassagne, "Technique chimico-physique de détection du polonium injecté dans les organes," *Comptes rendus* 178 (1924): 630–2.
7 E. Montel, "Sur la pénétration du polonium dans le plomb," *Journal de Physique* 10 (1929): 78–80.
8 We thank Christer Wijkstrom, Librarian, Royal Academy of Sciences, Centre for the History of Science, Stockholm, and Carl-Otto von Sydow, Keeper of Manuscripts, Uppsala Universitetsbibliotek, Uppsala, for the biographical information. Anne-Marie Kubanek, John Abbott College, Ste Anne-de-Bellevue, is thanked for the translation.

9 Eva Ramstedt, "Sur la solubilité de l'emanation du radium dans les liquides organiques," *Le Radium* 8 (1911): 253–6.
10 Eva Ramstedt, "Sur la solubilité du dépôt actif du radium," *Le Radium* 10 (1913): 159–65.
11 Ellen Gleditsch and Eva Ramstedt, *Radium og de radioaktive processer* (Kristana: Aschehoug, 1917).
12 Eva Ramstedt to Frederick Soddy, 29 November 1922, OUA.
13 Eva Ramstedt to Marie Curie, 11 August 1925, BN.
14 Eva Ramstedt to Lise Meitner, 3 September 1933, Meitner Collection, CCA.
15 *Svenska män och Kvinnor: Biografisk Upplagsbok, volume 6* (Stockholm: Albert Bonniers Forlag, 1949), 210; *Lärarmatrikeln 1934* (Stockholm: Utgivarnas Förlag), 395–6.
16 Inga Fischer-Hjalmars, "Women Scientists in Sweden," in Derek Richter, ed., *Women Scientists: The Road to Liberation* (London: MacMillan, 1982), 122.
17 Magda Fekete, Head, Department of Bibliography and Information, Library of the Hungarian Academy of Sciences, is thanked for discovering the biographical information on Götz.
18 I. Götz, "Über quantitative Bestimmung der Radiumemanation," *Chemiker Zeitung* 35 (1911): 724.
19 I. Götz to André Debierne, 16 September 1911, AIC.
20 J. Danysz and I. Götz, "Sur les rayons β de la radioactivité induite à évolution lente," *Le Radium* 9 (1912): 6.
21 After authoring several biochemical publications between 1914 and 1918, she returned to physical chemistry research in 1919. See Irene D. Götz, "Über die beim Verdünnen konzentrierter Lösungen bzw. beim Mischen zweier Flussigkeiten auftretenden Volumänderungen," *Zeitschrift für Physikalische Chemie* 94 (1920): 181–209. While working at the animal research institute she had managed to co-author one chemical publication: Julius Gróh and Irene D. Götz, "Stalagmometrische Bestimmung kleiner Hydroxylionen- konzentrationen," *Biochemische Zeitschrift* 66 (1914): 165–72.
22 See chapter 20, below.
23 Jörg K. Hoensch, *A History of Modern Hungary: 1867–1986* (London: Longman, 1988), 98.
24 Irén Götz, *Az elemek átváltozása és a modern anyagfogalom* (Hungary: Korunk, 1926).

CHAPTER ELEVEN

1 This chapter is an updated, modified, and improved version of Marelene F. Rayner-Canham and Geoffrey W. Rayner-Canham, "Harriet

Brooks – Pioneer Nuclear Scientist," *American Journal of Physics* 57 (1989): 899–902. We thank the American Journal of Physics for permission to reproduce the material. A fuller account of her life is described in Marelene F. Rayner-Canham and Geoffrey W. Rayner-Canham, *Harriet Brooks: Pioneer Nuclear Scientist* (Montreal: McGill-Queen's University Press, 1992). We thank Paul Brooks Pitcher and Cicely Grinling (son and niece of Harriet Brooks) for their help and encouragement.

2 Barbara M. Solomon, *In the Company of Educated Women* (New Haven, CT: Yale University Press, 1985), 63.

3 Phebe Chartrand, Archivist, McGill University, is thanked for information on Brooks's time at McGill.

4 D. Suzanne Cross, "The Neglected Majority: The Changing Role of Women in 19th Century Montreal," *Histoire Sociale/Social History* 6 (1973): 202–23.

5 Harriet Brooks, "Damping of Electrical Oscillations," *Royal Society of Canada, Transactions,* section 3 (1899): 13–15.

6 Harriet Brooks, "Damping of the Oscillations in the Discharge of a Leyden Jar," MA thesis, McGill University, 1901.

7 P.L. Kapitza, "Recollections of Lord Rutherford," *Royal Society (Great Britain), Proceedings,* series A, A294 (1966): 123–37.

8 W. Peterson to H. Brooks, 22 March 1901, McGill University Archives.

9 E. Rutherford, "A Radioactive Substance Emitted from Thorium Compounds," *Philosophical Magazine,* series 5, vol. 49 (1900): 1–14.

10 Ernest Rutherford and Harriet T. Brooks, "The New Gas from Radium," *Royal Society of Canada, Transactions,* section 3 (1901): 21–5.

11 We thank Dr Montague Cohen, McGill University, for pointing out to us the significance of this work.

12 Marjorie Malley, "The Discovery of Atomic Transmutation: Scientific Styles and Philosophies in France and Britain," *Isis* 70 (1979): 213–23.

13 Ernest Rutherford and Harriet T. Brooks, "Comparison of the Radiations from Radioactive Substances," *Philosophical Magazine,* series 6, vol. 4 (1902): 1–23.

14 Caroline Rittenhouse, Archivist, Bryn Mawr College, is thanked for information on Brooks's time at Bryn Mawr.

15 H. Brooks to E. Rutherford, 8 December 1901, CUA. A.E.B. Owen, Archivist, Cambridge University, is thanked for copies of the correspondence.

16 Sonya (Sophia) Kovalesky (1850–91) was one of the great mathematicians. See Ann Hibner Koblitz, *A Convergence of Lives: Sofia Kovalevskaia, Scientist, Writer, Revolutionary* (Cambridge, MA: Birkhäuser, 1983); Don H. Kennedy, *Little Sparrow: A Portrait of Sophia Kovalevsky* (Athens, OH: Ohio University Press, 1983); and Roger Cook, *The Mathematics of Sonya Kovalevskaya* (New York: Springer-Verlag, 1984).

17 M. Cary Thomas, "Present Tendencies in Women's College and University Education," *Educational Review* 35 (1908): 64–85.
18 H. Brooks to E. Rutherford, 18 March 1902, CUA.
19 Joyce Antler, "'After College, What?': New Graduates and the Family Claim," *American Quarterly* 32 (Fall 1980): 409–34.
20 J.J. Thomson to E. Rutherford, 13 May 1902, CUA.
21 H. Brooks to E. Rutherford, Spring 1903 (n.d.), CUA.
22 A. Stewart Eve, "Some Scientific Centres. VIII The MacDonald Physics Building, McGill University, Montreal," *Nature* 74 (1906): 272–5.
23 Her life at the Cavendish is discussed more fully in Rayner-Canham and Rayner-Canham, *Harriet Brooks: Pioneer Nuclear Scientist*, 32–8.
24 Harriet Brooks, "A Volatile Product from Radium," *Nature* 70 (1904): 270.
25 Ernest Rutherford, "The Succession of Changes in Radioactive Bodies," *Philosophical Transactions of the Royal Society*, series A, 204 (1904):169–219.
26 Thaddeus J. Trenn, "Rutherford and Recoil Atoms: The Metamorphosis and Success of a Once Stillborn Theory," *Historical Studies in Physical Sciences* 6 (1975): 513–47.
27 Trenn, "Rutherford and Recoil Atoms," 513–47.
28 E. Rutherford to O. Hahn, 22 December 1908, CUA.
29 Otto Hahn, *Otto Hahn: A Scientific Autobiography*, trans. and ed. Willy Ley (New York: Charles Scribner's Sons, 1966), 268.
30 Harriet Brooks, "The Decay of the Excited Radioactivity from Thorium, Radium, and Actinium," *Philosophical Magazine*, series 6, vol. 8 (1904): 373–84.
31 Rutherford, "The Succession of Changes in Radioactive Bodies," 169–219.
32 Patricia K. Ballou and Lucinda Manning, Archivists, Barnard College, are thanked for copies of the relevant correspondence at Barnard College.
33 Harold W. Webb, "Bergen Davis," in *Biographical Memoirs*, vol. 34 (New York: Columbia University Press, 1960), 65–82.
34 L. Gill to H. Brooks, 12 July 1906, BCA.
35 H. Brooks to L. Gill, 18 July 1906, BCA.
36 M. Maltby to L. Gill, 24 July 1906, BCA.
37 L. Gill to M. Maltby, 30 July 1906, BCA.
38 E. Rutherford to O. Hahn, 20 August 1906, CUA.
39 Maxim Gorky, *Letters* (Moscow: Progress Publishers, 1966), 49. L.P. Bykovtseva, Gorky Museum, is thanked for information concerning Brooks's travels with Gorky and Andreyeva.
40 Alexander Kaun, *Maxim Gorky and His Russia* (New York: Benjamin Blom, 1968), 573.
41 Monique Bordry, Curie Institute, is thanked for copies of the Brooks-Curie correspondence.

42 Maria F. Andreeva, *Perepiska Vospominaniia* (Moscow: Iskusstvo, 1968), 147.

43 André Debierne, "Sur le coefficient de diffusion dans l'air de l'émination de l'actinium," *Le Radium* 4 (1907): 213–218; André Debierne, "Sur le dépôt de la radioactivité induite du radium," *Le Radium* 6 (1909): 97–106; and L. Blanquies, "Comparaison entre les rayons a produits par différentes substances radioactives," *Le Radium* 6 (1909): 230–2.

44 E. Rutherford to A. Schuster, 25 March 1907, Royal Society, London.

45 Margaret W. Rossiter, *Women Scientists in America* (Baltimore: Johns Hopkins University Press, 1982), 195.

46 Quoted in Roberta Frankfort, *Collegiate Women* (New York: New York University Press, 1977), 33.

47 Rayner-Canham and Rayner-Canham, *Harriet Brooks*, 72–82.

48 John Martin and Prestonia Mann Martin, *Feminism: Its Fallacies and Follies* (New York: Dodd, Mead, 1917), 73.

49 M. Rutherford to H. Brooks, 28 November 1907, H. Brooks files of Paul Brooks Pitcher.

50 Rayner-Canham and Rayner-Canham, *Harriet Brooks*, 92–102.

51 Elizabeth Fee, "Critiques of Modern Science: The Relationships of Feminism to Other Radical Epistimologies," in Ruth Bleier, ed., *Feminist Approaches to Science* (New York: Pergamon Press, 1986), 45.

52 Lewis Pyenson, "The Incomplete Transmission of a European Image: Physics at Greater Buenos Aires and Montreal, 1890–1920," *American Philosophical Society, Proceedings* 122 (1978): 92–114.

53 E. Rutherford to A.S. Eve, 6 May 1933. We thank Montague Cohen, McGill University, for a copy of this letter.

54 David Wilson, *Rutherford* (Cambridge, MA: MIT Press, 1983); A. Stewart Eve, *Rutherford* (Cambridge, UK: Cambridge University Press, 1939); Daniil S. Danin, *Rezerford* (Moscow: Molodaya gvardiya, 1967).

55 Eugene Garfield, "Premature Discovery or Delayed Recognition – Why?," in *Essays of an Information Scientist* (Philadelphia: ISI Press, 1981), 4: 488–93.

CHAPTER TWELVE

1 The biographical details for Gates have been mostly assembled from the following sources: James McKeen Cattel and D.R. Brimhall, eds., *American Men of Science*, 3d ed. (Lancaster, PA: Science Press, 1921), 250; Margaret E. Maltby, *History of the Fellowships Awarded by the American Association of University Women, 1888–1929* (Washington: American Association of University Women, n.d.), 17–18; the *Grinnell College Bulletin, 1914–1915* (we thank Anne Kintner, College Archivist, for this item); assorted documents from the Northwestern University Archives (Patrick M. Quinn, University Archivist, is thanked for

providing these); and some information from Maureen A. Harp, Archives and Manuscripts Assistant, University of Chicago Library. For certain years, the sources contain contradictory information; however, we believe that we have pieced together a reasonably accurate version of Gates's travels and appointments.

2 Frederick Rudolph, *The American College and University: A History* (Athens: University of Georgia Press, 1990), 323. Rudolph comments: "At Northwestern the [increase in] enrollment of women students so threatened the coeducational nature of the institution that an engineering course was added to bolster the dwindling male forces."

3 Roberta Wein, "Women's Colleges and Domesticity, 1875–1918," *History of Education Quarterly* 14 (1974): 31–48.

4 Margaret W. Rossiter, *Women Scientists in America: Struggles and Strategies to 1940* (Baltimore: Johns Hopkins University Press, 1982), 38.

5 Rossiter, *Women Scientists in America,* 40.

6 Women's College to Gates, 6 August 1898, GCA. Ms Sydney Roby, Archivist, Goucher College, is thanked for copies of this correspondence.

7 Gates to Goucher, 22 June 1899, GCA.

8 E.H. Moores to F.C. Gates, 14 May 1901, GCA.

9 David Wilson, *Rutherford: Simple Genius* (Cambridge, MA: MIT Press, 1983), 183.

10 Marelene F. Rayner-Canham and Geoffrey W. Rayner-Canham, *Harriet Brooks: Pioneer Nuclear Scientist* (Montreal: McGill-Queen's University Press, 1992).

11 E. Rutherford to J.J. Thomson, 26 December 1902, Rutherford Collection, CUA.

12 Fanny Cook Gates, "Effect of Heat on Excited Radioactivity," *Physical Reviews* 16 (1903): 300–5.

13 E. Rutherford, "Einfluss der Temperatur auf die 'Emanationen' radioaktiver Substanzen," *Physikalische Zeitschrift* 2 (1901): 429–31.

14 Lawrence Badash, "Radioactivity before the Curies," *American Journal of Physics* 33 (1965): 128–35.

15 See John Heilbron, "Physics at McGill in Rutherford's Time," in Mario Bunge and William R. Shea, *Rutherford and Physics at the Turn of the Century* (New York: Dawson & Science History Publications, 1979), 62.

16 Fanny Cook Gates, "On the Nature of Certain Radiations from the Sulphate of Quinine," *Physical Reviews* 17 (1903): 499–501.

17 Mary J. Nye, "N-rays: An Episode in the History and Psychology of Science," *Historical Studies in the Physical Sciences* 11 (1980): 125–46.

18 Mary J. Nye "Gustave Le Bon's Black Light: A Study in Physics and Philosophy in France at the Turn of the Century," *Historical Studies in the Physical Sciences* 4 (1974): 163–95.

19 F.C. Gates to E. Rutherford, 21 July 1903, CUA.
20 F.C. Gates to E. Rutherford, 9 April 1904, CUA.
21 "Miss Gates Honored in France," *Baltimore Sun*, 31 March 1906.
22 Anna Heubeck Knipp and Thaddeus P. Thomas, *The History of Goucher College* (Baltimore: Goucher College, 1938), 80. Unfortunately, the setting up of the sorority is the only reference to Gates in this work.
23 C.M. Stuart to F.C. Gates, 14 May 1901, GCA.
24 F.C. Gates to E. Rutherford,? April 1904, CUA.
25 F.C. Gates to E. Rutherford, 20 July 1904, CUA.
26 J.J. Thomson to E. Rutherford, 1 October 1905, CUA.
27 S.B. Sinclair, "J.J. Thomson and Radioactivity: Part II," *Ambix* 35 (1988): 113–26.
28 *Science*, 22 (1905): 574.
29 Rossiter, *Women Scientists in America*, 71.
30 This summary of her time at Illinois is taken from Mary Lou Filbey, "Deans of Women of the University of Illinois" (unpublished manuscript), 70–80. We thank Maynard Brichford, University Archivist, University of Illinois, for a copy of this work.
31 Filbey, "Deans of Women," 83.
32 R.A. Millikan to F.C. Gates, 3 May 1901, GCA.
33 Filbey, "Deans of Women," 90.
34 *The Scarlet and Black* (Grinnell College), 25 February 1931, 91 (curiously, this obituary claims that she obtained her PhD at the University of Leipzig for work with Dr William Conrad Roentgen, though there is no other evidence of this); and *Chicago Tribune*, 25 February 1931.
35 Elizabeth Róna, "Laboratory Contamination in the Early Period of Radiation Research," *Health Physics* 37 (1979): 723–7.
36 For example, the physicist Margaret Maltby had followed a similar career path into a woman's college (Barnard) where she stayed for a "safe" but active life. See Shirley W. Harrison, "Margaret Eliza Maltby (1860–1944)," in Louise S. Grinstein et al., eds., *Women in Chemistry and Physics: A Biobibliographic Sourcebook* (Westport, CT: Greenwood Press, 1993), 354–60.

CHAPTER THIRTEEN

1 She spelled her name "Jadwiga Szmidt" on her English-language research papers. However, the contemporary conversion of the letters of her names from the Cyrillic alphabet would be "Yadviga Shmidt."
2 We thank T.I. Nazarovskaya, Archivist, Academy of Sciences of the USSR, Leningrad, and N.A. Sheshina, Gorky Science Library, Leningrad University, for this information. Michael Newton, Department of

Religious Studies, Sir Wilfred Grenfell College, is thanked for the translation of the letters from Russian.

3 Sharon B. McGrayne, *Nobel Prize Women in Science* (New York: Birch Lane Press, 1993), 6.

4 M.C. Golonka, J. Róziewicz, J. Starosta, and K.G. Tokhadze, "Jadwiga Szmidt (1889–1940): A Pioneer Woman in Nuclear and Electrotechnical Sciences," *American Journal of Physics* 62 (1994): 947–8.

5 Ann H. Koblitz, "Science, Women, and the Russian Intelligensia," *Isis* 79 (1988): 208–26.

6 J. Szmidt to Marie Curie, 3 October 1923, BN. We thank Stephanie Weinsberg-Tekel for a translation of this letter from Polish.

7 From the records of the Laboratoire Curie. Monique Bordry, Chargée des Archives, Institut Curie, is thanked for permission to view these files.

8 H.R. Robinson, "Rutherford: Life and Work to the Year 1919, with Personal Reminiscences of the Manchester Period," in J.B. Birks, ed., *Rutherford at Manchester* (New York: W.A. Benjamin, 1963), 73.

9 Jadwiga Szmidt, "On the Distribution of Energy in the Different Types of Gamma Rays emitted from Certain Types of Radioactive Substances," *Philosophical Magazine*, series 6, vol. 28 (1914): 527–39.

10 Jadwiga Szmidt, "Note on the Excitation of Gamma Rays by β rays," *Philosophical Magazine*, series 6, vol. 30 (1915): 220–4.

11 E.N. da C. Andrade, "Rutherford at Manchester, 1913–14," in Birks, ed., *Rutherford at Manchester*, 33.

12 Ernest Rutherford to Bertram Boltwood, 20 June 1914, CUA.

13 E. Gleditsch to B. Boltwood, 17 July 1914, in Thorleiv Kronen and Alexis C. Pappas, *Ellen Gleditsch. Et liv i forskning og medmenneskelighet* (Oslo: Aventura Forlag, 1987), 223.

14 Robinson, "Rutherford: Life and Work to the Year 1919," 76.

15 J. Szmidt to Marie Curie, 3 October 1923, BN.

16 J. Szmidt to Ernest Rutherford, 8 September 1915, CUA.

17 J. Szmidt to Ernest Rutherford, n.d. (October 1915?), CUA.

18 E. Rutherford to Ellen Gleditsch, 10 November 1915, Oslo University Archives. Anne-Marie Weidler-Kubanek is thanked for this reference.

19 J. Szmidt to Ernest Rutherford, 29 December 1915, CUA.

20 See A.T. Grigorian, "Abram Fedorovich Ioffe," in C.C. Gillespie, ed., *Dictionary of Scientific Biography*, vol. 15, Supp. 1 (New York: Charles Scribner's, 1978), 251–2.

21 It is customary in Russian to address an individual by the first name followed by the first name of their father, which was Ryszard, or Richard in its anglicized form.

22 A.F. Ioffe to A.P. Ioffe, 1916, cited in M.S. Sominski, *Abram Fedorovich Ioffe* (Moscow-Leningrad: n.p., 1964), 1900. T[di] Nazarovskaya is

thanked for this information. Dr Michael Newton is thanked for the translation.

23 Paul R. Josephson, *Physics and Politics in Revolutionary Russia* (Berkeley: University of California Press, 1991).

24 J. Szmidt to E. Rutherford, n.d. (October 1920?), CUA. As the address is clipped from the original letter, it is likely that Rutherford replied to Szmidt's pleading, though the contents of his reply are not known.

25 We thank T.I. Nazarovskaya, Archivist, Academy of Sciences of the USSR, Leningrad, for this and some of the subsequent information about her life.

26 J.W. Boag, P.E. Rubenin, and D. Shoenberg, *Kapitza in Cambridge and Moscow: Life and Letters of a Russian Physicist* (Amsterdam: North-Holland, 1990), 9.

27 Joyce Antler, "'After College, What?': New Graduates and the Family Claim," *American Quarterly* 32 (1980): 409–34.

28 J. Szmidt to Marie Curie, 3 October 1923, BN. We thank Ms Stephanie Weinsberg-Tekel for a translation of the letter.

29 L. Razet (secretary to M. Curie) to J. Szmidt, 20 October 1923, BN.

30 Josephson, *Physics and Politics in Revolutionary Russia*, 88.

31 J.R. Szmidt-Tshernyshev and A.A. Tshernyshev, "A Long-distance Vision Apparatus," *Collection of Works in Applied Physics* (1926): 13–18 (in Russian).

32 A.A. Tshernyshev and J.R. Szmidt-Tshernyshev, Patent 2463 (1927).

33 Josephson, *Physics and Politics*, 7.

34 N. Semenoff, *Chemical Kinetics and Chain Reactions* (Oxford: Clarendon Press, 1935), translated by Professor Frenkel and Mrs Jadviga Szmidt-Chernysheff.

35 Golonka, Róziewicz, Starosta, and Tokhadze, "Jadwiga Szmidt."

CHAPTER FOURTEEN

1 The other women who worked with Soddy were Winifred Moller Beilby (his spouse) and Ruth Pirret. See chapter 15 below.

2 We thank Lesley A. Richards, Deputy Archivist, University of Glasgow, for the biographical information.

3 The remainder of Hitchins's biography was assembled from an undated reference (probably about 1927) written by Frederick Soddy, and from a letter to the Private Secretary (Appointments), Colonial Office, London, from Soddy, 18 March 1927. Both of these items are in the Soddy Collection, OUA.

4 Thaddeus J. Trenn, *The Self-Splitting Atom: The History of the Rutherford-Soddy Collaboration* (London: Taylor & Francis, 1977).

5 George B. Kauffman, ed., *Frederick Soddy (1877–1956): Early Pioneer in Radiochemistry* (Dortrecht: D. Reidel, 1986).
6 Bertram B. Boltwood, "Note on a New Radioactive Element," *American Journal of Science* 24 (1907): 370–2.
7 Frederick Soddy and Ada F.R. Hitchins, "The Relation between Uranium and Radium. Part VI. The Life-period of Ionium," *Philosophical Magazine*, series 6, vol. 30 (1915): 209–19.
8 We thank Colin A. McLaren, University Archivist, Aberdeen University Library, for this information.
9 *The Aberdeen University Calendar for the Year 1915–1916* (Aberdeen: The University Press, 1915): 174–80.
10 F. Soddy and J.A. Cranston, "The Parent of Actinium," *Proceedings of the Royal Society*, Part A, 94 (1918): 385–405.
11 John A. Cranston, "Frederick Soddy, The Pioneer," *Aberdeen University Review* 39 (1961): 20–7.
12 Frederick Soddy, "The Relation between Uranium and Radium. Part VII," *Philosophical Magazine*, series 6, vol. 38 (1919): 483–8.
13 Soddy wrote "According to analyses by Miss A.F.R. Hitchins and myself," in Frederick Soddy, "The Atomic Weight of Thorium Lead," *Nature* 98 (1917): 469.
14 See chapter 17, above.
15 George B. Kaufmann, "The Atomic Weight of Lead of Radioactive Origin," *Journal of Chemical Education*, 59 (1982): 3–8, 119–23.
16 See also chapter 5 above.
17 Kenneth R. Page, "Frederick Soddy: The Aberdeen Interlude," *Aberdeen University Review* 48 (1979): 127–48.
18 Margaret W. Rossiter, *Women Scientists in America: Struggles and Strategies to 1940* (Baltimore: Johns Hopkins Press, 1982), 118.
19 A.D. Cruickshank, "Soddy at Oxford," *British Journal for the History of Science* 12 (1979): 277–88.
20 File 201, Papers of Professor Frederick Soddy, OUA.
21 *Oxford University Gazette*, 13 June 1923, 673. We thank F.J.C. Rossotti, Sub Faculty of Chemistry, Oxford University, for this information and for a fruitless search of the administrative records.
22 Frederick Soddy and Ada F.R. Hitchins, "The Relation between Uranium and Radium. Part VIII. The Period of Ionium and the Ionium-Thorium Ratio in Colorado Carnotite and Joachimsthal Pitchblende." *Philosophical Magazine*, series 6, vol. 47 (1924): 1148–58.
23 Lawrence Badash, "The Suicidal Success of Radiochemistry," *British Journal of the History of Science* 12 (1979): 245–56.
24 Michael I. Freedman, "Frederick Soddy and the Practical Significance of Radioactive Matter," *British Journal for the History of Science* 12 (1979): 257–60.

25 Files 202, 204, and 208, Papers of Professor Frederick Soddy, OUA.

26 Frederick Soddy, n.d., OUA. This reference was probably provided to Hitchins for use in obtaining a position in Kenya.

27 Frederick Soddy to the Private Secretary (Appointments), Colonial Office, 18 March 1927, OUA.

28 Membership Records, Royal Institute of Chemistry. We thank F. Boyle, Senior Information Scientist, Royal Society of Chemistry, for this information.

29 We thank Mrs S.A. Akhaabi, Kenya National Archives and Documentation Service, for this information.

CHAPTER FIFTEEN

1 Reginald J. Stephenson, "The Scientific Career of Charles Glover Barkla," *American Journal of Physics* 35 (1966): 140–52.

2 H.S. Allen, "Charles Glover Barkla: 1877–1944," *Obituary Notices of Fellows of the Royal Society* 5 (1947): 341–66.

3 C.G. Barkla and Janette G. Dunlop, "Scattering of x-rays and Atomic Structure," *Philosophical Magazine*, series 6, vol. 31 (1916): 222–32.

4 "Janette Gilchrist Dunlop," *University of Edinburgh Journal* 25 (1971–72): 152.

5 "Margaret Pirie Dunbar (née White)," *University of Edinburgh Journal* 26 (1973–74): 69.

6 C.G. Barkla and Margaret P. White, "Absorption and Scattering of x-rays and the Characteristic Radiation of the J Series," *Philosophical Magazine*, series 6, vol. 34 (1917): 270–85. For a discussion of the J phenomenon, see Brian Wynne, "C.G. Barkla and the J Phenomenon," *Physics Education* 14 (1979): 52–5.

7 We thank Mrs Jo Currie, Assistant Librarian, Special Collections, Edinburgh University Library, for the information on Dunlop and White.

8 Otto Hahn, *Otto Hahn: My Life*, trans. Ernst Kaiser and Eithne Wilkins (New York: Herder and Herder, 1970), 65. Hahn added, however, that his relationship with Meitner had been formal to the point where they never socialized outside of the laboratory.

9 We thank Ann Phillips, Archivist, Newnham College, Cambridge, for biographical information.

10 Rita McWilliams-Tulberg, *Women at Cambridge* (London: Victor Gollancz, 1975).

11 J.M.W. Slater, "On the Emission of Negative Electricity by Radium and Thorium Emanation," *Philosophical Magazine*, , series 6, vol. 10 (1905): 460–6.

12 J.M.W. Slater, "On the Excited Activity of Thorium," *Philosophical Magazine*, series 6, vol. 9 (1905): 628–44.

13 Winifred I. Vardy, *King Edward VI High School for Girls, Birmingham, 1883–1925* (London: Ernest Benn, 1928).
14 Hilda M. Beilby (wife of Hubert Beilby, Winifred's brother) to Muriel Howorth, 30 October 1957, Soddy Collection, OUA.
15 "Chapter 15: The Perfect Marriage," in Muriel Howorth, *Pioneer Research on the Atom: The Life Story of Frederick Soddy* (London: New World Publications, 1958). Ernest Rutherford commented to Bertram Boltwood, 14 October 1906: "He [Soddy] writes me that he is engaged to a Miss Beilby of Glasgow." Rutherford Collection, CUA.
16 Mr and Mrs Soddy and Russell, "The Question of the Homogeneity of γ Rays," *Philosophical Magazine*, series 6, vol. 19 (1910): 725–57.
17 Howorth, *Pioneer Research on the Atom*, 168.
18 Letter from unknown author, Somerville College, Oxford, to M. Howorth, n.d., OUA.
19 Howorth, *Pioneer Research on the Atom*, 274.
20 We thank Venora Skelly, Assistant Archivist, University of Glasgow, for the biographical information.
21 Lord Fleck, "Early Work in the Radioactive Elements," *Proceedings of the Chemical Society* (1963): 330.
22 Frederick Soddy and Ruth Pirret, "The Ratio between Uranium and Radium in Minerals," *Philosophical Magazine*, series 6, vol. 20 (1910): 345–9.
23 Ruth Pirret and Frederick Soddy, "The Ratio between Uranium and Radium in Minerals II," *Philosophical Magazine*, series 6, vol. 21 (1911): 652–8. It is noteworthy that in this second paper, Soddy gave Pirret's name priority.

CHAPTER SIXTEEN

1 For a fuller biography of Meitner, see Ruth Lewin Sime, *Lise Meitner. A Life in Physics* (Berkeley: University of California Press, 1996).
2 Walter Moore, *Schrödinger* (Cambridge, UK: Cambridge University Press, 1989), 40.
3 Lise Meitner to Alfred Meitner, March 24, 1940. Folder 5/12, Meitner Collection, CCA.
4 Berta Karlik, "Lise Meitner, Nachruf," *Sonderabdruck aus dem Almanach der Österreichischen Akademie der Wissenschaften*, 119 Jahrgang (1969), Wien, 1970.
5 O.R. Frisch, "Lise Meitner," *Biographical Memoirs of Fellows of the Royal Society* 16 (1970): 405–20.
6 I am indebted to Ruth Lewin Sime for information about Lise Meitner's baptism. Sime located a baptismal certificate at CCA, dated 30 September 1908.

7 Information on the adult careers of the Meitner children comes principally from interviews with surviving relatives and family friends, and letters, obituary notices, and similar documents in the Meitner Collection at CCA.
8 Otto Frisch, *What Little I Remember* (Cambridge: Cambridge University Press, 1979), 1.
9 Charlotte Kerner, *Lise, Atomphysikerin, Die Lebensgeschichte der Lise Meitner* (Weinheim und Basel: Beltz Verlag, 1986), 11.
10 Phyllis Stock, *Better Than Rubies: A History of Women's Education* (New York: Capricorn Books, G. P. Putnam's Sons, 1978), 133.
11 Auguste Dick, 1981. Private communication.
12 O.R. Frisch, "A Nuclear Pioneer Lecture Honouring Dr Lise Meitner," Miami Beach, Florida, June 1973 (Society of Nuclear Medicine Audio Visual Education Program # NF 36).
13 "Hi Jinx Woman of the Week Program,"presented by Frida Meitner Frischauer, National Broadcasting Corporation, 13 February 1949.
14 Frisch, "A Nuclear Pioneer Lecture."
15 It has been said that when "speaking to people she will at times 'employ a friendly sympathetic smile that begins quickly about the mouth and fades slowly and reluctantly from her brown eyes.'" N.a., *Current Biography* (New York: H.W. Wilson Company, 1945), 395.
16 Frisch, "Lise Meitner," 405.
17 L. Meitner, "Looking Back," *Bulletin of the Atomic Scientists* 20 (1964): 2–7.
18 Auguste Dick, 1981. Private communication.
19 Meitner, "Looking Back," 2.
20 Frisch, "Lise Meitner," 405.
21 Meitner, "Looking Back," 3.
22 Hans Benndorf, *Physikalische Zeitschrift* 28 (1927): 397–409.
23 Moore, *Schrödinger*, 70.
24 Karlik, "Lise Meitner, Nachruf," 406.
25 Meitner, "Looking Back," 2.
26 Ibid.
27 Christel Allers, 1987. Private communication.
28 Martin Klein, *Paul Ehrenfest: The Making of a Theoretical Physicist* (New York: American Elsevier Publishers, 1970), 49.
29 Information cited here regarding the approval of Meitner's dissertation and the awarding of her PhD degree was obtained from the Archives of the University of Vienna.
30 Meitner, "Looking Back," 3.
31 Fritz Krafft, "Lise Meitner: Her Life and Times – On the Centenary of the Great Scientist's Birth," *Angewandte Chemie (International Edition)* 17 (1978): 826–42.
32 N.a., *Current Biography*, 393.

33 Lise Meitner, "Uber die Absorption der Alpha- und Beta-Strahlen," *Physikalische Zeitschrift* 7 (1906): 588.
34 Meitner, "Looking Back," 3.
35 Moore, *Schrödinger*, 58.
36 J.L. Heilbron, *The Dilemmas of an Upright Man* (Berkeley: University of California Press, 1986), 38.
37 Frisch, "Lise Meitner," 406.
38 Meitner, "Looking Back," 4.
39 Heilbron, *The Dilemmas of an Upright Man*, 39.
40 Stock, *Better Than Rubies*, 139.
41 Ibid., 140.
42 Krafft, "Lise Meitner," 826.
43 Meitner, "Looking Back," 5.
44 Jonathan Tennenbaum, "Fission and the Breakthrough of Women in Fundamental Scientific Research," *21st Century Science & Technology* 4 (Spring 1991): 29.
45 Frisch, "A Nuclear Pioneer Lecture."
46 Otto Hahn, *Otto Hahn: A Scientific Autobiography*, trans. and ed. Willy Ley (New York: Charles Scribner's Sons, 1966), 66.
47 Heilbron, *The Dilemmas of an Upright Man*, 42.
48 E. M. Wellisch to L. Meitner, 23 September 1911, CCA.
49 Hahn, *A Scientific Autobiography*, 65.
50 Otto Hahn, *Otto Hahn: My Life*, trans. by Ernst Kaiser and Eithne Wilkins (New York: Herder and Herder, 1970), 85.
51 Hahn, *My Life*, 61. Information on personal events in the early life of Otto Hahn is derived principally from this reference and from reference 58.
52 Meitner, "Looking Back," 5.
53 Krafft, "Lise Meitner," 834.
54 Werner Heisenberg, "Gedenkworte für Otto Hahn und Lise Meitner," *Orden pour le merite fur Wissenschaften und Kunste: Reden und Gedenkworte* 9 (1968/69): 113, as quoted in Krafft, "Lise Meitner."
55 Hahn, *My Life*, 42, 43.
56 Ibid., 88.
57 Ruth Lewin Sime, "The Discovery of Protactinium," *Journal of Chemical Education*, 63 (1986): 653–57.
58 Hahn, *My Life*, 89.
59 Armin Hermann, *The New Physics* (Bonn/Bad Godesberg: Inter Nations, 1979), 41, as quoted in William R. Shea, ed., *Otto Hahn and the Rise of Nuclear Physics* (Dordrecht: D. Reidel, 1983), 11.
60 Lise Meitner, "Über die Absorption der Alpha- und Beta-Strahlen," *Physikalische Zeitschrift* 7 (1906): 588, and "Über die Zerstreuung der Alpha-Strahlen," *Physikalische Zeitschrift* 8 (1907): 489–91.

61 Meitner, "Looking Back," 5.
62 Ibid.
63 Ibid., 4.
64 Ibid.
65 Ibid.
66 For further discussion of this work, see Sallie A. Watkins, "Lise Meitner and the Beta-ray Energy Controversy: An Historical Perspective," *American Journal of Physics*, 51 (1983): 551–3.
67 Hahn, *My Life*, 61.
68 Ibid. The literature shows that Harriet Brooks was the first to observe radioactive deposits on the inside of testing vessels in her work at McGill University. She published an account of this observation in 1904, attributing the phenomenon to a volatility of the decay product (Harriet Brooks, "A Volatile Product from Radium," *Nature* 70 [1904]: 270.) For a detailed discussion of the question of attribution in the discovery of radioactive recoil, see Marelene F. Rayner-Canham and Geoffrey W. Rayner-Canham, *Harriet Brooks, Pioneer Nuclear Scientist* (Montreal: McGill-Queen's University Press, 1992), 40, 41.
69 Meitner, "Looking Back," 6.
70 Hahn, *My Life*, 98.
71 Ibid., 106; Heilbron, *The Dilemmas of an Upright Man*, 39.
72 Hahn, *My Life*, 106.
73 Ibid., 137, 138.
74 Meitner, "Looking Back," 6.
75 Sime, "The Discovery of Protactinium."
76 Ibid., 653.
77 Ibid., 656.
78 Otto Hahn and Lise Meitner, "Die Muttersubstanz des Actiniums, ein neues radioaktives Element von langer Lebensdauer," *Physikalische Zeitschrift*, 19 (1918): 208–18, 438.
79 Sime, "The Discovery of Protactinium," 657.
80 L. Meitner and J. Franck, "Über radioaktive Ionen," *Verhandlungen der Deutschen Physikalischen Gesellschaft* 13: 671.
81 Meitner, "Looking Back," 5.
82 P. Curie, "Action du champ magnétique sur les rayons de Becquerel. Rayons déviés et rayons non déviésé," *Comptes rendus* 130 (1900): 73–6.
83 H. Becquerel, "Déviation du rayonnement du radium dans un champ électrique," *Comptes rendus* 130 (1900): 809–15.
84 For a detailed discussion of this issue, see Marjorie Malley, "The Discovery of the Beta Particle," *American Journal of Physics* 39 (1971): 1454–61.
85 Hahn, *A Scientific Autobiography*, 62–3.

86 L. Meitner, "Über Entstehung der Beta-Strahl-Spektren radioaktiver Substanzen," *Zeitschrift für Physik* 9 (1922): 131–44.
87 J. Chadwick, "Intensitätsverteilung im magnetischen Spektrum der α-Strahlen von Radium B + C," *Berichte der Deutschen Physikalischen Gesellschaft* 12 (1914): 383–91.
88 C.D. Ellis and W. A. Wooster, "The Average Energy of Disintegration of Radium E," *Proceedings of the Royal Society,* series A, vol. 117 (1927): 109–23.
89 L. Meitner and W. Orthmann, "Über eine absolute Bestimmung der Energie der primären Beta-Strahlen von Radium E," *Zeitschrift für Physik* 60 (1930): 143–55.
90 W. Pauli to "Dear Radioactive Ladies and Gentlemen," c/o Hans Geiger and Lise Meitner, 4 December 1930, *Collected Scientific Papers, Volume* 2 (New York: Interscience, 1964), 1313. See also Laurie M. Brown, "The Idea of the Neutrino," *Physics Today* 31 (1978): 23–8.
91 Lise Meitner, "Die gamma-Strahlung der Actiniumreihe und der Nachweis, daß die gamma-Strahlen erst nach erfolgtem Atomzerfall emittiert werden," *Zeitschrift für Physik* 34 (1925): 807–18.
92 Charles Ellis to Lise Meitner, 8 December 1925, CCA.
93 L. Meitner, "Über den Zusammenhang zwischen beta- und gamma-Strahlen," *Zeitschrift für Physik* 9 (1922): 145–52; L. Meitner, "Das β-Strahlenspektrum von UX1 und seine Deutung," *Zeitschrift für Physik* 17 (1923): 54–66.
94 R. Sietmann, "False Attribution," *Physics Bulletin* 39 (1988): 316–17.
95 James Chadwick to Lise Meitner, 7 February 1928, CCA.
96 C.D. Ellis to Lise Meitner, 8 February 1928, CCA.
97 Lise Meitner to J. Chadwick, 14 February 1928, CCA.
98 Lise Meitner to C.D. Ellis, 14 February 1928, CCA.
99 J. Chadwick to L. Meitner, 11 July 1928, CCA.
100 L. Meitner to J. Chadwick, 4 August 1928, CCA.
101 Lise Meitner and H.H. Hupfeld, "Über die Prüfung der Streuungsformel von Klein und Nishina an kurzwelliger gamma Strahlung," *Naturwissenschaften* 18 (1930): 534–5.
102 Dietrich Hahn, ed., *Otto Hahn, Erlebnisse und Erkenntnisse* (Dusseldorf: Econ Verlag, 1975), 43.
103 David C. Cassidy, *Uncertainty: The Life and Science of Werner Heisenberg* (New York: W.H. Freeman and Company, 1992), 218. By way of note, Cassidy comments that only "Planck was missing; he was away on his annual spring vacation" (see p. 587)
104 Lise Meitner, "Right and Wrong Roads to the Discovery of Nuclear Energy," *Advancement of Science* (January 1963): 364; Enrico Fermi, "Possible Production of Elements of Atomic Number Higher than 92," *Nature* 133 (1934): 898–9.

105 Fritz Krafft, "Internal and External Conditions," in Shea, ed., *Otto Hahn*, 136.
106 Meitner, "Right and Wrong Roads," 363.
107 Fritz Strassmann, "Kernspaltung – Berlin Dezember 1938," Privately Published, Mainz (1978), 23; reprinted in Fritz Krafft, *Im Schatten der Sensation: Leben und Wirken von Fritz Strassmann* (Weinheim: Verlag Chemie, 1981), chapter 2.
108 This presentation was published as Lise Meitner, "Atomkern und periodisches System der Elemente," *Naturwissenschaften* 22 (1934): 733–9.
109 Ida Noddack, "Über das Element 93," *Angewandte Chemie* 47 (1934): 653–5.
110 Tennenbaum, "Fission and the Breakthrough of Women," 36.
111 O.R. Frisch, "How It All Began," *Physics Today* 20 (1967): 47–52.
112 Hahn, *A Scientific Autobiography*, 140.
113 Hahn, *My Life*, 147, 148. Hahn credits von Grosse's assertion with initiating the Meitner-Hahn collaborative work on uranium bombardment. Meitner's recollections of the events leading to their renewed joint work are more detailed and more plausible; see Meitner, "Right and Wrong Roads." See also Meitner to Laue, 4 September 1944, CCA.
114 O. Hahn and L. Meitner, "Über die küntsliche Umwandlung des Urans durch Neutronen," *Naturwissenschaften* 23 (1935): 37–8, and Über die küntsliche Umwandlung des Urans durch Neutronen. II. Mitteilung," ibid., 230–1; O. Hahn, L. Meitner and F. Strassmann, "Einege weitere Bemerkungen über die küntslichen Umwandlungen beim Uran," ibid., 544–5; L. Meitner and O. Hahn, "Neue Umwandlungsprozesse bei Bestrahlung des Urans mit Neutronen," ibid., 24 (1936): 158–9; L. Meitner, O. Hahn, and F. Strassmann, "Über die Umwandlungsreihen des Urans, die durch Neutronenbestrahlung erzeugt werden," *Zeitschrift für Physik* 106 (1937): 249–70; L. Meitner, "Über die beta- und gamma-Strahlen der Transurane," *Annalen der Physik* 29 (1937): 246–50; O. Hahn, L. Meitner, and F. Strassmann, "Über die Trans-Urane und ihr chemisches Verhalten," *Berichte de Deutschen Chemischen Gesellschaft* 70B (1937): 1374–92, and "Ein neues langlebiges Umwandlungsprodukt in den Trans-Uranreihen," *Naturwissenschaften* 26 (1938): 475; O. Hahn and L. Meitner, "Trans-Urane als künstliche radioaktive Umwandlungsprodukte des Urans," *Scientia* 63 (1938): 12–15.
115 Meitner, "Right and Wrong Roads," 363.
116 Ute Frevert, *Women in German History* (Oxford: Berg, 1989): 197; Alan D. Beyerchen, *Scientists under Hitler: Politics and the Physics Community in the Third Reich* (New Haven: Yale University Press, 1977).
117 Roger H. Stuewer, *Nuclear Physics in Retrospect: Proceedings of a Symposium on the 1930s* (Minneapolis: University of Minnesota Press, 1979),

84. Maurice Goldhaber recalls that in the early 1930s, he "got interested in nuclear physics ... partly through a stimulating course given by Lise Meitner. She was very enthusiastic about her field." See Maurice Goldhaber, "With Chadwick at the Cavendish," *Bulletin of the Atomic Scientists* (December 1982): 12, 13.

118 Heilbron, *The Dilemmas of an Upright Man*, 152.

119 This remark, attributed to Kurt Hess, a National Socialist who headed the guest department at the institute, is quoted in Krafft, "Lise Meitner," 835.

120 Ibid.

121 For a minutely detailed account of Meitner's last months in Germany and her flight to Holland, see Ruth Lewin Sime, "Lise Meitner's Escape from Germany," *American Journal of Physics* 58 (1990): 262–7.

122 Hahn, *My Life*, 149.

123 Sime, "Lise Meitner's Escape from Germany," 266.

124 Lise Meitner to O. Hahn, 4 March 1951, MPG.

125 Lise Meitner to Otto Hahn, 23 October 1938, MPG.

126 Johannes Meitner to Lise Meitner, 24 September 1939, CCA; Gisela Lion Meitner to Lise Meitner, 1 October 1939, CCA; Walter Meitner to Lise Meitner, 3 October 1939, CCA; Johannes Meitner to Lise Meitner, 26 October 1939, CCA; Johannes Meitner to Lise Meitner, 20 November 1939, CCA; Illa Lion (Meitner) to Lise Meitner, 16 December 1939, CCA; Frida Lion to Lise Meitner, 18 December 1939, CCA; Illa Lion (Meitner) to Lise Meitner, 12 February 1940, CCA; Illa Lion (Meitner) to Lise Meitner, 15 March 1940, CCA; Illa Lion (Meitner) to Lise Meitner, 1 April 1940, CCA; O. R. Frisch to Lise Meitner, 11 October 1942, CCA; Lise Meitner to O. R. Frisch, 19 November 1942, CCA; Lise Meitner to Frida Frischauer-Meitner, 27 December 1942, CCA.

127 Lise Meitner to Otto Hahn, 24 August 1938, MPG.

128 Lise Meitner to Otto Hahn, 21 September 1938, MPG.

129 Krafft, "Lise Meitner," 839. Austria was by that time a part of the German Reich.

130 L. Meitner to O. Hahn, 13 August 1919 (postcard), and 19 August 1919. I am indebted to Ruth Lewin Sime for copies and translations of correspondence from Meitner to Hahn that came to light only in 1983. Having been in Hahn's possession for many years, the letters disappeared during World War II but were recovered in a collection of personal papers eventually incorporated into the Otto Hahn Nachlass at the MPG.

131 L. Meitner to O. Hahn, 24 April, 4, 12, and 18 May 1921, MPG.

132 L. Meitner to O. Hahn, 31 March 1938, MPG.

133 Ibid.

134 L. Meitner to O. Hahn, 3 September 1938, MPG.

135 L. Meitner to O. Hahn, 15 October 1938, MPG.

136 R. L. Sime, "Belated Recognition: Lise Meitner's Role in the Discovery of Fission," *Journal of Radioanalytical and Nuclear Chemistry* 142 (1990): 13–26.

137 O. Hahn to L. Meitner, 25 October 1938, CCA.

138 L. Meitner to O. Hahn, 20 October 1939, MPG.

139 F. Krafft, *Im Schatten der Sensation, Leben und Wirken von Fritz Strassmann* (Weinheim: Verlag Chemie, 1981), 84.

140 O. Hahn to L. Meitner, 19 December 1938, MPG.

141 L. Meitner to O. Hahn, 21 December 1938, MPG.

142 See, for example, the following: Oral history interview with Charles Weiner, 3 May 1967, American Institute of Physics Center for History of Physics, New York, NY, 33–5; "The Interest Is Focussing on the Atomic Nucleus," in S. Rozental, ed., *Niels Bohr: His Life and Work as Seen by His Friends and Colleagues* (New York: Wiley, 1967), 144–5; Frisch, "How It All Began," 43–8; Otto R. Frisch, "A Walk in the Snow," *New Scientist* 60 (1973): 833; Otto Robert Frisch, "The Origin of Nuclear Fission," undated manuscript, Frisch Papers, Trinity College Archives, Cambridge University, Cambridge, 2; Frisch, "A Nuclear Pioneer Lecture," 9–10; Otto Robert Frisch, "Experimental Work with Nuclei: Hamburg, London, Copenhagen," in Roger H. Stuewer, ed., *Nuclear Physics in Retrospect*, 71; Frisch, *What Little I Remember*, 114–17; Roger H. Stuewer, "Bringing the News of Fission to America," *Physics Today* 38 (1985): 3–10.

143 Meitner, "Right and Wrong Roads," *Advancement of Science* (January 1963): 364.

144 Ibid.

145 For a detailed study of the history of the liquid-drop model of the nucleus, see Roger H. Stuewer, "The Origin of the Liquid-Drop Model of the Nucleus," *Perspectives in Science* 2 (1994): 76–129.

146 Meitner, "Right and Wrong Roads," 364–5.

147 Stuewer, "Bringing the News of Fission to America," 3–10.

148 O. Hahn and F. Strassmann, "Über den Nachweis und das Verhalten der bei der Bestrahlung des Urans mittels Neutronen entstehenden Erdalkalimetalle," *Naturwissenschaften* 27 (1939): 11–15.

149 L. Meitner and O. Frisch, "Disintegration of Uranium by Neutrons: A New Type of Nuclear Reaction," *Nature* 143 (1939): 239.

150 This is described in more detail in Sharon B. McGrayne, *Nobel Prize Women in Science* (New York: Birch Lane Press, 1993), 60–2.

151 Elizabeth Crawford, J.L. Heibron, and Rebecca Ullrich, *The Nobel Population 1901–1937: A Census of the Nominators and Nominees for the Prizes in Physics and Chemistry* (Berkeley: Office for History of Science and Technology, 1989).

152 L. Meitner to O. Hahn, 1 January 1939, MPG.
153 O. Frisch to O. Hahn, 4 January 1939, MPG.
154 Frisch, "Lise Meitner," 414.
155 Meitner, "Right and Wrong Roads," 365.
156 Frisch, "Lise Meitner," 413.
157 W.L. Bragg to Manne Siegbahn, 23 August 1938; J. D. Cockroft to Otto Robert Frisch, 30 August 1938, CCA.
158 W.L. Bragg to L. Meitner, 17 January 1939, CCA.
159 Program announcement, CCA.
160 AAAS Program announcement, CCA.
161 Meitner Archives, Department of Physics, CU.
162 Program announcement, CU.
163 Lecture schedule, CU.
164 Krafft, "Lise Meitner," 839.
165 L. Meitner to O. Hahn, 4 March 1951, MPG.
166 L. Meitner to O. Hahn, 15 October 1953, MPG.
167 Walter Michels to L. Meitner, 19 March 1958, Meitner Archives, BMC.
168 W. Michels to Alvin C. Eurich, 7 January 1959, BMC.
169 Meitner to Berta Karlik, 13 July 1959, quoted in ref. 3.
170 W. Michels to O. Frisch, 25 May 1959, BMC.
171 Ibid. This letter also reveals that it was Frisch who suggested that Meitner be invited to Bryn Mawr, and who provided encouragement when it seemed doubtful whether she could make the trip.
172 This talk was published as "Looking Back" (ref. 17).
173 Ulla Frisch, 1981. Private communication.
174 Karlik, "Lise Meitner, Nachruf," 353.
175 Author unknown, "Atomic Age Mona Lisa: Dr Lise Meitner," 1959, BMC.
176 Meitner, "Looking Back," 2.

CHAPTER SEVENTEEN

1 We thank Karl Muhlberger, University Archivist, University of Vienna, for this biographical information. Jytte Selno, German Department, Memorial University, is thanked for translations.
2 Bruno Böttcher and Stephanie Horovitz, "Über die Umlagerung von Chinin durch Schwefelsäure," *Monatshefte für Chemie und Verwandte Teile Anderer Wissenschaften* 32 (1913): 793–6; Bruno Böttcher and Stephanie Horovitz, "Über die Umlagerung von Chinin durch Schwefelsäure II," *Monatshefte für Chemie und Verwandte Teile Anderer Wissenschaften* 33 (1914): 567–82.
3 E. Zintl, "Otto Hönigschmid zum 60. Geburtstag," *Zeitschrift für anorganische und allgemeine Chemie* 236 (1938): 3–11.

4 The support for women researchers by Otto Hönigschmid was confirmed by his last woman student, Dr. Luitgard Görnhardt. Personal Communication, Dr Wolfgang Hönigschmid, nephew of Otto Hönigschmid, 24 March 1991.
5 Appendix: Verzeichnis der Mitarbeiter (1910–1950) in *Festschrift des Institutes für Radiumforschung: Anlässlich seines 40Jährigen Bestandes (1910–1950)* (Wien, 1950).
6 Otto Hönigschmid to Max Lembert, 13 June 1914, Deutsche Staatsbibliothek.
7 Ralph E. Oesper, "Otto Hönigschmid," *Journal of Chemical Education* 18 (1941): 562.
8 George B. Kaufmann, "The Atomic Weight of Lead of Radioactive Origin," *Journal of Chemical Education* 59 (1982): 3–8, 119–23; Lawrence Badash, "The Suicidal Success of Radiochemistry," *British Journal of the History of Science* 12 (1979): 245–56.
9 A. Fleck, "Early Work in the Radioactive Elements," *Proceedings of the Chemical Society* (1963): 330.
10 F.A. Paneth, "St Joachimstal and the History of the Chemical Elements," in Herbert Dingle and G.R. Martin, eds., *Chemistry and Beyond: A Selection from the Writings of the Late Professor F.A. Paneth* (New York: Interscience, 1964) 20–8.
11 O. Hönigschmid and St Horovitz, "Sur le poids atomique du plomb de la pechblende," *Comptes rendus* 158 (1914): 1796–8; O. Hönigschmid and St Horovitz, "Über das Atomgewicht des 'Uranbleis,'" *Monatshefte für Chemie* 35 (1914): 1557–60.
12 Henry M. Leicester, ed., *Source Book in Chemistry 1900–1950* (Cambridge: Harvard University Press, 1968), 82–4.
13 Lawrence Badash, *Radioactivity in America: Growth and Decay of a Science* (Baltimore: Johns Hopkins University Press, 1979), 206.
14 O. Hönigschmid and St Horovitz, "Über das Atomgewicht des 'Uranbleis' II," *Monatshefte für Chemie* 36 (1915): 335–80.
15 Otto Hönigschmid to Lise Meitner, 27 June 1914, Meitner Collection, CCA.
16 Otto Hönigschmid and Stefanie Horovitz, "Zur Kenntnis des Atomgewichtes des Urans," *Monatshefte für Chemie* 37 (1916): 185–90; "Zur Kenntnis des Atomgewichtes des Ioniums," *Monatshefte für Chemie* 37 (1916): 305–34; and "Revision des Atomgewichtes des Thoriums. Analyse des Thoriumbromids," *Monatshefte für Chemie* 37 (1916): 335–45.
17 According to Dr Wolfgang Hönigschmid, nephew of Otto Hönigschmid, all of his uncle's papers were destroyed during an air raid in 1944. Personal communication, 24 March 1991.
18 Kasimir Fajans to Elizabeth Róna, 3 January 1963, BHL.

19 Elizabeth Róna to Kasimir Fajans, 6 January 1963, BHL.
20 Kasimir Fajans to Elizabeth Róna, 31 August 1963, BHL.
21 See, for example, Frederick Soddy, "The Origins of the Conceptions of Isotopes," in *Nobel Lectures: Chemistry 1901–1921* (Amsterdam: Elsevier, 1966), 396.
22 Grete Ronge, "Otto Hönigschmid," in C.C. Gillispie, ed., *Dictionary of Scientific Biography* (New York: Charles Scribner's Sons, 1972), 6: 481.

CHAPTER EIGHTEEN

1 Michael W. Friedlander, *Cosmic Rays* (Cambridge, MA: Harvard University Press, 1989), 19–21; Yataro Sekido and Harry Elliot, eds., *Early History of Cosmic Ray Studies* (Dortrecht: D. Reidel, 1989), 28, 210.
2 Herta R. Leng, "Pioneer Woman in Atomic Science," *American Journal of Physics* 59 (1991): 584.
3 Most of this account of Blau's life was derived from conversations with Blau herself between 1953 and 1961. As well, a considerable quantity of research has been undertaken to verify the nature of the Blau-Wambacher controversy described in this chapter. [Editor's note: L.E. Halpern came to the US on a Fulbright fellowship to work with Dr Herta Leng, who introduced him to Blau. Blau and Halpern discovered that their families in Vienna were acquainted. When Blau needed help she turned to Halpern she could trust and who had contacts in atomic science.]
4 Additional information was obtained from Wolfgang L. Reiter, *Emigration und Exil Österreichischer Wissenschaft* (Wien-München: Verlag Jugend und Volk, 1988), 720–2.
5 This work was published as M. Blau, "Über die Absorption divergenter g-Strahlung," *Sitzungberichte Akademie der Wissenschaften Wein Math-naturwissenschaften,* abt. 2A 127 (1918): 1253–79.
6 See, for example, M. Blau and K. Altenburger, "Über einige Wirkungen von Strahlen II," *Zeitschrift für Physik* 12 (1923): 315–29.
7 Marietta Blau and Elizabeth Rona, "Weitere Beiträge zur Ionisation durch H-Partikeln," *Sitzungberichte Akademie der Wissenschaften Wein Math-naturwissenschaften,* abt. 2A 138 (1929): 717–31.
8 Mlle Marietta Blau, "La méthode photographique et les problèms de désintégration artificielle des atomes," *Journal de physique et le Radium,* series 7, 5 (1934): 61–6. Blau closes the article with the comment "I express my thanks to Mme Pierre Curie for the interest that she has shown concerning my work and for the samples of polonium that she has put at my disposal. As well, I thank M. and Mme Joliot for their valuable advice."
9 Blau to Leopold Halpern, personal communication.

10 Qiaozhen Xu and Laurie M. Brown, "The Early History of Cosmic Ray Research," *American Journal of Physics* 55 (1987): 23–33; M. de Maria, M.G. Ianniello, and A. Russo, "The Discovery of Cosmic Rays: Rivalries and Controversies between Europe and the United States," *Historical Studies in the Physical and Biological Sciences* 22 (1991): 165–92.
11 Laurie M. Brown and Helmut Rechenberg, "Quantum Field Theories, Nuclear Forces, and the Cosmic Rays (1934–1938)," *American Journal of Physics* 59 (1991: 595–605.
12 M. Blau and H. Wambacher, "Die photographische Methode in der Atomforschung," *Photographische Korrespondenz* 74 (1938): 2–6, 23–9.
13 M. Blau and H. Wambacher, "Disintegration Processes by Cosmic Rays with the Simultaneous Emission of Several Heavy Particles," *Nature* 140 (1937): 585.
14 M. Blau, "Bericht über die Entdeckung der durch kosmische Strahlung erzeugten 'Sterne' in photographischen Emulsionen," *Festschrift des Institutes für Radiumforschung Anlässlich seines 40Jährigen bestandes (1910–1950)*, in *Sitzungberichte Akademie der Wissenschaften Wein Math-naturwissenschaften*, abt. 2A, 159 (1950): 53–7.
15 Dr Peter Galison, Stanford University, is thanked for information relating to Stetter and Wambacher.
16 M.M. Shapiro, "Tracks of Nuclear Particles in Photographic Emulsions," *Reviews of Modern Physics* 11 (1941): 58–71.
17 Walter Moore, *Schrödinger, Life and Thought* (Cambridge, UK: Cambridge University Press, 1989), 335.
18 M. Blau, "Photographic Tracks from Cosmic Rays," *Nature* 142 (1938): 613.
19 M. Blau, "Über das Vorkommen von alpha-Teilchen mit Reichweiten zwichen 1.2 und 2 cm in einer Samariumlösung," *Archiv for Mathematik og Naturvidenskab* 42(4) (1939): 1–10. At the end of this paper, Blau gives effusive thanks to Ellen Gleditsch and E Föyn for their hospitality.
20 A. Einstein to Kathryn McHale, 18 April 1938, cited in E.L. Horwitz and D. Sponder, "Letters from Dr. Einstein," *AAUW Outlook* 85 (1992): 46. Ms Herta Leng is thanked for providing a copy of this article.
21 A. Einstein to G. Bucky, 14 Febuary 1938, Albert Einstein Archives, Jewish National and University Library, Jerusalem, Israel. Prof. P. Havas is thanked for copies of the Einstein letters relating to Blau.
22 M. Blau "Algunas Investigaciones Sobre Radiactividad Llevadas a Cabo en Mexico," *Ciencia* 5 (1944): 12–17.
23 M. Blau and B. Dreyfus "Multiplier Photo-Tube in Radioactive Measurements," *Review of Scientific Instruments* 16 (1945): 245–8.
24 M. Blau and J.R. Carlin "Industrial Applications of Radioactivity," *Electronics* 21(4) (1948): 78–82.

25 M. Blau and J.E. Smith, "Beta-ray Measurements and Units," *Nucleonics* 2(6) (1948): 67–74.
26 M. Blau and J.A. De Felice, "Development of Thick Emulsions by a Two-Bath Method," *Physical Review* 74 (1948): 1198.
27 M. Blau, M.M. Block, and J.F. Nafe, "Heavy Particles in Cosmic-Ray Stars," *Physical Review* 76 (1949): 860–1.
28 M. Blau, "Semiautomatic Device for Analyzing Events in Nuclear Emulsions," *Review of Scientific Instruments* 21 (1950): 978–85.
29 See, for example, M. Blau, "Hyperfragments and Slow κ-Mesons in Stars Produced by 3-Bev Protons," *Physical Review* 102 (1956): 495–501.
30 M. Blau, S.C. Bloch, C.F. Carter, and A. Perlmutter, "Studies of Ionization Parameters in Nuclear Emulsions," *Review of Scientific Instruments* 31 (1960): 289–97.
31 Marietta Blau, "1.7 Photographic Emulsions" in L.C.L. Yuan and C-S. Wu, eds., *Methods of Experimental Physics: Nuclear Physics Part A* (New York: Academic Press, 1961): 208–63; "2.2.1.1.5 Momentum Measurement in Nuclear Emulsions," ibid., 388–408; "2.2.3.8 Detection and Measurement of Gamma Rays in Photographic Emulsions," ibid., 676–82; "2.3.5 Determination of Mass of Nucleons in Emulsions," in L.C.L. Yuan and C-S. Wu, eds., *Methods of Experimental Physics: Nuclear Physics Part B* (New York: Academic Press, 1963): 37–44.
32 M. Blau to B. Karlik, 24 March 1956, quoted in Wolfgang L. Reiter, *Emigration und Exil Österreichischer Wissenschaft* (Wien-München: Verlag Jugend und Volk, 1988), 722.
33 E. Schrödinger to Max Born, 24 October 1960. Cited in Walter Moore, *Schrödinger, Life and Thought* (Cambridge, UK: Cambridge University Press, 1989), 479.
34 It was in 1962 that she submitted her last research paper, co-authored with her former colleagues at the University of Miami. M. Blau, C.F. Carter, and A. Perlmutter, "An Example of Hyperfragment Decay in the π^+ Mode and Other Interactions of κ-Mesons and Hyperons in Emulsion," *Nuovo Cimento* 27 (1963): 774-85.
35 It is of note that Stretter and Hans Thirring co-authored a laudatory obituary describing the work of Herta Wambacher: G. Stetter and H. Thirring, "Herta Wambacher," *Acta Physica Austraica* 4 (1950): 318-20. Professor C. Quigg is thanked for bringing this article to the author's attention. As well, Moore, in his life of Schrödinger, discusses the questions relating to rehiring of the Nazi scientists Stetter and G. Orthner by Thirring. See Moore, *Schrödinger, Life and Thought.*
36 In his memoirs, Frisch comments: "Coarse tracks formed by alpha particles had been seen early in the century, and attempts to improve the emulsions for getting finer detail had been pursued in the twenties by a group of Viennese women scientists (mainly Kara

Michailowa, Marietta Blau and Berta Karlik)." Otto R. Frisch, *What Little I Remember* (Cambridge, UK: Cambridge University Press, 1979), 186.

CHAPTER NINETEEN

1 Many of the biographical details were obtained from an article on Karamihailova in *Eminent Bulgarian Physicists* (n.p.), written by one of her former students, Hristo Hristov. The book was published in 1981 as part of the communist "celebration" of the 1500th anniversary of the establishment of the Bulgarian state. The reason why E.K. was mentioned (although she was obviously suppressed and isolated by the regime) was to promote a favourable image of Bulgarian science in the field of the history of physics, and its achievements under communism. Some additional information was provided by Wolfgang Reiter, Bundesministerium für Wissenschaft und Forschung, Vienna, Austria.

2 A DPhil was not equivalent to a modern PhD degree but was at a higher academic level than a modern MSc. Details of Karamihailova's early academic career were obtained from the *Girton College Register 1869-1946* (Cambridge, UK: Cambridge University Press, 1948), 686. Ms Kate Perry, Archivist, Girton College, Cambridge is thanked for this information.

3 Elizabeth Kara-Michailova, "Elektrische Figuren auf verschiedenen Materialien, insbesondere auf Krystallen," *Sitzungberichte Akademie der Wissenschaften Wein Math-naturwissenschaften*, abt. 2A, 131 (1922): 155-69.

4 Elizabeth Kara-Michailova and Hans Pettersson, "The Brightness of Scintillations from H-particles and from α-particles," *Nature* 113 (1924): 715.

5 Berta Karlik and Elizabeth Kara-Michailova, "Über die durch α-Strahlen erregte Lumineszenz und deren Zusammenhang mit der Teilchenenergie," *Sitzungberichte Akademie der Wissenschaften Wein Math-naturwissenschaften*, abt. 2A, 137 (1928): 363-80; Berta Karlik and Elizabeth Kara-Michailova, "Zur Kenntnis der Szintillationsmethode," *Zeitschrift für Physik* 48 (1928): 765-83.

6 Marietta Blau and Elizabeth Kara-Michailova, "Über die durchdringende Strahlung des Poloniums," *Sitzungberichte Akademie der Wissenschaften Wein Math-naturwissenschaften*, abt. 2A, 140 (1931): 615-22.

7 E. Föyn, E. Kara-Michailova, and E. Róna, "Zur Frage der künstlichen Umwandlung des Thoriums durch Neutronen," *Naturwissenschaften* 23 (1935): 391-2.

8 Ann Bishop and Marjorie Bawden, "The Yarrow Scientific Awards," unpublished. A copy of this document was obtained from Ms Kate Perry.

9 Miss Elizabeth Kara-Michailova, "The Total Energy of the g-radiation emitted from the Active Deposit of Actinium," *Proceedings of the Cambridge Philosophical Society*, 34 (1938): 429-34.

10 Elizabeth Kara-Michailova and D.E. Lea, "The Interpretation of Ionization Measurements in Gases at High Pressures," *Proceedings of the Cambridge Philosophical Society* 36 (1940): 101-26.

11 Ms Kate Perry is thanked for obtaining this quote in a personal conversation, June 1991.

12 E. Róna, "Laboratory Contamination in the Early Period of Radiation Research," *Health Physics* 37 (1979): 723-7.

13 Hristov, *Eminent Bulgarian Physicists.*

14 E. Karamichailova, Khr. Kamburov, K. Nikolov, V. Marinov, and L. Manolov, "The Radioactivity of Rainwater and Atmospheric Aerosols in Sophia" (in Bulgarian), *Bulgarska Akademie Na Naukite Sofia Fizicheski Institut Izvestiia* 8 (1960): 53-78; E. Karamichailova, K. Nikolov, and K. Doichinova, "Radioactive Studies of Thermal and Cold Subterranean Waters in the Valley of Chepino and Its Immediate Environs" (in Bulgarian), *Bulgarska Akademie Na Naukite Sofia Fizicheski Institut Izvestiia* 9 (1962): 91-8.

15 E. Karamihajlova (deceased) and Zh. T. Zhelev, "On the Radioactivity of the Bulgarian Spring Waters" (in Bulgarian), *Iadrena Energiia* 2 (1975): 95-105.

16 Hristov, in his biographical notes on Karamikhailova, claims that a series of three of her papers were published jointly with papers of her colleagues posthumously in 1980. However, there is no record of them in *Chemical Abstracts.*

17 Hristov, *Eminent Bulgarian Physicists.*

CHAPTER TWENTY

1 An outline of her life is given in Wolfgang L. Reiter, *Emigration und Exil Österreichischer Wissenschaft* (Wein-München: Verlag Jugend und Volk, 1988), 718-20.

2 Unless stated otherwise, the details of her life were obtained from Elizabeth Róna, *How It Came About* (Oak Ridge: Oak Ridge Associated Universities, 1978). This work is very vague about dates and, written late in life, cannot always be relied on factually. However, her correspondence to Kasimir Fajans, FA, includes an undated small piece of paper listing the precise dates of her different appointments.

3 Elizabeth Róna, "Über die Geschwindigkeit der gegenseitigen Wirkung des Broms und der einwertigen, gesättigten aliphatischen

Alkohole in wässeriger Lösung," *Zeitschrift für Physikalische Chemie* 82 (1913): 225-48.

4 Elizabeth Róna, "I. Über die Reduktion des Zimtaldehyds durch Hefe. II. Vergärung von Benzylbrenztraubensäure," *Biochemische Zeitschrift* 67 (1914): 137-42.

5 Róna to K. Fajans, 29 September 1963, FA.

6 Her first in a series of publications being Georg von Hevesy and Elizabeth Róna, "Die Lösungsgeschwindigkeit molekularer Schichten," *Zeitschrift für Physikalische Chemie* 89 (1915): 294-305.

7 Elizabeth Róna, "Diffusionsgrösse und Atomdurchmesse der Radiumemanation," *Zeitschrift für Physikalische Chemie* 92 (1917): 213-18.

8 Reiter, *Emigration und Exil*, 718.

9 Róna, *How It Came About*, 8.

10 For information on Ramstedt, see chapter 10, above.

11 Elizabeth Róna, "Über die Wirksamkeit der Fermente unter abnormen Bedingungen und über die angebliche Aldehydnatur der Enzyme," *Biochemische Zeitschrift* 109 (1920): 279-89.

12 Such as Irén Götz; see chapter 10, above.

13 Róna, *How It Came About*, 12-13.

14 Elizabeth Róna, "Über den Ionium-Gehalt in Radium-Rückständen," *Berichte der Deutschen Chemischen Gesellschaft* 55B (1922): 294-301.

15 Róna, *How It Came About*, 14.

16 Ibid., 15.

17 Elizabeth Róna und Ewald A.W.Schmidt, "Eine Methode zur Herstellung von hochkonzentrierten Poloniumpräparaten," *Sitzungberichte Akademie der Wissenschaften Wein Math-naturwissenschaften*, abt. 2A, 137 (1928): 103-15.

18 Róna, *How It Came About*, 24-5.

19 Ibid., 25.

20 Elizabeth Róna, "Verdampfungsversuche an Polonium," *Sitzungberichte Akademie der Wissenschaften Wein Math-naturwissenschaften*, abt. 2A, 141 (1932): 533-7.

21 For information on Karlik, see chapter 22, below.

22 Berta Karlik und Elizabeth Róna, "Untersuchungen der Reichweite von α-Strahlen des Actinium X und seiner Folgeprodukte mit der Lumineszenzmethode," *Sitzungberichte Akademie der Wissenschaften Wein Math-naturwissenschaften*, abt. 2A, 142 (1933): 121-6; and Berta Karlik und Elisabeth Róna, "Untersuchungen über die Reichweite der Polonium-α-Strahlen in Abhängigkeit von Intensität der Strahlung, Alter der Präparate und Art der Unterlage," *Sitzungberichte Akademie der Wissenschaften Wein Math-naturwissenschaften*, abt. 2A, 143 (1934): 127-221.

23 Elizabeth Róna, "Laboratory Contamination in the Early Period of Radiation Research," *Health Physics*, 37 (1979): 723-7.

24 Róna, *How It Came About*, 33.
25 See chapter 19, above.
26 Róna, *How It Came About*, 37.
27 Ibid.
28 N.a., "Pioneer in isotope research," *Oak Ridger* (28 July 1981): 1,7.
29 Reiter, *Emigration und Exil*, 718.
30 Ellen Gleditsch und Elisabeth Róna, "Contribution to the Study of the Exchange of Ions between Salts and Saturated Solutions," *Archiv for Mathematik og Naturvidenskab* 44 (1941): 53-62.
31 Elizabeth Róna to Kasimir Fajans, 26 September 1965, FA. This is contrary to Meitner's opinion as she felt that Hahn was far too accomodating of the Nazi authorities.
32 Elizabeth Róna to Kasimir Fajans, 29 September 1963, FA.
33 Elizabeth Róna and William D. Urry, "Radioactivity of Ocean Sediments. VIII. Radium and Uranium Content of Ocean and River Waters," *American Journal of Science* 250 (1952): 241-62.
34 Elizabeth Róna, "Exchange Reactions of Uranium Ions in Solution," *Journal of the American Chemical Society* 72 (1950): 4339-43.
35 Elizabeth Róna, "A Method to Determine the Isotopic Ratio of Thorium-232 to Thorium-230 in Minerals," *American Geophysical Union, Transactions* 38 (1957): 754-9.
36 Edith Wilson to Herta Leng, 24 January 1992. Dr Leng is thanked for a copy of this letter.
37 Ernst Föyn, Berta Karlik, Hans Pettersson, and Elizabeth Róna, "Radioactivity of Sea Water," *Nature* 143 (1939): 275-6.
38 Elizabeth Róna, Donald W. Hood, Lowell Muse, and Benjamin Buglio, "Activation Analysis of Manganese and Zinc in Sea Water," *Limnology and Oceanography* 7 (1962): 201-6.
39 N.a., "Pioneer in Isotope Research," *Oak Ridger* (28 July 1981): 1,7.
40 Elizabeth Róna to Kasimir Fajans, 26 September 1965, FA.
41 E. Róna and B.L. Brandau, "Absolute Dating of Ocean Sediments by Use of Th-230/Pa-231 Ratio," *Bulletin Volcanologique* 32 (1969): 509-14; E. Róna and C. Emiliani, "Absolute Dating of Caribbean Cores P6304-8 and P6304-9," *Science* 163 (169): 66-8.
42 Elizabeth Róna to Kasimir Fajans, 29 August 1974, FA.
43 Edith Wilson to Herta Leng, 24 January 1992.
44 N.a., "Pioneer in Isotope Research," *Oak Ridger* (28 July 1981): 1,7.
45 Edith Wilson to Herta Leng, 24 January 1992.

CHAPTER TWENTY-ONE

1 Otto Hahn, *Otto Hahn: A Scientific Autobiography*, trans. and ed. Willy Ley (New York: Charles Scribner's Sons, 1966), 140.

2 Ida Noddack, personal communications.
3 D. Holde and I. Tacke, "Anhydrides of the Higher Fatty Acids," *Chemische Zeitung* 45 (1921): 949-50, 954-6.
4 For an account of Moseley's work, see: J.L. Heilbron, *H.G.J. Moseley* (Berkeley: University of California Press, 1974). In fact, Moseley tentatively identified element 43 as being canadium. An account of the claims for this spurious element can be found in G.W. Rayner-Canham, "The Curious Case of Canadium," *Canadian Chemical Education* 8 (1973): 10-11.
5 See, for example, "Two New Elements of the Manganese Group," *Nature* 116 (1925): 54-5; Walter Noddack and Ida Tacke, "Eka- and Dvi Manganese," *Chemical News* 131 (1925): 84-7 (a translation into English of the original paper in *Naturwissenschaften*).
6 Mary E. Weeks and Henry M. Leicester, *Discovery of the Elements*, 7th ed. (Easton, PA: Journal of Chemical Education, 1968), 824.
7 The discovery of element 43 is reviewed in G.W. Rayner-Canham and G. Pike, "The Search for the Elusive Element 43," *Education in Chemistry* 30 (1993): 12-14.
8 B.T. Kenna, "The Search for Technetium in Nature," *Journal of Chemical Education* 39 (1962): 436-42.
9 P. Van Assche, "The Ignored Discovery of the Element Z=43," *Nuclear Physics* A480 (1988): 205-14.
10 P.K. Kuroda, "A Note on the Discovery of Technetium," *Nuclear Physics* A503 (1989): 178-82.
11 J. Newton Friend, *Man and the Chemical Elements*, 2d ed. (London: Charles Griffin, 1961): 251.
12 Aaron J. Ihde, *The Development of Modern Chemistry* (New York: Harper & Row, 1964), 591.
13 Ute Frevert, *Women in German History* (Oxford: Berg, 1989), 197.
14 In fact, Nazi women were divided between the modernist camp, which considered women could have certain roles outside the family, and the more numerous traditionalist wing, which believed women's sole role was motherhood. See Claudia Koonz, "Nazi Women before 1933: Rebels against Emancipation," *Social Science Quarterly* 56 (1976): 553-63.
15 Alan D. Beyerchen, *Scientists under Hitler* (New Haven: Yale University Press, 1977).
16 Ibid., 66-7.
17 I. Noddack, "The Periodic System of the Elements and Its Gaps," *Angewandte Chemie* 47 (1934): 301-5.
18 E. Fermi, "Possible Production of Elements of Atomic Number Higher than 92," *Nature* 133 (1934): 898-9.
19 I. Noddack, "Über das Element 93," *Angewandte Chemie* 47 (1934): 653-6.

20 William R. Shea, ed., *Otto Hahn and the Rise of Nuclear Physics* (Dortrecht: D. Reidel, 1983).
21 O. Hahn and F. Strassmann, "Nachweis der Entstehung activer Bariumisotope aus Uran und Thorium durch Neutronenbestrahlung," *Naturwissenschaften* 27 (1939): 11-15.
22 D. Hahn, *Otto Hahn Begründer des Atomzeitalters* (Munich: List Verlag, 1979).
23 Robert Jungk, *Brighter than a Thousand Suns* (New York: Harcourt, Brace, Jovanovich, 1958), 62.
24 I. Noddack, "Bererkung zu den Untersuchungen von O. Hahn, L. Meitner und F. Strassman über die Produkte, die bei der Bestrahlung von Uran mit Neutronen entstehen," *Naturwissenschaften* 27 (1939): 212-13.
25 Teri Hopper, "She Was Ignored: Ida Noddack and the Discovery of Nuclear Fission," Master's thesis, Palo Alto, Stanford University, 1990.
26 See, for example, Fritz Krafft, "Internal and External Conditions," in Shea, ed., *Otto Hahn*, 149. "Indeed, she neither reiterated her criticism nor made any experimental investigation, although she was a chemical expert on rare earths and together with her husband was the codiscoverer of the element rhenium itself. Nor did anybody else take in earnest her objections, perhaps also because she and her husband insisted on their detection of element 43 (called 'masurium' by them) which nobody was able to reproduce and which undermined considerably their scientific credibility."
27 Elizabeth Crawford, J.L. Heibron, and Rebecca Ullrich, *The Nobel Population 1901-1937: A Census of the Nominators and Nominees for the Prizes in Physics and Chemistry* (Berkeley: Office for History of Science and Technology, 1989).
28 F.A. Paneth, "The Making of the Missing Chemical Elements," *Nature* 159 (1947): 8-10.

CHAPTER TWENTY-TWO

1 We thank Dr Klaus-Peter Hoepke, University Archivist, University of Karlsruhe, for this information and Jytte Selno, German Department, Memorial University, for the translation.
2 Kasimir Fajans and Helene Towara, "Über ein neues langlebiges Glied der Wismutplejade," *Naturwissenschaften* 2 (1914): 685-6.
3 Kasimir Fajans to Elizabeth Róna, 15 September 1974, FA.
4 G.v. Hevesy and Hilde Levi, "Radiopotassium and Other Artificial Radio-elements," *Nature* 135 (1935): 580; "Artificial Radioactivity of Dysprosium and Other Rare-Earth Elements," *Nature* 136 (1935): 103; and "Action of Slow Neutrons on Rare Earth Elements," *Nature* 137 (1935): 185.

5 D. Lockner, "George de Hevesy – 80 years," *International Journal of Applied Radiation and Isotopes* 16 (1965): 513-16.

6 There is an entry for Karlik in John Turkevich and Ludmilla B. Turkevich, "Austria," *Prominent Scientists of Continental Europe* (New York: American Elsevier, 1968), 2.

7 Franz Patzer to the Magistrate's Office of the City of Vienna, 6 December 1988. We thank Herwig Würtz, Director of the Vienna Stadt- und Landesbibliothek for a copy of these biographical notes. Also, we thank Dr Wolfgang L. Reiter, Bundesministerium für Wissenschaft und Forschung, Vienna, for additional information on Karlik.

8 B. Karlik, "Über die Abhängigkeit des Szintillationen von der Beschaffenheit des Zinksulfide und das Wesen des Szintillationsvorgangs," *Sitzungberichte Akademie der Wissenschaften Wein Math-naturwissenschaften*, abt. 2A 136 (1928): 531-62.

9 Helen Simpson Gilchrist and Berta Karlik, "Separation of Normal Long-chain Hydrocarbons by Fractional Distillation in High Vacuum," *Journal of the Chemical Society* (1932): 1992-5.

10 Isabel Ellie Knaggs and Berta Karlik, *Tables of Cubic Crystal Structure of Elements and Compounds* (London: A. Hilger, 1932).

11 Ernst Föyn, Berta Karlik, Hans Pettersson, and Elizabeth Róna, "Radioactivity of Sea Water," *Nature* 143 (1939): 275-6.

12 Karlik's lengthy and diverse list of publications (only three of which are in English) are detailed in *J.C. Poggendorff Biographisch-Literarisches Handwörterbuch der Exacten naturwissenschaften,* Band VIIa, Teil 2:F-K, 1932-1953 (Berlin: Akademie-Verlag, 1958), 691-2.

13 Berta Karlik and Traude Bernert, "Über eine vermutete β-Strahlung des Radium A und die natürliche Existenz des Elementes 85," *Naturwissenschaften* 44/45 (1942): 685-6; and "Das Element 85 in den natürlichen Zerfallsreihen," *Zeitschrift für Physik* 123 (1944): 51-72; B. Karlik, "Unsere heutigen Kenntnisse über das Element 85 (Ekajod)," *Monatschäfte für Chemie* 77 (1947): 348-51.

14 Berta Karlik, "1938 bis 1950," in *Festschrift des Institutes für Radiumforschung anlässlich seines 40Jährigen bestandes (1910-1950)* (Vienna, 1950): 35-41.

15 Kart Lintner, "Nachruf – Berta Karlik," *Österreichischen Hochschulzeitung* 4 (1990): 21. Dr Kurt Mühlberger, University Archivist, University of Vienna, is thanked for a copy of this obituary.

CHAPTER TWENTY-THREE

1 "Mary E. Weeks and Henry M. Leicester, *Discovery of the Elements*, 7th ed. (Easton, PA: Journal of Chemical Education, 1968).

2 M. Alic, *Hypatia's Heritage: A History of Women in Science from Antiquity through the Nineteenth Century* (Boston: Beacon Press, 1986).

3 We thank Elaine D. Trehub, Archivist, Mount Holyoke College, and Lorrett Treese, Archivist, Bryn Mawr College, for this information.
4 Marelene F. Rayner-Canham and Geoffrey W. Rayner-Canham, "Women in Chemistry: Participation during the Early Twentieth Century," *Journal of Chemical Education* 73 (1996): 203-5.
5 Betty M. Vetter, "Changing Patterns of Recruitment and Employment," in V.B. Haas and C.C. Perrucci, *Women in Scientific and Engineering Professions* (Ann Arbor: University of Michigan Press, 1984), 59.
6 Vera Kistiakowsky, "Women in Physics: Unnecessary, Injurous and Out of Place?" *Physics Today* 33 (1980): 32-40.
7 Vetter, 59.
8 Jessie Barnard, *Academic Women* (New York: Meridian, 1964), 62.
9 E. Roberts, *Women's Work 1840-1940* (London: Macmillan Education, 1988), 67.
10 Margaret Rossiter, *Women Scientists in America* (Baltimore: Johns Hopkins Press, 1982), 118.
11 Victoria de Grazia, *How Fascism Ruled Women: Italy 1922-1945* (Berkeley: University of California Press, 1992), 161.
12 Veronica Strong-Boag, *The New Day Recalled: Lives of Girls and Women in English Canada, 1919-1939* (Toronto: Copp Clark Pittman, 1988), 25.
13 Ibid.
14 Susan Faludi, *Backlash* (New York: Doubleday, 1991), 50.
15 Jacques R. Pauwels, *Women, Nazis, and Universities: Female University Students in the Third Reich, 1933-1945* (Westport, CT: Greenwood Press, 1984), 33-48.
16 Miss G.I. Harper and Miss E. Salaman, "Measurements on the Ranges of α-particles", *Proceedings of the Royal Society* 127 (1930): 175-85.
17 Kathleen A. Davis, "Katharine Blodgett and Thin Films," *Journal of Chemical Education* 61 (1984): 437-9; K. Thomas Finley and Patricia J. Siegel, "Katharine Burr Blodgett (1898-1979)," in Louise S. Grinstein, Rose K. Rose, and Miriam H. Rafailovich, eds., *Women in Chemistry and Physics: A Biobibliographic Sourcebook* (Westport, CT: Greenwood Press, 1993), 65-71.
18 George B. Kauffman and J.P. Adloff, "Marguerite Catherine Perey (1909-1975)," in Grinstein, Rose, and Rafailovich, eds., *Women in Chemistry and Physics*, 470-5.
19 Lorella M. Jones, "Intellectual Contributions of Women to Physics – Yvette Cauchois," in G. Kass-Simon and Patricia Farnes, eds., *Women of Science: Righting the Record* (Bloomington: Indiana University Press, 1990), 203-5.
20 Mary R.S. Creese, "British Women of the Nineteenth and Early Twentieth Centuries Who Contributed to Research in the Chemical Sciences," *British Journal for the History of Science* 24 (1991): 275–305.

21 Lawrence Badash, "The Suicidal Success of Radiochemistry," *British Journal for the History of Science,"* 12 (1979): 245-56.
22 Roger L. Geiger, *To Advance Knowledge: The Growth of American Research Universities, 1900-1940* (New York: Oxford University Press, 1986), 233-40; Bernadette Bensaude-Vincent and Isabelle Stengers, *Histoire de la chimie* (Paris: Éditions de la Découverte, 1993), 309.
23 John C. Slater, "Quantum Physics between the Wars," *Physics Today* (1968): 43-51.
24 Daniel J. Kevles, *The Physicists: The History of a Scientific Community* (Cambridge, MA: Harvard University Press, 1987), 207.
25 Joan A. Dash, *A Life of One's Own* (New York: Harper & Row, 1973), 277.
26 Caroline L. Herzenberg and Ruth H. Howes, "Women of the Manhatten Project," *Technology Review,* 36 (Nov./Dec. 1993): 32-40; Ruth H. Howes, "Leona Woods Marshall Libby (1919-1986)," in Grinstein, Rose, and Rafailovich, eds., *Women in Chemistry and Physics,* 320-8.
27 Joan Freeman, *A Passion for Physics: The Story of a Woman Physicist* (Bristol: Adam Hilger, 1991), 115.
28 Sharon Traweek, "High-Energy Physics: A Male Preserve," *Technology Review* 87 (1984): 42-3.

Index